FOREST POLICIES AND SOCIAL CHANGE IN ENGLAND

WORLD FORESTS

Series Editors

World Forests Description

As forests stay high on the global political agenda, and forest-related industries diversify, cutting edge research into the issues facing forests has become more and more transdisciplinary. With this is mind, Springer's *World Forests* series has been established to provide a key forum for research-based syntheses of globally relevant issues on the interrelations between forests, society and the environment.

The series is intended for a wide range of readers including national and international entities concerned with forest, environmental and related policy issues; advanced students and researchers; business professionals, non-governmental organizations and the environmental and economic media.

Volumes published in the series will include both multidisciplinary studies with a broad range of coverage, as well as more focused in-depth analyses of a particular issue in the forest and related sectors. Themes range from globalization processes and international policies to comparative analyses of regions and countries.

Forest Policies and Social Change in England

SYLVIE NAIL
University of Paris III, Sorbonne Nouvelle, France

Springer

Dr. Sylvie Nail
University of Paris III
Sorbonne Nouvelle
France

ISBN 978-1-4020-8364-8 e-ISBN 978-1-4020-8365-5

Library of Congress Control Number: 2008925104

Cover illustration: New social uses for the woods. Photograph by Sylvie Nail.

Printed on acid-free paper

9 8 7 6 5 4 3 2 1

springer.com

In memory of Pam Jackson, the dearest of friends,
always with me.

Acknowledgements

Over the past few years, during the preparation of this book, I have been very lucky to meet many wonderful people involved in forestry, all of whom are dedicated to trees but more than anything else to people. In particular, I would like to thank all my interviewees, whether members of the general public, members of charitable organisations or of institutions like the Forestry Commission and the Community Forests. My gratitude goes especially to Paul Hill-Tout, Director of the Forestry Commission England, and to the Mersey Forest Team – especially Paul Nolan, its director, Lee Dudley, Jo Sayers and Dave Perry – who gave freely of their time for interviews, who were ever helpful and provided numerous illustrations.

I am also grateful to Eleanor Harland, librarian at the Forestry Commission at Alice Holt Research Station for her help and patience, and to Claudie Petit at the Laboratoire d'anthropologie urbaine (LAU) of the Centre national de la recherche scientifique (CNRS) in Paris, who made sure I received the documents I needed. I am greatly indebted to the LAU for its support which made many field trips possible in all seasons.

Many thanks to all my correspondents who gave me permission to reproduce their photographs. For documents with Crown copyright, a PSI licence has been obtained (number C2007002067).

At Springer, I would like to acknowledge the encouragement and help of Catherine Cotton and Ria Kanters, as well as the perceptive comments of Jussi Uusivuori and Matti Palo.

Special thanks are due to my friend Janet Rossiter, not only on account of her hospitality and of her constant endeavours to help me in my fieldwork by providing numerous contacts, but also for being brave enough to take on the proofreading of the book. Sarah Pickard was kind enough to check the last version of the manuscript. The mistakes that remain are mine, of course.

If the protracted work involved in research and publication tests lifelong friendships, it comes with the added bonus of new acquaintances, some of which turn into friendships. Cecil Konijnendijk and Mark Johnston are two very special people I was lucky to meet during this work, and I want to thank them for their support, comments and enthusiasm.

I am especially indebted to Jose Luis, my husband, whose patience and listening capacities I tested to the utmost, and in whom I always found a source of support and of relevant advice. He and Théo have had to live with this book, and they deserve my heartfelt thanks for this.

Contents

List of Figures

List of Tables

Introduction

Forests lie at the heart of intersecting topical issues at the beginning of the third millennium. With the advent of the post-industrial, post-productivist era in many countries of the Western world, the forestry sector has undergone dramatic changes. For instance in England, the price of timber decreased dramatically in the late 1990s, but indirect market benefits and non-market benefits, including recreation, health and regeneration, have grown. These now loom large in the equation of forest profitability.

One of these benefits relates to a wider international agenda, that of the preservation of biodiversity, which has important consequences in terms of management policies and practices. Traditional resource management has led to criticisms and to forestry practices being questioned over the last decades, while sustainable ecosystem-based forest management has developed. What Kennedy et al. describe as the transition from a machine model to an organic model, hinging on relationships [Kennedy et al. 1998], has led to another characteristic, the inclusion of the socio-cultural environment of forestry.

What follows from this is two-fold. First, forests are now perceived as cultural landscapes, that is to say intentional as well as unintentional expressions of history, environment and cultural identity combined. These social constructions reflect the values of society, understood as contingent concepts of worth informed by 'ethical and moral judgements' [O'Brien 2003: 5]. Conversely, forests feed societies' dreams, myths and memories. The second corollary is the widening of the forestry agenda to include such social issues as the fight against social deprivation and exclusion, especially in urban areas. Indeed, urbanisation has been a driving force, not only in Europe, but also in developing countries, as the subtitle of the UN World Environment Day 2005, held in San Francisco, emphasised: 'Green Cities: where the future lives'. Forestry in urban areas, a recent development as will be seen, has undoubtedly become a way of delivering sustainable development: international projects exemplify this strategic approach to urban forestry in Europe.

All in all, the stakes have become increasingly complex with the advent of multipurpose forestry.

These issues, together with the rise of new academic disciplines (environmental history, landscape ecology, cultural geography, environmental psychology, among them), have led to copious research in relation to wooded landscapes in many

S. Nail, *Forest Policies and Social Change in England.*

countries, focusing on history, politics, aesthetics, biodiversity, cultural meanings, recreational functions or economic value. More often than not, however, such aspects are dealt with in isolation, cut off from one another.

Similarly, urban and rural issues are often dealt with separately, which in many cases is no longer relevant. Forests, like any other landscape, are more than the sum of their parts, and their analysis must be based on both their material forms and on the economic and ideological context in which they were construed. Policies and perceptions are frequently dealt with in different books, written by specialists of different disciplines, and the connection between the two is lost. Yet, in order to understand the applications and implications of policies, it is imperative, not only to address economics, leisure, heritage and symbolism, but to address them jointly so as to see how they interact, in order to form a picture of the place forests occupy, physically and symbolically, today. Interactions are fundamental, and policies can fail if they do not take perceptions into account.

The main thrust of this book is thus to explore the multifarious uses, functions and values of forests, whether rural, peri-urban or urban, at the beginning of the 21st century in England. This entails, first, analysing British governments' responses,[1] in so far as values underlie public policies and determine to a certain extent the means deemed most appropriate to reach the public good. It also entails relating these courses of action to social change, that is, to the transformations in the functioning, structure and culture of society in England.

International agreements and European policies obviously frame British policies and influence perceptions to a certain extent. They are presented whenever relevant, as well as examples of initiatives in other countries. Yet, it is argued that England offers a particularly relevant illustration of the issues mentioned above, and of original policies implemented to respond to the new challenges of forestry today.

The definition of woodland in the UK is slightly different from that in other countries. It refers to land under stands of trees with a canopy cover of, or the potential to reach, at least 20%, but with no minimum height to form a mature woodland, whereas the international definition is based on a 10% canopy cover and a minimum height of 5 metres at maturity. There is no precise definition of 'forest' in the UK, and no minimum size, either (although statistics sometimes make a distinction between woodlands of more than 2 hectares and those of less than 2 hectares). As will be shown, the term 'forest' has been used with different meanings over the centuries; today, it tends to be associated mostly, but not exclusively, with large planted areas. Forests and woodlands are part and parcel of the English culture and identity, strongly marked by class and power relationships throughout history. As such, they are invaluable elements of analysis of the English society. Furthermore, England is at the leading edge of urban and social forestry on the international scene.

[1] Until the devolution process of the late 1990s, which is explained in Chapter 3, the Forestry Commission operated at the level of Great Britain (i.e. England, Scotland and Wales), hence the need to analyse forest policies at the British level.

In 2007, English woodlands covered 1.124.000 hectares, that is to say 8.6% of the total land area, a smaller percentage of the land than in most other European countries [Forestry Commission 2007: table 1.2]. Yet, they have over the last 15 years become the objects of an amazing array of initiatives, programmes, grant schemes and, more than anything else, of copious discourse on their value and multiple benefits. This attention bears no relation to their modest contribution to the national economy, which raises the question of their real missions, material and symbolical. England therefore offers a privileged environment in which to observe how perceptions shape policies, and conversely how policies can modify perceptions, as well as how both are grafted onto to the national cultural heritage.

Public policies such as the *England Forestry Strategy* (1998) jointly address all the issues related to woodland. The purpose of this book is to deal jointly with all these topics, too, through a social sciences perspective, in order to assess public policies in relation to people's perceptions of woodlands, and to show, in a mirror effect, the place and value of woodland and what these reveal of the English society in the 21st century, of its problems, aspirations and vision of itself. It makes use of existing literature and of the numerous reports on multifarious aspects of English woodland. More than anything else, it is based on a close scrutiny of official publications and abundant fieldwork with numerous interviews.

Through this book, it is hoped to contribute to improving the knowledge of issues which may be peripheral to arboricultural and forestry practices, but are felt by professionals working in these fields, especially in urban forestry, to be lacking from their training. The readers may include specialists of public policies in relation to open spaces and (urban) forests, including those who devise and implement them (among them professionals of arboriculture), as well as social science researchers, human geographers and forest historians. The book is also intended as a resource to higher education, hoping that students of forestry/arboriculture, landscape architecture, ecology, environmental studies and urban/rural planning will find in it tools to reflect on practices. Because England is a leader in the field of urban and social forestry, this research may be of interest to academics and practitioners in other countries, too. The wide public of tree-lovers will hopefully find here fresh insight into the stakes behind tree policies.

The thematic organisation of the book reflects its wide-ranging scope.

The preliminary chapter is designed to provide the historical landmarks in the woodland history of England, i.e. the background necessary to understand the present situation. It constitutes the starting-point of the questioning on the roles and functions of today's woodland. It reviews the main characteristics of English woodland through time, so as to enhance the political, economic, social and cultural traits indispensable to an understanding and assessment of present trends: the natural data, the links of woodland with political power and the traditional economic and symbolical functions of woodland in pre-industrial England. Furthermore, it illustrates the two basic principles that underlie forestry throughout history: in order for woodlands to exist, there has to be no better use of the land, and forestry has to be profitable. This raises the question which informs the whole book: what is the rationale of woodland today? Do the old criteria still exist? If so, are they still valid, or are they present in a different form?

The first part focuses on the institutionalisation of forestry since 1919, a turning-point in the destiny of forests and woodlands through the creation of the Forestry Commission. A lot has been written on the first 50 years of the Forestry Commission, and this chapter avoids repeating previous publications on the subject. Rather, it wants to attract attention to some characteristics which were dogma for several decades, not only in Britain but also on the international scene, and have shaped today's forestry, for example in terms of landscapes and economics. The chapter then explores the changes in emphasis following the economic failure of British forestry and presents the contents and impacts of institutional changes, within the Forestry Commission and also in the UK at large, due to devolution at the end of the 20th century. It also presents the new focus of public policies through an analysis of the 1998 *England Forestry Strategy*.

The next chapter demonstrates how today's new forestry paradigm stems directly from two main influences: the failure of the productivist ethos and a change in perceptions and priorities on the international scene. One of the dominant features of the last three decades, in England as in many other countries, has been the rise of urban forestry, the definition, remit, tools and applications of which are discussed in the last chapter of this first part.

To a certain extent, the second and the third parts reflect the official classifications of woodland benefits: market benefits and non-market benefits. But the purpose goes beyond following this classification, which does not do justice to the width and breadth of woodland issues, in particular to the underlying values of woodlands or to the power of representations. Yet, the latter can promote or hinder public policies, especially in the case of non-market benefits. Typical examples of this phenomenon are presented in connection with programmes related to education, health, community participation and access, where values and perceptions, when they are ignored, can clearly stand in the way of public policies being effective. Furthermore, the advent of urban forestry, especially since the 1980s as far as England is concerned, has changed the agenda and put an end to traditional categories (rural/urban, market/non-market). Instead, two visions of woodland emerge which apply to both cities and the countryside.

The second part reviews the concept of multi-purpose forestry and its consequences on woodlands. Although financial returns are given a low priority today, the calculations made to assess the benefits of forestry nevertheless rely more often than not on financial valuation, which leads to a certain ambiguity. The first chapter of this part focuses on the direct market and non-market benefits, with new outlets and new emphases. Another aspect of multi-purpose forestry very much stressed in official literature is its role in rural and urban regeneration, which is the object of the next chapter. This particular area relates to the economy, in so far as there are obvious financial returns expected from regeneration (e.g. inward investment), but it also reflects changes in the economy that call for changes in land use. The discourse on regeneration is scrutinised, so as to elicit what missions forestry is supposed to fulfil, then compared with incentives and programmes, in particular in the 12 Community Forests, showing that action does not always match rhetoric or social needs.

Arguably the first function of 21st century forestry is linked to the environment, pollution and climate change in particular. This is what the following chapter focuses on, bearing in mind the context of the production of scientific knowledge, in order to review the recognised scientific facts, and expose the fantasies and the exploitation of the tree image to enlist support and generate profits. This chapter analyses these issues and shows that, all in all, the image of the tree is exploited beyond what can reasonably be expected of trees and woodlands. The latest agenda of forestry, i.e. social issues, is the object of the last chapter of this part. Health and education, which are part of this agenda, are presented together because they have a lot in common in terms of objectives and of strategies. As in the previous chapters of this part, it is shown here that economic benefits are not absent from the agenda (for example, in terms of savings in the health sector or better trained youths). Moreover, official programmes, as well as those implemented by charitable organisations, reveal new definitions of health and education and a new conception of the missions of the State in these areas, to which social forestry can contribute.

The third part of the study moves into the realm of perceptions and politics. It explores how policies may differ from people's expectations and what woodlands, old or new, mean to people. The first chapter revolves around the controversial issue of amenity and access. All the policies described in the previous chapters rely for their success on regular contact with woodland, yet this is far from being simple, with the structure of land ownership in England making access/ibility problematic. Moreover, the issue of access/ibility is a culturally rooted one and its treatment today is indicative of social relationships. The discourse on the benefits of urban and community forestry emphasises the importance of people's participation as part of sustainable community building. The next chapter thus explores the theoretical framework behind participation and how it is interpreted by the Community Forests teams. This is supported by fieldwork and other studies showing, first, that woodland represents a minority interest and, secondly, that there remain obstacles to social inclusion, let alone empowerment, in English society.

Last but not least, the final theme explored in this book focuses on questions of conservation and argues that behind what may appear as motivated solely by issues of biodiversity/landscape heritage, there lies in the woodland a wealth of representations and identity markers which contribute to forming and informing people's perceptions of woodlands in the 21st century.

References

Forestry Commission. 2007. *Forestry Statistics 2007. Woodland Areas and Planting*. http://www.forestry.gov.uk/website/forstats2007.nsf/LUContents/4E46614169475C868025735 D00353CC8. Accessed 20 November 2007.

Kennedy, J., M. Dombeck and N. Koch. 1998. Values, beliefs and management of public forests in the Western world at the close of the twentieth century. *Unasylva* 49 (192): 16–26.

O'Brien, E. 2003. Human values and their importance to the development of forestry policy in Britain: a literature review. *Forestry* 76 (1): 3–17.

Chapter 1
Preliminary Chapter: Woodlands as Landscapes of Power

The organisation of space is considered as a reflection of class relations

[Darby 2000: 103]

The rich history of British woodlands is well documented. Yet, since the object of the study is to elicit connections that can then be transposed to other contexts, it is worth taking up the main tenets of forestry and the main cultural traits associated with woodlands in England. This will show how they inform present practices and debates which will be analysed and assessed further on. This introductory chapter will first explore the natural characteristics of woodlands, to throw light on the discourse about the conservation of native and ancient woodlands discussed in the last chapter of the book. Then, it will elicit the use value as well as the cultural values of the woodlands in England, as these frame and feed today's conscious – and at times unconscious – reactions and policies to woodlands.

1.1 British Woodlands, from Nature to Culture

Environmental historians and geographers agree to say that the British Isles were covered with ice until about 10,000 years ago, with a glacial maximum around 18,000 years ago. The glaciers shaped the Isles into their present landforms, which can roughly be divided into two main regions: the Highlands and the Lowlands. The Highland Zone, in the north and the east of the Isles, is dominated by mountains, thin acidic soils and a harsh climate, while the Lowland Zone, with lower altitudes, benefits from a milder climate and more fertile soils.

1.1.1 Natural Woodland

While few trees can have lived in this adverse climate, apart possibly from patchy instances of birch, willow and juniper, they began to advance as the glaciers retreated when the present period, known as the Holocene, began about 10,000

S. Nail, *Forest Policies and Social Change in England.*

years ago. With the warming climate, an open tundra landscape was colonized and the first forests of Britain appeared.

It is now commonly accepted that native tree species are those 'natural colonists' [Brown 1997: 191] which are assumed (as it is a rather complex task to detect which species came naturally and which were introduced by humans) to have migrated from mainland Europe to the British Isles without human aid, in the millennia between the beginning of the Holocene and 5,500 B.C. Around that time, as the sea-level rose, a flat marshy bridge between Lincolnshire and the north of Germany which was the connection with the rest of Europe was severed for good. Therefore, tree species crossing the Channel after this period had to be brought into the British Isles by boat and are not considered as native to Britain.

Palinology (which reveals the history of the spread of different tree species through pollen data), dendrochronology (using the precise counting of tree growth-rings to date trees) and carbon dating have made it easier to visualise the landscape of the British Isles as far back as 10,000 years ago. These methods have thus reduced the part of fantasy and contributed to a more informed debate on the controversial question of native versus exotic which, as we shall see, has been an important issue in the discourse on tree-planting in the late 20th century. With the help of these tools, we can reconstruct a picture of these first forests of Britain at their peak, between 5,500 and 3,000 B.C. They were in all probability composed first of juniper, then birch practically everywhere, pine in the north and the mountains, alder in damp places, ash with hazel, elm and maple on limestone. Lime trees were prevalent on lower-lying fertile soils throughout most of England, while oak dominated the more acidic and less fertile soils. Silver fir, spruce, beech and hornbeam were later additions to the mostly mixed deciduous woodland that grew on land up to at least 800 metres above sea level [Reed 1990: 29–31].

It is worth noting that, apart from a few rare exceptions present in Britain more than 10,000 years ago, these tree species 'are all recent invaders from Europe', as N. Brown rather provocatively puts it [Brown 1997: 193]. On the whole, Britain is considered to be poor in native tree species, at around 35 [Peterken 2001: 31], or as many as 70 if shrubs are included [Rackham 2001: 4]. The analysis of pollens also reveals how clearance of the woods on the chalk downlands created the shallow-soiled open grassland [Bettey 2000: 30].

The hunter-gatherers and fisherfolk of the late upper Palaeolithic and the Mesolithic gave way to farmers around 6,000 B.C., with important consequences for the natural forest. Indeed, human occupation went hand in hand with clearance for tillage, but also with the exploitation of woodland resources. From then on, the destiny of the English woods has been closely associated with that of the inhabitants of the country, who have constantly altered it, hence the debate on the concept of 'naturalness' in the environment. As environmental historian Ian Simmons summarizes,

> to talk of the 'natural environment' in terms of plants and animals is to attempt a reconstruction of some stable interval in the past. But perhaps there have been no periods of equilibrium between climate, soils and biota that could attract the title of '*the* natural environment of Britain', or the last 10,000 years at any rate [Simmons 2001: 1].

Furthermore, the concept of a 'natural environment' disregards 'natural' changes such as climate, for instance, which affect the range of plant species that can grow in a particular area. In other words, there may be no such thing as a natural pattern. Finally, in relation to the species that had not colonised Britain before the Mesolithic, it is impossible to tell which of them might have finally arrived there without human assistance, and how they might have spread. Few ecological certainties can substantiate the vision of 'native' and 'climax' plant communities which would inevitably develop, even less so the 'equilibrium paradigm' which dominated ecology until the 1970s [Steward et al. 1992]. The assumption underlying this paradigm was that, provided no human intervention came to disrupt it, this state of equilibrium might yet be reached if humankind left nature to its own devices.

Recently, the climax theory has been replaced by the belief that succession is chaotic rather than deterministic [Worster 1993]. However, the deterministic interpretation, disregarding the force of history of which disturbance is part and parcel, has often been convoked to this day to justify the planting of new woodlands. For instance, the objectives of the Forestry Authority in the 1990s included the creation of new native woodland in Britain, in order to create 'the kind of climax vegetation that would develop... if succession was allowed to take its full course' [Rodwell and Patterson 1994, in Brown 1997: 193]. This interpretation disregards the fact that history is neither predictable, nor can it be repeated, so that managing nature in order to maintain a *status quo* means at best reinforcing 'an ecological myth' [Brown 1997: 194].

1.1.2 The Beginnings of Cultivation

As it was, human control played an important part in the scenario. Botanist and landscape historian Oliver Rackham claims that half of England was no longer 'wildwood' by the early Iron Age. Hunter gatherers and fisherfolk lived in symbiosis with nature, in the sense that their gathering of mushrooms, fruit and nuts and collecting of honey from wild hives did not exert enough pressure to alter the ecology of the woodland. By the middle of the 5th millennium B.C., however, they were being pushed northwards and replaced by several waves of immigration coming from the east and bringing with them their knowledge of cultural systems. They also exploited the organic produce of the woods with techniques that have continued to be used well into the 19th century. Contrary to what has been asserted concerning the damaging effects of these practices on the extent of the woodlands, it now appears that, far from diminishing the area of woodland, the vital importance of the wood resource ensured that it was managed so as to keep productivity high.

What may have contributed to the misunderstanding of the effects of the management of woodlands is that another phenomenon was taking place simultaneously which did entail the disappearance of woodlands, namely the cultivation of crops and the raising of domestic animals. The function of 'land banks' of the woodlands increased as populations grew: as Ian Simmons puts it, 'agriculture was, then, responsible for hewing a humanised landscape out of a largely forested matrix' [Simmons 2001: 58]. Woodland was cleared from the 4th millennium B.C. onwards, either to be converted permanently into fields, or to be cultivated, then abandoned to return to scrub and tree cover after a while. The natural 'wild-wood' was progressively cleared for agriculture and pastoral farming: trees were felled, woodlands were burnt, pines disappeared from England and Wales and limes were much reduced in number. The extension of farmland was such that by 2,500 B.C., the agricultural lifestyle dominated in most lowland areas, and that by about 500 B.C., more than half the country had ceased to be forest [Rackham 1998: 141]. The clearance of woodland was certainly the greatest environmental change of the period stretching from the Neolithic to the arrival of the Normans in England in 1066.

Agriculture implies preparing the ground for sowing the seeds (first wheat and barley, to which were later added rye and oats), but also protecting the crops from animals, and waiting for the crops to be ready to be harvested. This entails permanent settlement, at least until the harvest, and indeed from the Bronze Age onwards, evidence survives of farming activities in the form of settlement sites, farmsteads and field systems and boundaries. Archaeological evidence points to the fact that the downland had been substantially altered and was intensively used before the Roman conquest in 43 A.D. The Romans contributed a lot to the extension of the agricultural landscape and it is even estimated that during the Roman occupation, 'the downland landscape was more extensively used for arable farming than at any subsequent period until the 20th century' [Bettey 2000: 33]. Extension in agricultural lands at the expense of the woods, as well as improved farming methods, made it possible to increase production to such an extent that it could both feed the army and export foodstuffs to the rest of the Empire. Moreover, the Romans cleared numerous areas of woodland in Wales and on the Scottish borders for military reasons.

The Roman occupation of 360 years took its toll on British forests, but it also meant continued management of the remaining woods in order to yield renewable supplies of fuel for brickmaking and iron-smelting industries, as well as to heat the Roman baths. Paradoxically, the Romans' main contribution to the English landscape was not in the cultivated surfaces, but in the interstices between them, namely the communications system. The Roman road system, parts of which remain to this day, is significant first because the roads cut through the woods, but also because according to landscape historian W. Hoskins, these roads made it easier for the Anglo-Saxon colonists to penetrate into the country [Hoskins 1975: 34].

It is hard to assess the extent of the forests in England when the Romans left in 409 A.D., and that of natural regeneration that took place subsequently. During the 'Dark Ages' (roughly from the late 5th century to 1000), considerable clearing of the woodland, draining of the marshes and reclamation of the heathlands went on. The Angles, the Saxons and the Jutes who arrived from the 5th century A.D. continued clearing the woods, in particular in areas like Middlesex and Warwickshire which still had substantial woodlands, to turn them into broad tracts of cultivated fields and pasture. It is estimated that the rate of woodland decline during the Anglo-Saxon period was only equalled in the 20th century: the Anglo-Saxons aptly termed the ploughman 'the grey-haired enemy of the wood' [Darby 1951: 379]. Clearing the woodlands contributed to the creation of farms and villages which remain to this day, and many place names ending in -ley, -hurst and -holt testify to the existence of woods or clearings in woods. The Saxons were followed by the Norse from the end of the 8th century A.D., who intensified wood clearance and established new village communities and hamlets.

By the time William the Conqueror set foot on the English coast of Sussex in 1066, the total population of the country may have been 1,250,000 people, 10% of whom would have lived in the boroughs. Medieval England was one of the least wooded countries in North-West Europe, with about 15% woodland cover in 1086, but it was irregular, as was population density. Most of the primeval wildwood was gone, except for a small area at the centre of the New Forest. Thick oak and ash woodlands were to be found on clays, and beech woodlands on chalk and limestone uplands, with birch on the hills, while in some areas of the Midlands woodlands would not account for more than 5% [Schama 1995: 142; Simmons 2001: 94; Hoskins 1975: 86].

This brief summary of over 10,000 years might give the impression that landscape change was solely directed at utilitarian use. Yet, other forces were at work, with at least as much influence on the amount, distribution and use of the woodland.

1.2 Woodlands as Spaces of Exclusion

Forest landscapes also operate as emblems of political power, therefore they have been contested arenas throughout history. The political dimension of forest landscapes and its sometimes mythical interpretation is an essential component of today's perception of ancient woodlands which will be analysed later on.

1.2.1 The Forest Law

Royal Forests first appeared in France under the Carolingian Empire in the 9th century. After the Norman Conquest, William I created Royal Forests in England, which were maintained and extended by his successors in the following 500 years. The term 'forest' in this context – from the Latin *foris*, meaning outside – is legal, not botanical. It designates some lands under direct royal control, set aside as deer farms and royal game preserves, and in which Forest Law was superimposed on Common Law. In the words of Manwood in his *Forest Laws* of 1598, a forest was

> a certain territory of woody grounds and fruitful pastures, privileged for wild beasts and foules of forrest, chase and warren, to rest and abide in, in the safe protection of the King for his princely delight and pleasure [Brown 1883: 12].

A set of Forest Laws known as the *Constitutiones de Foresta* or *Charter of the Forest* were later defined at the Assize of Woodstock in 1184, 'the first code of forest law that applied to the kingdom as a whole' [James 1981: 10]. According to it, *vert* (the trees, undergrowth and grass, and their produce) and *venison* (the beasts, essentially the red and fallow deer) were protected by foresters under the supervision a Chief Justice of all the Royal Forests of England. The task was divided into two Chief Justices in 1236, with the river Trent acting as a dividing line between their territories (these two justiceships functioned until they were abolished in 1817) [Grant 1991: 76,88]. Contrary to what their name seems to indicate, the lands classified as Royal Forests were not necessarily the king's property, although they were managed for the benefit of the Crown: some were part of the royal demesne (*boscus dominicus regis*), while others belonged to the clergy and nobility (*boscus baro, boscus priori, boscus mili*) [Serovayskay 2000: 254]. They were not necessarily woodlands, either, but included cultivated land, villages and commons. Waltham Forest in Essex, for instance, covered 60,000 acres, three-quarters of which were ordinary countryside [Coates 2005: 46–47]. Indeed, more than trees, the characteristic feature of Royal Forests was the animals they contained, which were reserved for the king's hunts. The boundaries of these areas of unenclosed land were roads, valleys, streams, banks or ditches. The three characteristics of Royal Forests, concerning legal status, land use and purpose, are to be kept in mind in relation to the new forests of the 21st century that we shall turn to later.

According to *The Rhyme of King William*, a poem included in the *Peterborough Chronicle* of 1087, William I is said to have loved the deer as if he were their father. This benign representation of the king as a – rather ambiguous – father figure hid a more practical political reality. It is undeniable, as king after king clearly realised, that there were advantages to be gained from exerting considerable powers over wide expanses of the national territory. Royal Forests are said to have covered between one fifth and one third of England at their maximum extent under Henry II (1154–1189), a surface far greater than what was strictly needed for hunting.

First of all, independently of who owned the land, the king had control of the hunting, which represented a source of fresh meat for the king and his court wherever they went. Forest Law thus forbade encroachment as well as the damaging – and sometimes felling – of trees. The control of the land also permitted bestowing gifts on those whom the sovereign wished to reward, in a subtle game of inclusion in exchange for services or money [Tsouvalis 2000]. Thus, parts of the forest were given to members of the Royal family or grand magnates, as under Henry III; timber was offered to the Church to build cathedrals; concessions were granted, among which hunting rights, warrants to create assarts (i.e. converting forest land into agricultural land by felling the trees) or liberty to erect forges. The monarch also exercised control over industries within the Forests, such as forges and tanneries. Thus, Royal Forests acted as important sources of revenue for the Crown. Hugh de Neville, Justice of the Forest of King John, thus collected vast sums of money for the king, in the form of Forest amercements (fines for offences against forest law), proceeds of sales and leasing of assarts [Grant 1991: 90; Reed 1990: 123]. Royal Forests turned into a lucrative business in kind and in cash, or occasionally, as Simon Schama puts it, into 'sylvan gangsterism' [Schama 1995: 148].

If financial needs became too desperate, the asset could be sold. Examples of this abound. Richard I, in search of funds for the forthcoming crusade, disafforested – i.e.

removed from Forest Law territory – large areas in return for money, thereby giving an impetus to settlement and agriculture. King John disafforested Dartmoor to its commoners in order to fund his war against France, and so did the Plantagenets. These are but a few instances of a continuous stream of demands made upon the Royal Forests over the centuries to relieve the sovereigns' financial difficulties. A tract of Royal Forest which passed from the Crown into private hands became a chase or a park, depending on the size, and could be enclosed provided royal license was given. Again it was the hunting which was preserved, but this time through Common Law rather than Forest Law and in the interest of the private owner. Over 20 chases and about 2,000 parks are known to have existed in medieval England [Reed 1990: 124].

With the rapid economic development of the Tudor era, demand for timber grew. The dissolution of the monasteries provided Henry VIII with opportunities to sell lands, the timber of which was processed to build ships and forges, etc. It is interesting to note that the forest administration set up by the Normans started to decline precisely at that time [Short 2000: 140]. In spite of the fact that Henry VIII was an ardent follower of the chase, as was his daughter Elizabeth I, both of them sold parts of the Royal Forests in exchange for ready money. This may mean that financial priorities took over from the pleasure of the hunt. It may also signify that their function of 'landscapes of power', emblems of political power, had been superseded – with the help of political stability – by more impersonal management of woodland resources for economic purposes.

Even though few monarchs resisted the lure of woodland revenue by disafforesting Royal Forests, the discourse was nevertheless always one of care and conservation. The history of Royal Forests reads like the swing of a pendulum between selling them off for financial reasons, and trying to reassert/reinforce their existence for political reasons, in contexts of political unrest (under Edward IV for instance) or of restoration of royal power (Henry VII). The Stuart dynasty provides a good illustration of this. James I tried to restore the Forest system as part of his royal prerogative, declaring soon after his accession that he would enforce the Forest Laws 'which were as ancient and authentic as the Great Charter' of 1215 [Grant 1991: 187]. This reference to *Magna Carta* conveniently omitted a reference to *Carta de Foresta* granted by Henry III in 1217 and 1225, which had temporarily brought the territories of the Royal Forests back to their area before king Henry II, a century before, in exchange for the barons' allegiance to the 9-year-old king [Schama 1995: 148]. Charles I also remembered the symbolic association of power with the Royal Forests when he had his portrait painted by Van Dyck under an emblematic oak tree.

But both had to find the means to finance their expenses. James I, in spite of his keenness for hunting, allowed the felling of 1,800 oaks in and around the New Forest in 1611 to provide the Royal Navy with timber [Grant 1991: 187]. Charles I did not hesitate to raise funds by selling off vast tracts of the Forest of Dean to entrepreneurs who cut them down, or to revive Forest eyres (tribunals that heard the most serious cases of attacks on vert and venison) to fill the royal coffers with the produce of fines, in order to finance his disastrous foreign policy [Schama 1995:

156]. After the Restoration, Charles II continued to sell off some of the remaining Royal Forests. It was during his reign that the last Forest Court took place, interestingly enough not to defend the king's rights to the hunt, but to preserve and promote the production of timber. But even before this last instance, Forest courts were found to be a cumbersome procedure and were often replaced by Common Law courts in prosecutions.

Thereafter, the Surveyors-General of Woods and Forests were responsible for enforcing forest policies in the Royal Forests, with little success if one judges from the reports of the Royal Commission on Crown Woods and Forests which reported between 1787 and 1793: 'the general picture was one of neglect and decline' [Grant 1991: 206]. Such a commentary might equally well apply to the royal powers and prerogatives at the time. The commissioners therefore recommended the 'disafforestment, enclosure and 'improvement' of forest wastes' [Grant 1991: 210].

1.2.2 Hunting as an Aristocratic Privilege

The original function of Royal Forests was to serve as deer reserves and hunting grounds for the monarchs and their lessees. Among the occupations of the aristocracy, hunting loomed large as a favourite pastime. Hunting served two main purposes. It helped to display and maintain social order, and it served as a daily practice for war, both in the strategy developed and the handling of weapons. In a nutshell,

> planning the chase was a mental challenge, taking part in it was physically and emotionally satisfying, and the spoils provided delight for the stomach [Cummins 1988: 2].

Although the strategic function of hunting as a practice for war progressively disappeared, the social one remained and the royal hunt acted as 'the most important blood ritual through which the hierarchy of status and honour around the king was ordered' [Schama 1995: 144]. Henry VIII, for instance, who hunted the deer with passion, was the last king to create a Royal Forest, the Forest of the Honour at Hampton Court in 1539. As Royal Forests dwindled, parks became the sites of royal hunts, down to Queen Anne, and including Oliver Cromwell [Rackham 2004: 3].

Following the decline of royal power in the second half of the 17th century, the social ritual of hunting was adapted to suit the needs of the aristocrats living in their country seats, and of the newly rich looking for social status by adopting a rural lifestyle. The warring aspect of hunting and the search for contact with primitive nature were less in keeping with the new landscapes of enclosures, but the hunt however remained a potent status symbol, linked to exclusion and social hierarchy. The 1671 *Game Law* enshrined the principle of private property by making hunting the privilege, no longer of the king, but of gentlemen owning freehold property of over £100. Poaching thus became an act of felony. This law survived until 1831. Many Georgian parks, as well as being landscaped for aesthetic purposes, were still deer-farms where hunting often took place. Kings George I, II and III, for example, hunted at Windsor [Rackham 2004: 4].

New ways of pursuing the sport were being devised, linked to the changes in the landscape and in the social order. On the one hand, fox-hunting took off as a fashionable sport between 1750 and 1800, whereas it had since medieval times been a way of controlling a pest, and was therefore practised by local populations. It grew to 'encapsulate all the virtues of the English: fortitude, physical prowess and liberty' [Finch 2004: 42]. By the end of the 18th century, it eclipsed other forms of hunting as the exclusive one, meant to 'cement' the social relationships of the well-off, as the rates of subscription and other expenses (expensive outfits among them) testify. On the other hand, grouse shooting on the moorlands, and pheasant shooting in the disused coppiced woodlands, also developed as a sport appealing to the newly rich in search of status as well as to landowners in search of income for their unproductive lands.

The development of deer-hunting, fox-hunting and grouse-shooting testifies to the link between landscape and social order. These sports exemplify how 'landscapes of exclusion' were devised in close connection with the political, economic and geographic realities.

1.2.3 Reactions and Controversy

Due to these connections, woodlands are also spaces where beliefs and myths are strongly rooted. In a paradoxical juxtaposition, they act both as hiding-places for outlaws and bandits and as refuges for people in search of justice or identity. For the former, Richard Siward in the early 13th century in the Forest of Windsor, robbers and thieves in the Wirral Forest in the 14th century, The Waltham Blacks or Dick Turpin in Epping Forest in the 17th and 18th centuries all serve as examples. They attract or frighten, acting as emblems of order and disorder. Simon Schama traces the descent of the English 'mythic memory of greenwood freedom' into the 19th-century historical novel, of which *Ivanhoe* is the epitome [Schama 1995: 140]. According to that myth, pre-Norman greenwood was supposed to have been a place of social harmony which was trampled underfoot by the invader and only temporarily restored by the Charter of 1217. The belief passed on from generation to generation that, as the archdeacon of Oxford, Walter Map, put it,

> the Conqueror took away much land from God and men and converted it for the use of wild beast and the sport of his dogs for which he demolished thirty-six churches and exterminated the inhabitants [Fitzgerald 1966: 79].

Alexander Pope suggests the same thing concerning the creation of the Forest of Windsor in his famous poem, *Windsor Forest*, written at the beginning of the 18th century, even though there is no proof that William I depopulated the area at all. So the greenwood memory was passed on and inflated until in the 18th century it was thought that

> the Conqueror and his heirs had been so determined to swathe Old England in woods populated only by boar and by buck that they had gone to the length of planting good arable fields with trees [Schama 1995: 140].

The invader apparently paid for this: the dynasty of the founder of the Forest Law seems to have been cursed for its attempts against 'greenwood liberty'. William the Conqueror's son Richard was gored to death by a stag during a hunt. His other son, Rufus, was shot by an arrow in the New Forest in 1100, and his death was interpreted in popular folklore as a punishment for reneging on his promise to restore the common rights over vert and venison in the Royal Forests outside the royal demesne. The Conqueror's grandson, Henry, died when he became caught up in the branch of a tree. But beyond strict reality, the greenwood myth served as an imaginary escape for the victims of feudal oppression.

Complaints addressed two distinct, but parallel, issues. On the one hand, they echoed the belief that civilisation can only be achieved at the cost of removing the primitive forest. Opposing the advance of this progress meant going against God's laws. On the other hand, the accusations stressed the cruelty of the Norman kings to the local populations. More pragmatically than the actual cruelty of the foreign kings maybe, for the many people involved in making charcoal, stripping bark for tanning, smelting lead, working iron, making glass, collecting wood for fuel, fattening their pigs on acorns and beech mast or collecting berries, Forest Law 'was the hallmark of despotism' [Schama 1995: 144].

The memory which passed by word of mouth on the establishment of Royal Forests and the negative changes they entailed for most of the population have remained powerful. Even to this day, some historians perpetuate ideas which have been proved wrong but which reflect a cultural inheritance, a token of the identity violated by the foreign invader. For instance, some historians stress the supposed forgery around the *Constitutiones de Foresta*. According to them, William I made a forgery in order to demonstrate that this document was drawn, not by himself but by King Canute at Winchester in 1016 [James 1981: 9]. This allegedly allowed William to claim continuity with his predecessors and undermine resistance to the new regime. By insisting on the forgery rather than on the claim to continuity, the argument misses the point. Indeed, Canute was himself a foreigner, and continuity with the Danish king could hardly serve the Norman's interests, or even be counterproductive.

On the whole historians agree to say that the English kings before the Normans had already claimed hunting-rights on tracts of country set aside as forests, and that a lot of William's Royal Forests were situated on older, Anglo-Saxon, royal hunting-grounds. Moreover, under Edward the Confessor (1042–1066), there already existed an embryo of Forest administration. The two main changes, then, seem to lie in the introduction of Forest Law and in the fact that the Normans' Royal Forests could include lands outside the royal demesne, which was the main bone of contention. Indeed, these lands were sometimes held back from cultivation and strongly devalued as a result. Furthermore, in urban areas close to forests (Nottingham to Sherwood Forest or Lancaster to Lancaster Forest, for example), access to woodland resources was also denied to the inhabitants of these towns [Short 2000: 135].

Unsurprisingly, complaints abounded: *The Anglo-Saxon Chronicle* [Whitelock 1965] or forestry statutes such as Sir John Manwood's *Treatise of the Laws of the Forest* [Manwood 1598] bear witness to this. Depredations, even confrontations between poachers and foresters, were numerous in spite of threats and repression, under Henry III for example. Various local studies detail these offences, sometimes taking the form of 'organised sabotage' [Serovayskay 2000], other times of attacks against the forest officials, in particular the foresters in charge of preserving vert and venison [Freeman 1996]. However, the accounts of forest eyres provide a different point of view. The punishments inflicted upon trespassers (from fines to mutilation and death) denounced in the grievances against Royal Forests, in reality often boiled down to fines, which was more in line with the state of the sovereigns' coffers. When they were harsher, the penalties were comparable to those for similar property crimes outside the Forest. Corporal punishment stopped after 1217, and even before that, there are good reasons to think that fines were by far the most frequent form of punishment.

The most virulent hostility to Royal Forests came from the nobility and the Church, their frustration stemming from the interdiction to hunt on their lands and the damage occasioned by the deer. These grievances led to the *Carta de Foresta* of 1217 and 1225 under Henry III, but the generosity of the king was short-lived and he soon went back on his promise. A number of tales, from medieval times onwards, take up the grievances of the victims of persecution in Royal Forests, one of them being *The Outlaw's Song* of 1305 [Hinde 1985: 221]. But the most famous, of course, concerns Robin Hood. It is not clear who the original model for Robin Hood was, but according to recent historians, he seems to have lived in the 1260s–1280s, might have been a supporter of Simon de Montfort and might have taken to the woods as a result of the latter's defeat. William Langland refers to him in his *Piers Plowman* in 1377, but the fact that the first edition of the *Lytell Geste of Robyn Hode* appeared at the end of the 15th century, during or just after the War of the Roses makes his story 'a product of a time of usurpation and chronic rebellion' [Schama 1995: 149], and his character

> a passionate and nostalgic conservative who yearns for the restoration of a just, personal monarchy and who wants a social order dislocated by rogues and parvenus to be set right in its proper ranks, stations and portions [Schama 1995: 150].

Royal Forests were also a source of grievances during the build-up to the Civil War. Charles I's 'Personal Rule' triggered off widespread discontent with regard to forests, the motives of which are echoed in *The Grand Remonstrance* in 1641:

> 21. The enlargements of forests, contrary to *Carta de Foresta*, and the composition thereupon. [...]
> 25. The general destruction of the King's timber, especially that in the Forest of Deane, sold to Papists, which was the best store-house of this kingdom for the maintenance of our shipping.

The plaintiffs demanded that

> 134. The forests [be] by a good law reduced to their right bounds [Gardiner n.d.].

It is probably as much the symbolic as the material space of the Royal Forests that the common people invaded when these landscapes of exclusion were abolished during the Civil War and Forest Law fell into abeyance. Manifestations of discontent did not stop, however. The disafforestation of the last Royal Forests went together with the enclosure of the lands divided between lords of the manor and commoners. These enclosures deprived a lot of local people of access to the woods to collect firewood or to use the pasture for their sheep and pigs, and many riots are documented, from Feckenham Forest (Worcestershire) in 1631–1632 to the Forest of Dean in 1831, among others [Large 2002]. Furthermore, the 18th century game laws provoked resentment by reserving the right to hunt to landowning aristocrats.

The epilogue to the Royal Forests created by the Normans as part of their prerogatives came after nine centuries of disputed policies and at times bitter conflicts. In 1923, the *Forestry (Transfer of Woods) Act* transferred their property to the newly-created Forestry Commission, and in 1971, the *Wild Creatures and Forest Laws Act* abolished the sovereign's rights over wild creatures and revoked the Forest law.

A study of landscape preservation campaigns at the turn of the 20th century, devotes a passage to the New Forest in the 1890s, and the arguments used are worth noting, all the more so as the New Forest represents the Royal Forest *par excellence*. They refer mostly, not to the landscape or amenity value of the site, but to the historical and heritage value of the Forest, stating that the New Forest provided a glimpse of 'the England that was and ceased to be', the 'England of the outlaw, or the singer of ballads, of the lover of the greenwood life' [Readman 2001: 71]. This vision of the New Forest as a 'national inheritance', providing a 'connection with the Saxon origins of modern England', ignores the history of the contested creation of the Royal Forest by William the Conqueror in 1079 and all the ensuing conflicts. Several reasons can be given for this surprising reading of the New Forest as a landscape emblematic of national English identity. First, a small part of the New Forest is admittedly a remnant of the primeval wildwood, so a reference to a distant national past. Secondly, it can hark back to the image inherited from the time of Henry VIII, when Royal Forests stopped being the portentous landscapes of power they had been in the previous dynasties, to become in the Robin Hood tales of the time the background to a 'chivalrous, merry greenwood England' [Schama 1995: 153]. Furthermore, a Whig interpretation of history 'tracing the happy and largely unbroken continuities of the ancient constitution and (seeing) the present as a natural, organic outgrowth of the past' [Readman 2001: 73] contributes to making of the New Forest a landscape of national inclusion far from its real history. Finally, and maybe decisively, the New Forest by the end of the 19th century was no longer a place of production and had become a place of recreation for a largely urban population; the stakes were therefore much less contentious. The case of the New Forest provides an instance of the mythical discourse on forests, which in turn provides the cultural backdrop to the discourse on ancient woodlands at the beginning of the 21st century.

1.3 The Economic Balance Between Agriculture and Forestry

French historian Jacques Le Goff stresses the fact that in the early Middle Ages, depending on what culture one deals with, the desert, the mountain or the forest was perceived as the negative side of civilisation and social order. In European history, agriculture and forestry often represent these antagonistic poles, with trees the enemies of grass and plants in terms of land use [Delort and Walter 2001: 273]. This dichotomy goes back to when Neolithic man started to cultivate the land. Forests were a negative sign in the landscape, lands of darkness and hostility, and legends of the Middle Ages focused on the conversion of forests into fields [Bechmann 1990: 289]. In fact, European landscape history often reads like a pendulum between agriculture and forestry, down to the 21st century, as will be shown further on.

1.3.1 Woodlands as Residual Land Use

Between 550 and 1700 A.D., agriculture underwent a dramatic change. In the 12th and 13th centuries, in a context of demographic growth, 'land hunger' more often than not led to the conversion of more land for cultivation. Arable land and pasture were extended at the expense of woodlands, heaths, marshes and fens. The wetlands of Somerset Levels, Pevensey and Romney Marsh, as well as parts of the Fens of East Anglia, were reclaimed for cultivation in these centuries. However, the most readily available source of land for conversion to arable was the woodlands, which could easily be cleared by axe or fire. Furthermore, once the roots had been grubbed out, they provided good soil for cultivation. The progress of mankind was thus seen as a passage from the forest to the field [Thomas 1984: 195]. The openfield system was established by individual peasants and by monasteries in many of the lower slopes on the downs by the 12th century and became the dominant cultivation system before the enclosures overtook it. This contributed a lot to the changes in the medieval landscape.

Aesthetic preferences reinforced the choice for the extension of agricultural land. For the Cistercian monks who cleared the Yorkshire Dales for sheep-farming in the 12th century, aesthetics and use went hand in hand: a beautiful landscape was before anything else a fertile one [Muir 2005: 138]. Gregory King, often dubbed the first great economic statistician, calculated at the end of the 17th century that out of 39 million acres of land in the country, *only* about half were under cultivation, while there were still ten million acres of heaths, mountains and barren lands, plus another three million of forests, parks and commons [Thomas 1984: 255]. Keith Thomas deems the latter figures an exaggeration, but they testify to a state of mind. A cultivated landscape was also the epitome of beauty for the 'improvers' of the 18th century. Urban Mrs Elizabeth Carter, writing in 1769, thought that the country was 'disgraced by … tracts of uncultivated land', and Arthur Young, writing on agriculture and economics at the end of the 18th century, dreamed of bringing 'the waste lands of the kingdom into culture' and to 'cover them with turnips, corn and clover'.

Gardener Samuel Collins, summarizing the current trend of thought, declared in 1717 that 'the best of all flowers was the cauliflower' [Thomas 1984: 256].

This desire to see the land cultivated can be put down to the cultural background mentioned before, but it was also linked to the religious context: domination over nature, from the Greek Stoics onwards, was justified by the fact that humans possessed reason and a soul. The protestant ethos, after the Reformation of Henry VIII, associated work (and its visible results) with holiness. According to Francis Bacon (1561–1626), through work, the arts and sciences, control over nature could be reached and the effects of the Fall reduced [Simmons 2001: 118–119]. Woodlands, like mountains until the mid-17th century, were perceived as obstacles to human progress, as tracts of land yet to be socialised, so the battle against nature had to be waged again and again. This was all the more important as, even after centuries of struggle, it was far from being won, and every halt in the conquest had significant consequences. Indeed, the fact that cultivated land had reverted to scrub after the Romans' departure was but a foretaste of the more dramatic consequences of the plague that swept across Europe in the mid-14th century. It is estimated that on the eve of the Black Death of 1348, there may have been four million people in England, about three times the population of 1086, and vast quantities of land had been won for cultivation and pasture between these two dates. And yet the bubonic plague which smote the country from the summer of 1348 and claimed the lives of between a quarter and half the population of England over 30 years, had tremendous environmental repercussions as well as human ones. With the decrease in population, the cultivated area shrank as a result of the surrender of marginal lands, and between 1,300 and 1,400 villages were deserted. Landlords had to look for tenants instead of the contrary, and as a consequence, they often converted the arable strips of the openfields into less labour-intensive enclosed pasture for cattle and sheep, especially in the Midlands. Demographic recovery at the beginning of the 16th century led to land hunger by the end of the century, which stimulated renewed interest in marginal lands (North York Moors, Dartmoor) for the first time in 200 years [Simmons 2001: 75].

Thus, when the Elizabethans spoke of a 'wilderness' or a 'desert' they meant, not a barren waste, but a dense, uncultivated wood, as appears in Shakespeare's Forest of Arden in *As You Like It* (Act II, scene 1):

> […] this desert inaccessible,
> Under the shade of melancholy boughs.

The woods were homes for animals, not for men. The poet William Browne could describe wild beasts as 'forest citizens'. John Locke (1632–1704) contrasted the 'civil and rational' inhabitants of cities with the 'irrational untaught' denizens of woods and forests. Literary convention as well as actual experience thus informed the 16th and 17th century commonplace that forest-dwellers tended to be lawless squatters, poverty-stricken, stubborn and uncivil. Of course everyone agreed that timber reserves were needed and that wood was wanted for fuel and other purposes. But it should be grown on inferior land as a crop, in coppices or high timber woods which were regularly cut and harvested. For woodland of any other kind there was no room.

The fluctuations between agricultural prices and the prices of timber provide another example of the superiority of agricultural land over timber in the landowners' minds. Examples are numerous of periods of plantations of trees at times of low agricultural prices. The most famous of them occurred between 1660 and 1760. The plantations undoubtedly had patriotic connotations as well as aesthetic ones (we shall come back to them later), but they were far from being devoid of economic advantages. Indeed the prices of wool and corn had fallen at the end of the 17th century [Miles 1967: 31–32], while the price of oak timber rose from 1690 to 1800 [Seymour 1998: 117]. Conversely, when the price of corn rose again above the price of oak timber between 1790 and 1812, in the context of the Napoleonic wars, the lands were reclaimed for arable and the trees cut down [Daniels 1988: 48]. This alternation between agriculture and forestry was rather shamelessly summed up by a Restoration peer who reputedly said that wood was 'an excrescence of the earth provided by God for the payment of debts' [Miles 1967: 26]. This comment disregarded the fact that, on top of being land banks, woodlands were a fundamental part of the economy on account of the numerous uses they could be put to.

1.3.2 The Uses of Wood and Timber

From times immemorial, tall trees (standards) have been used for timber, mostly for building purposes, while the underwood (shrubs and small trees) has been used for fuel (firewood and charcoal) and managed so as to obtain poles of various lengths for different uses. Oliver Rackham classifies traditional land-use involving trees into four main categories: first woodland, that is, 'areas of naturally-occurring trees which have been managed'; secondly, wood-pasture; thirdly non-woodland trees, such as hedges or free-standing trees in fields; and finally plantations, 'usually unrelated to the natural vegetation' [Rackham 1998: 139]. He adds that the first three of these land-uses are ancient, while plantations appeared in the 16th century, but developed mostly in the 19th century. In early topographical writing, the word 'woodland' sometimes refers to one or the other of these land-uses [Dyer 2000: 97–98]. To these should be added the use of trees as boundary markers.

As the term wood-pasture suggests, woodland around villages served as grazing areas for animals, providing oak mast and beech mast for pigs ('pannage') and 'herbage' for other animals. Then, timber was essential for building houses, churches and cathedrals, bridges and castles. It is estimated that between them, the roofs and walls of the Old Court of Corpus Christi College in Cambridge (14th century), contained about 1,400 oaks [Rackham 2001: 74–75]. Wood was an essential resource for domestic fuel and also served to repair constructions and to build wagons, carts and ploughs. These uses were recognised in the granting of rights to the local population, respectively called firebote, haybote and housebote. Poles, branches and rods resulting from coppicing and pollarding the trees were put to all sorts of uses, from handles to hurdles and from furniture to clogs. A lot of trees also provided medicines and were renowned for their therapeutic effects. During the

Fig. 1.1 As many species of trees, willow coppices, sending up shoots from the stump, which can be cut down at regular intervals to be put to all sorts of domestic uses (Photo Sylvie Nail)

Middle Ages, wood pasture was progressively replaced by coppiced woodland, which offered the advantage of producing more timber. On the other hand, these areas could not accommodate grazing animals, who would have eaten the young shoots springing from the stools.

Apart from these domestic uses, wood and timber played a fundamental part in the development of industry: furnaces, forges, stud-farms, were among examples of proto industries. Industries using wood or charcoal were essential in pre-industrial Britain. As the Romans had used wood to heat their baths, make bricks and smelt and forge iron, tanning and glassmaking among many other industries required a constant supply of wood, and mines (lead, copper and tin) needed pit-props. Finally, shipbuilding developed, stimulated by demographic growth and by the development of the fishing industry and commerce. No doubt the dissolution of the monasteries under Henry VIII and the willingness of gentlemen to turn into entrepreneurs helped to develop industries with the aid of wood and charcoal.

In the meantime, concern was growing over the disappearance of woods from grazing, cutting and converting into arable.

1.3.3 The Beginnings of Plantations

Natural woodlands have been used and managed since the Neolithic period in order to provide for the population's various needs. For a long time, regeneration was natural, not the result of planting. Young shoots had to be protected from grazing animals if regeneration was to take place, but measures were not always taken. Hence, as a result of the growing domestic and industrial needs over the centuries, woodland shrank progressively. Both Henry III and Edward I issued royal edicts against certain forms of timber sales in 1258 and 1290 respectively. An Act was then voted in 1482 to allow enclosure for 7 years in order to allow natural regeneration after a wood had been felled, and the office of Purveyor of the King's Timber was created. But the turning-point came in the mid-16th century.

Henry VIII, a keen supporter of the expansion of the navy, took a deep interest in knowing what the woodland asset was, with a view to preserving it as a source of timber for the shipyards. Under growing pressure on reserves and as a reaction to the wasteful felling of trees when coppice could have been used, the first *Timber Preservation Act* was passed in 1543, also known as the *Statute of Woods*. It stated that when a wood was cut, 12 timber trees were to be left to the acre (0.4 hectare), and that the coppices were to be enclosed after cutting in order to protect new growth from grazing. The novelty was that it was a compulsory Act as opposed to the permissive one of 1482. In 1547, two Masters and two Surveyors of the Woods were appointed, one for either side of the Trent. In 1558, under Elizabeth I, an Act was passed forbidding the use of timber in the iron industry 'for the avoiding of destruction and wasting of timber', and protecting ship timber within 22 kilometres of navigable waterways [James 1981: 127]. By the late 16th century, the idea of plantations, originating on the continent, had taken hold in England. The earliest English oak plantation, 13 acres planted with acorns, is said to have been made by Burghley near Cranbourne Lodge in Windsor Great Park during 1580. Soon, it became common for conifers to be used. From the 17th century onwards, periods of neglect alternated with periods of plantation. James I took a great interest in the forests and tried to encourage planting, but Charles I did not. In the second half of the 17th century, fear for the supplies of wood and timber occasioned by the excesses of the Civil War, as well as by the enclosures of agricultural lands, led to great concern. What the wooded landscape looked like when Charles II was restored to the throne in 1660 is a moot point. Some argue that it was 'a scene of unparalleled desolation' [Schama 1995: 159], while others argue that the situation was made to look worse than it really was, in order to urge the king to act [Williamson 2000: 114]. In 1688, Gregory King estimated that there were three million acres of woodland in England, compared to at least four at the beginning of the 16th century. On Charles II's accession there were still 68 Royal Forests, in a bad state of neglect but with fine timber in them [Miles 1967: 27]. The situation was supposed to be serious enough to justify action, but awareness did not necessarily lead to more rapid and efficient intervention than in the previous century. Only one measure was significant, namely the passing of an *Act for the Increase and Preservation of Timber within the Forest of Dean* in 1668. Thus, 11,000 acres were enclosed for planting, these being considered as the first plantations ever made by an English government. In the New Forest, 6,000 acres were enclosed in 1698 with a view to planting, following the Great Fire of London and the Dutch Wars, but that was not followed up efficiently enough to improve the situation significantly.

The New Forest was particularly exposed to depletions for the navy, since 'it lay virtually on the doorstep of the royal dockyards at Portsmouth' [Muir 2005: 163]. Seventeenth-century improvers' works calculated costs and returns in order to convince their readers of the interest of planting for the Navy, but this was not sufficient to reverse the trend, nor were declarations such as Captain John Smith's in 1670 convincing enough:

> there was a time when England had been overgrown with woods and it had been beneficial to grub them up. But that time was past [Thomas 1984: 198].

Plantations began under the last Stuarts. An Act of 1756 noted that previous attempts by kings, Henry VIII, Charles II and William III in particular, to encourage the production of timber had failed. It was left to the 18th century to really improve the state of the forests, for the needs of the mercantile nation that Britain was becoming. Indeed,

> the Enlightenment produced the theory that the natural and proper state of trees was growing close together in forests of tall trees, suitable for European ideas of good timber quality. This led to a separation of forestry from farmland and the loss of intermediate categories [Rackham 2000: 12].

Leading the way to this positivist conception of forestry was Germany. There, the organisation of rational silviculture was assisted by mathematics and geometry and the first forestry schools were created by Georg Ludwig Hartig in Hunden (Hesse) in 1789 and by Johann Heinrich Cotta in Tharandt (Saxony) in 1816 [Ciancio and Nocentini 2000: 47–49]. From Germany, the concept of scientific forest management was exported to many parts of the world, including to the British Empire. Contrary to the traditional management of woodlands, plantations mean that successor types (oak for example) are planted among pioneer species (such as birch, holly or, later, conifers) which give them shelter until they are no longer needed. The trees are kept close together so that the branches grow tall and straight, the weakest of the trees are thinned out and the best are allowed to reach full height and are cut down in one crop when they reach maturity. This new landscape involved regular forms: wood plantations, ruled Walter Blith in 1653, could be square, triangular, rectangular, oval or circular; but they should not be made 'rudely or confusedly'; similarly, a straight hedgerow was 'a thing of delight' [Thomas 1984: 256].

After centuries of unthinking exploitation, afforestation in the modern sense of establishing trees on land not previously forested, had started. Two different kinds of forestry coexisted henceforth: first, the traditional one, made up of coppice woodland and coppice with standards; and secondly, plantations. But the word 'forest' itself progressively became a synonym for plantations. This change of meaning from the legal Norman term bears witness to the change of role of wooded spaces and the difference in their perception, from spaces of power to spaces of production. On top of the past political and economic dimensions of woodland which have relevance to this day, albeit in a different guise, another historical reading gives important clues to 21st-century perceptions of English woodland, which concerns social and national identity.

1.4 Wooden Walls and Hearts of Oak

However unfounded the claim, some argue that the land of the British Isles would have been covered by oak forest had nature not been tampered with. It is important to read history in retrospect to understand where this discourse comes from. By so doing, we shall see the process of social construction of reality in action, and lay the bases for the perception of identity in forests today.

1.4.1 Political Identity

English forest history keeps the name of John Evelyn and his famous *Silva* as a turning-point [Evelyn 1786]. John Evelyn, 'scholar and court official' [Miles 1967: 25] was invited in October 1662 to deliver a conference to the Royal Society. It became his famous *Silva, or A Discourse of Forest Trees and the Propagation of Timber in His Majesty's Dominions*, first published in 1664, which ran into five editions until 1729 and was republished with annotations by Dr A. Hunter in 1776. That version was itself published five times, the last one in 1825. *Silva* was both a learned treaty on the management of trees and an instrument of propaganda for afforestation. However, its influence has been overestimated, not least by himself, since he boasted in the preface to the third edition in 1679 that thanks to its publication, 'many millions of timber-trees, besides infinite others, have been propagated and planted... at the instigation and sole direction of this work' [Miles 1967: 26]. It is estimated that in fact not until the last edition of Evelyn's book did mentalities change with regard to trees, in the context mentioned above. *Silva* was directly addressed to the king, and its clever presentation of history was meant to create a link between the republican politics of the Commonwealth, the disintegration of monarchy, and the dilapidation of the Royal Forests. What was implied was that the improvement of Royal Forests should be part and parcel of the Restoration of monarchy. Maybe this association between royalty and forests was the most important part of Evelyn's work, as it contributed greatly to a long-lasting association between monarchy and trees, the oak in particular.

A tradition of political imagery already existed connecting monarchs with oak trees, but it was diffuse. Edward II was said to have held his parliament in 1290 under an oak tree, and Queen Elizabeth I supposedly learnt that she was queen under an oak tree at Hatfield. In the Restoration context, conveying to the king the importance of his role as the royal carer of the forests was not too difficult for two reasons. First, the invasion of Royal Forests represented the breakdown of the previous social order, so putting the onus on the king to regenerate the Royal Forests was a way of visually restoring the monarchy, hence a way of turning the page of history. Secondly, there was a specific association between Charles II and oak trees. It was during the reign of his father, Charles I, thanks to the aforementioned portrait by Van Dyck, that the oak tree became the enduring symbol of the power of the monarchy, power that supposedly went hand in hand with the protection of the people and their 'greenwood liberties'. On top of being heir to his father's emblem, Charles II also had personal reasons for being associated with the oak. Indeed, he had hidden in an oak tree at Boscobel after the defeat of his army at Worcester in September1651 [Stamper 2002]. Allegedly, an oak tree had saved the king's life, therefore trees must be on the royalist side. At least, that was the gist of the narratives in the Restoration, such as Isaac Fuller's painting, *Charles II and Colonel Careless in the Royal Oak* (1660–1670) or Thomas Blount's *Boscobel, or the History of His Sacred Majesty's Most Miraculous Preservation*, published in 1660. These attracted a lot of visitors to the site of this landmark, to the detriment of the tree which had to be walled off to protect it from fetishist collectors. So it was easy by that twist of history to blame the Republicans for the spoliation of the woodland resource and to forget how the previous kings had dealt with their forest resources. A famous print, entitled 'The Royal Oake of Brittayne', shows Oliver Cromwell attending the cutting of an oak tree, emblem of the British Constitution, visually associating the disappearance of the woodland with that of the monarchy and the advent of anarchy.

THE BRAVE OLD OAK.

Fig. 1.2 Anonymous woodcut engraving, c.1830, showing crowned King Charles II amidst the foliage of the Boscobel Oak: a variation on the theme of political symbology in relation to oak trees (From document SVII.113g, copyright William Salt Library, Stafford)

The subtle change that occurred in the process is that trees, not beasts, became the focus of attention, in connection with the need for timber and the association of Charles II with the oak. Beasts became pests from which the trees had to be protected [Tsouvalis 2000: 298]. This ideological association of trees with the power of the monarchy served the king well and indeed he encouraged tree planting and re-established the Forest Laws. But discourse obviously went further than action, and while he urged tree planting and posed as the patron of the greenwood, he also let entrepreneurs like Sir John Winter strip the Forest of Dean of its oak trees in 1668. This double dealing of monarchs in relation to Royal Forests was not new, as was shown above. What the situation at the Restoration actually shows is that things had in fact not changed, or if they had, it was in speech only. But the conjuncture served the myth well between 1660 and 1760.

1.4.2 The Oak and the Navy

Indeed, it was no coincidence that John Evelyn's lecture was entitled 'Upon Occasion of certain Quaeries propounded to that Illustrious Assembly, by the Honourable the Principal Officers and Commissioners of the Navy'. One of the most preoccupying facts with regard to timber was the defence of the nation through its navy. Shipbuilding was a great consumer of timber: it took 1,000 oaks to build a ship the size of Nelson's *Victory*, and an estimated 2,000 oaks to build a 'third-rate' 74 gun warship [Hinde 1985: 15]. More than 40,000 cubic metres of oak are said to have been used annually by the six main British shipyards between 1730 and 1789 [Williamson 1995: 124]. During the Restoration, the supplies of hardwood for naval purposes in the Royal Forests centred on the Forest of Dean (Gloucestershire), and the New Forest and Alice Holt Forest (Hampshire). However, the Royal Forests almost never provided more than a tenth of the timber used for naval purposes, the rest coming from private estates and from imports; it was therefore essential to stimulate oak plantations by private landowners for defence purposes, especially in sites which were accessible for transport to the shipyards on the Thames. Even though it took more than oak to build a ship, the elm and beech (required to build hulls) were not as much part of the patriotic discourse, even less so the pine and spruce (necessary for masts and spars). The patriotic association of forests, and the idea of the threat to the country if they were not maintained in a good state of management, were not new. As John Evelyn made sure to stress in *Silva*, the Forest of Dean had reputedly been targeted by the Spanish Armada in 1588, to undermine English resistance:

> I have heard that in the great expedition of 1588 it was expressly enjoined the Spanish commanders of that signal Armada that if, when landed, they should not subdue the nation, and make good their conquest, they should yet be sure not to leave a tree standing in the Forest of Dean [Nisbet 1906: 447].

This version became adorned with more details as time went by. In 1875, Menzies described the episode thus:

> When the Spanish Armada was wrecked, in one of the ships, stranded on the coast of Devon, was found a paper of instructions to the admiral, telling him, above all things, to cut down and destroy the Forest of Dean, so that the navy of England might be effectually crippled. This paper fell into Lord Burghley's hands, and alarmed him and the whole nation. Among the projects submitted to him for providing timber for the navy in some inland locality, which could not be so easily reached as Dean Forest by an enemy, was a scheme to enclose thirteen acres in a corner of Cranbourne Chase, in Windsor Forest, and sow it with acorns as an experiment. Lord Burghley stretched his powers over the forest, so far as to get this done [Nisbet 1906: 446–447].

This interpretation of the first plantation speaks volumes of the associations of plantations with the salvation of the nation. It had antecedents under Charles I, whose opponents, as was said before, had accused him in the *Grand Remonstrance* of disposing of the Royal Forests in order to make the conquest of the country easier for the French, his allies. The great storm of 1703, which destroyed many ancient oaks, also fuelled the fear of penury. The needs for timber became even more pressing as the wars with France developed in the late 18th and early 19th centuries. The French were quicker at managing their woodland resources, especially since the famous *Ordonnance de Saint Germain en Laye*, promoted by Colbert in 1669. So the anecdote of Admiral Collingwood surreptitiously dropping acorns in his hosts' parks when he went ashore, so as to favour the growth of oak, is representative of a state of mind which Joseph Addison makes explicit when he talks of tree-planting in *The Spectator* of the 20 of August 1714:

> But I do not only recommend this Art to Men of Estates as a pleasing amusement, but it is a kind of Virtuous Employment, and may therefore be inculcated by moral Motives; particularly from the love which we ought to have for our Country, and the Regard which we ought to bear to our posterity. As for the first, I need only mention what is frequently observed by others, that the Increase of Forest-Trees does by no means bear a Proportion to the Destruction of them, insomuch that in a few Ages the Nation may be at a loss to supply it self with Timber sufficient for the Fleet of *England* [Gandrillon-Purcell 2003: 52].

New plantations were named after famous British admirals. The frontispiece to *The Modern Druid* by James Wheeler, published in 1747, bears witness to the patriotic connotations of oak planting: Britannia, holding an oak seedling, sits under an oak bearing a good crop of acorns, while in the background ships of the fleet are visible. The motto above the oak can be translated as 'The Glory and Protection of Britain'. The patriotic discourse on trees and the navy was not necessarily anchored in reality:

> in fact the navy never required more than a fraction of the oak taken by the tanning industry, and even the merchant navy, a far larger consumer of oak than the navy, probably took less than did tanning [Hinde 1985: 15].

Moreover, at the end of the 18th century in particular, oak and woodlands were far from being the only preoccupation of landowners. It was all the more important to praise the planting of trees as a lot of woodlands were grubbed up to sow wheat on arable land in the years 1790–1812, when agricultural prices rose. During the Napoleonic Wars, 'the political iconography' of corn-growing was quite strong in painting and poetry: 'a field of wheat proved as patriotic a spectacle as a grove of oaks and for some writers eclipsed the virtues of woodland' [Daniels 1988: 48]. Thus, it may have been felt that the landowners who resisted the lure of immediate profit in order to ensure more profit for future generations needed praising and encouraging. At least until the mid-18th century, vindicating the patriotic value of tree-planting could almost be seen as wishful thinking, or it could endeavour to give a guilty conscience to those landowners who did not opt for tree planting. The frequent occurrence of the word 'profit' in many book titles concerned with planting and managing trees in the 18th century points to economic benefits more than to patriotic ones.

Under the Tudors, especially after the split with the Church of Rome, the sovereigns began to have imperial aspirations for England. This of course called for the development of industry and commerce, which required wood and timber. Shipbuilding thus became associated with the empire not only in its political, but also in its economic sense of building ships both to engage in trade and to protect the merchant ships. This is what the Earl of Coventry wanted to convey when he coined the expression 'wooden walls' to speak of the protection of the country by ships in a speech in 1635. Larger ships began to be built to fill this double purpose. But again, it was mostly in the 18th century that the landowners' efforts to protect and make the country prosper, by planting trees, were endorsed and rewarded. From 1758 to 1835, the Royal Society for the Encouragement of Arts, Manufactures and Commerce awarded prizes to landowners who had planted great amounts of acorns and trees. Among the trees planted other than oak were alder, plane, elm and Lombardy poplar [James 1981: 168]. The efforts were beginning to bear fruit at the beginning of the 19th century: it is estimated that by 1835, well over 50 million trees had been planted.

The multifarious benefits of tree, and in particular of oak planting, are epitomised in an illustration published in a book by Percival Lewis on the New Forest:

> The shield and arms [hanging from an oak tree] indicate that it is a Royal Forest; timber production is shown by the timber feller with his axe and the felled trees; hunting is implied by the deer and two hounds while ship building is represented by the ship under construction [James 1981: 178].

In that very New Forest, changing times and needs were taking their toll on the deer whose very presence lay behind the creation of the Forest by William the Conqueror. In order to implement planting programmes with oak for the navy and pine for pit props in the first half of the 19th century, the *Deer Removal Act* was passed in 1851, ordering their destruction.

1.4.3 Hardwoods and Social Status

Although timber preservation had been, on and off, the order of the day since the 16th century, the publication of books testifies to the rapid rise of interest in woodland management and timber production in the second half of the 18th century.

Whereas only seven books focusing on 'woodlands and tree-growing' were published between 1700 and 1750, no less than 27 were in the second half of the 18th century [James 1981: 168].

For economic and cultural reasons, tree planting became a true fashion in the 18th century among landowners. As the power of the monarchy declined, and feudalism gave way to capitalism, so planting trees for national purposes was symbolically taken over by the landed aristocracy, who had obtained the political reins of the country and could muster the resources necessary to implement capitalist farming on their estates. This can explain the apparent paradox of trees being deemed emblems of loyalty to the Stuarts at the Restoration, as well as to the Hanovers during the period of threat of Jacobite rebellion in 1744–1746. The defence of patriotic values now lay with the members of the landed aristocracy – 'the great oaks which shade the country' in Edmund Burke's words – not the monarch.

Hitherto, private lands had been managed more efficiently than Royal Forests, unhindered by the limitations of Forest Laws. Furthermore, the economic conditions were particularly favourable to tree-planting between 1670 and 1760, so that it was not too difficult for landowners to appear patriotic. Planting trees in vast numbers was the result of the political stability that permitted long-term investments, and because trees take a long time to grow, this in turn generated a wish to protect the growing trees from profligate inheritors. Therefore, trees were protected by laws of strict settlement and tenants for life 'could be prevented from felling timber trees (usually oak, ash and elm) and constrained also from felling trees planted for shelter and amenity' [Daniels 1988: 44]. The *Black Act* of 1723, on top of punishing poaching,

> made it a capital offence to 'cut down or otherwise destroy any trees planted in any avenue, or growing in any garden, orchard or plantation, for ornament, shelter or profit' [Daniels 1988: 44].

Similarly, in 1766, an Act made wood gathering from plantations, copses and hedges a criminal act. Beyond trees and game, what was at stake in these laws was the Whig assertion of exclusive property rights, an emblem of 'class warfare' [Thompson 1975: 191].

The discourse since the Restoration had been powerful enough, given the proper circumstances, to be taken on board and converted into action. Tree-planting was connected to loyalty to the monarchy, it was also linked with Conservative policies and the analogy was obvious between sound politics and the good management of an estate. Finally, it acted as an emblem of aristocratic families. The oak claimed the richest metaphors of all trees. The adjectives applying to the oak could equally well qualify the typical landed family: 'venerable, patriarchal, stately, guardian and quintessentially English' [Daniels 1988: 48]. Vice versa, planting trees became emblematic of a gentlemen's estate, hence of power and status. The symbolism of the oak was progressively transferred from a national level to an individual one, the power of the country reflecting the worth of its families of high lineage, the 'hearts of oak' celebrated by Garrick in 1759, 'the perfect image - according to the poet Shenstone - in all respects of the manly character: the British one' [Thomas 1984: 220]. The associations of trees with strength could also be used with the opposite meaning, trees also being emblems of renewal in a revolutionary context, for instance in Tom Paine's *The Rights of Man* (1791–1792) or in William Blake's 'Preludium' of *America* (1793).

Patriotism was sometimes associated with coniferous planting, especially on poor soils, as William Pontey expressed it in 1800:

> As well as providing timber for ships to fight 'tyranny and injustice', the fir and the larch would increase the value and fertility or moorland and turn the once cheerless and solitary wilds into the social haunts of men [Daniels 1988: 51].

However, in his poem *The English Garden* (1772–1782), William Mason advised landscapers against exotic species of conifers and encouraged them to prefer native hardwoods, for better integration in the rural landscape. On the whole, the image of conifers tended to be viewed negatively. In his last published work, *Fragments on the Theory and Practice of Landscape Gardening* (1816), Humphry Repton shows two scenes supposedly representing 'improvements' to an old estate which is sold to an upstart by its aristocratic landowner. The 'after' scene, devoid of grace, displays a plantation of conifers enclosed by wooden palings, the symbol of the 'cultural decline' for the country as a whole. In a telling metaphor, Repton came to envision his *parvenu* clients as conifers among the hardwoods of the old aristocracy:

> How could I hope to suggest an idea to this man who shewed me what he called 'the LARGEST ACORN he had ever seen!' at the same time producing the CONE of a STONE PINE that grew near an oak and had fallen among the acorns! (fit emblem of him I thought who had fallen among gentlemen but could NOT be mistaken for one) [Daniels 1988: 69].

Conifers could be perceived as unworthy of the true gentlemen farmers. They could even be seen as anti-monarchical when they replaced hardwoods, and according to Uvedale Price, 'felling humble trees and planting conifers were equally examples of levelling' [Daniels 1988: 61].

Yet, against all odds, conifers were making headway into the English landscape. Although softwood planting may not have corresponded to the image of patriotic planting, it nevertheless corresponded to the reality of the country's needs, both for the navy and for building purposes, and therefore to the country landowners' economic interests; the overwhelming majority of wood used for building purposes was deal (Scots pine). As early as the 1770s and the 1780s, this reality was acknowledged by the Royal Society for the Encouragement of Arts, which rewarded owners of lands on which they had planted conifers (essentially Scots pine, spruce, silver fir and larch). Larches and pines were extensively planted, especially on thin soils. The larch (*Larix europea*) is one of the late-18th century success stories in forestry, and closely associated with the Dukes of Atholl. They believed so much in its capacity to replace oak in the navy, because it grew much more quickly than the latter, that they planted over 14 million larch trees in their Scottish estates between 1764 and 1826. Other than the Fourth Duke of Atholl, 'the Planting Duke', famous examples include the Duke of Portland, who nurtured his larches and pines 'with as much pride as [his] oaks' at Welbeck; Thomas Coke, nicknamed 'King Pine' for the thousands of conifers he planted at Holkham; and John Ingilby who recorded in his diary the numerous conifers he planted at Ripley.

The juxtaposition of hardwoods (in the discourse) and conifers (on the ground) testifies to the coexisting and complementary uses of trees: the economic one, in which trees are treated like a crop, and a symbolic one, made up of a growing sensitivity to old, native trees, increasingly valued for their ornamental and amenity value.

1.4.4 The Aesthetics of Tree-Planting

Aristocratic landowners, intent on praising a rural retreat as the best lifestyle one could ever wish for, and not as the result of political circumstances they were forced to suffer, were keen to surround themselves by artists who could diffuse this idea. In the poetic realm, *Ode on Solitude* by Alexander Pope or *Retirement* by William Cowper are examples of how a virtue could be made of necessity.

From the early 17th century onwards, plantings had begun to be integrated into the design of the grounds of country houses. 'Wildernesses' and avenues radiating from the house and extending over the countryside contributed to the visual power and stability of the landowning aristocracy in the country, 'putting a whole country into a nobleman's livery', in William Wordsworth's deprecating terms [Thomas 1984: 208]. The country houses were often surrounded by belts of woodland that not only provided shelter for animals (foxes and pheasants in particular) which could be hunted, shot and eaten, but also promenades on foot or on horseback. On top of that, they had the advantage of hiding the grounds from the outside, or of framing what was deemed aesthetically pleasing. The whole purpose of a belt of trees, explained Uvedale Price, was to conceal the extent of the property if it was small, and to display it if it was large. The idea of hiding is closely connected to that of excluding, as landscape designer Humphry Repton made it clear when he talked of 'appropriation… that charm which only belongs to ownership, the *exclusive right* of enjoyment, with the power of refusing that others should share our pleasure' [Daniels 1988: 45. Italics added]. Joseph Addison and Richard Steele also connected economy and landscape improvement, or profit and pleasure, when they asked the rhetorical question

> But why may not an estate be thrown into a garden by frequent plantations, that may turn as much to the profit as the pleasure of the owner? [Gandrillon-Purcell 2003: 63]

If for the early settlers, clearing the woodlands was a sign of the progress of mankind, the English gentlemen of the 18th century saw trees as pets, to be protected from pests (squirrels, deer, etc.) and enjoyed for profit and beauty. The pet-like vision of trees and the 18th century sensibility to them is clearly expressed by William Marsden in 1783 when he describes the feeling associated with the destruction of an aged tree: 'it appears a violation of nature, in the exercise of a too arbitrary right' [Thomas 1984: 213]. Never had trees been so domesticated when this prose was written, which makes it almost ironical. A rather biased historical vision made Addison declare in 1713 'the love of woods seems to be a passion implanted in our natures' [Thomas 1984: 212]. Similarly, William Gilpin, oblivious to the love-hate relationship of humankind with wooded landscapes, declared in a Romantic vein at the end of the 18th century that

> perhaps of all species of landscape, there is none, which so universally captivates mankind, as forest-scenery: and our prepossession in favour of it appears in nothing more than in this; that the inhabitants of bleak countries, totally destitute of wood, are generally considered, from the natural feelings of mankind, as the objects of pity [Short 2000: 122].

By that time indeed, under the influence of poets like John Milton and Alexander Pope, or philosophers like Jean-Jacques Rousseau, the appreciation of nature had become imbued with semi-religious overtones. If Nature was healing and a source of enlightenment, as was thought, then it was all the more so as it was unadorned. The influence of the Romantics and of the philosophy of Locke applied to landscape parks combined to make wooded landscapes scenes of meditation and sensory pleasures, as James Thomson's poem *The Seasons* (1730) exemplifies. This was all the truer as the trees were old, which was another argument for letting woodlands reach maturity.

Trees, used alone or in clumps, formed the backbone of the English landscape garden which Lancelot 'Capability' Brown brought to a climax. They gave it its beauty thanks to the arrangement of their colours, shapes and textures into a composition, so that aesthetics was added to the political, economic and social symbols of tree-planting. The palette available to the landscape garden designer was immensely richer than it had been before: according to J.C. Loudon, more than 500 new species of trees were introduced over the 18th century [Gandrillon-Purcell 2003: 73]. It took Charles Bridgeman between 60,000 and 70,000 trees to make the new park at Stowe in 1735 [Clarke 1985: 72]. To make their gardens appear older, therefore more beautiful and prestigious, some landowners did not hesitate to incorporate into them parts of previous hunting parks or trees from hedgerows [Rackham 2004].

At the same time as trees were becoming a major element of the designed landscape, they were becoming prominent in portrait painting from the 1720s onwards. According to Ronald Paulson, portrait painting after the Reformation – which forbade the painting of icons – had three main functions, 'political, moral and commemorative' [Paulson 1989: 259]. Due to the three-fold association of trees with patriotism, social rank and aesthetic value, trees formed an increasingly important and meaningful backdrop to the portraits of the landed aristocracy. Among many others, the composition of the famous painting by Thomas Gainsborough *Mr and Mrs Andrews* (1750) throws light on the values and the codes of representation of landowners at the time. The young man, poised on the root of an oak tree, is rooted in a genteel tradition. Mrs Andrews' position is a metaphor of her function: just in front of the tree trunk, under a main branch which divides into two, she is symbolically in charge on continuing the family tree by giving an heir to their estate, the Auberies, near Sudbury in Suffolk. To the right of the newly married couple, a ripe wheat field testifies to the prosperity of the estate, as do the grazing sheep. The enclosed field and the presence of trees growing in rows in the background are sure signs of Mr Andrews' sound management of his estate after the lessons of improvement. The genealogical metaphor of the tree is even more obvious in portraits of aristocratic families – which is not the case of Mr Andrews – such as George Stubbs' depiction of the Milbanke and Melbourne families in 1769. Patricians naturally gifted to govern are accompanied by the oak, the 'patrician tree' *par excellence*, as Abraham Cowley called it in his poem *Of Solitude* in 1688 [Gandrillon-Purcell 2003: 223]. Similarly, the growing fashion of 'conversation pieces' as a genre in 18th painting echoes the lifestyle and values of the 'polite society'. Among others, Philip Mercier's or William Hogarth's paintings stage groups of 'polite' people amidst a backdrop of trees which act as a metonymy of the taste of the landed elite both for natural settings and for socialising.

Catherine Gandrillon-Purcell argues that it is the use of the oak tree that makes these portraits specifically British. The cultural dimension of the oak tree had become such by the 18th century that, in spite of the revival of classical culture and the preference of the Royal Academy for the great Italian masters, British artists – whose patrons were often wealthy landowners – chose not to depict trees inherited from the Greek and Roman mythologies, but to use the powerful evocative power of the oak: 'the motives for aristocratic planting were thus a complex mixture of social assertiveness, aesthetic sense, patriotism and long-term profit' [Thomas 1984: 209].

This historical overview lays the basis for a lot of the questions that have remained valid to this day, concerning the ambivalent status and use of forests. Thanks to literary associations going back to antiquity, and to specific elements of British history, the oak developed into the emblem of the British nation after the Restoration. Also of great bearing today is the appearance in the late 17th century of the dichotomy between old woodlands, to be appreciated for their historic as well as aesthetic values, and plantations, necessary for the welfare of the economy. Interestingly, to celebrate its 250th anniversary in 2004, the Royal Society for the Encouragement of Arts, Manufactures and Commerce launched a tree-planting initiative. For a donation of £15, a tree was planted by the RSA and its partners, to be maintained until maturity on one of 28 sites. A 10-hectare site, named Anniversary Wood, was provided by Forest Enterprise. In October 2007, the number of trees planted under the scheme was 27,346. Although not a negligible figure in contemporary terms, this stands in sharp contrast with the 50 million trees planted in 77 years under the encouragements of the Royal Society at the end of the 18th century.

Fig. 1.3 Anniversary Wood is part of Tunnel Woods (Derbyshire), a 34-hectare new wood planted as part of the National Forest with native tree species, with the contribution of the RSA (Photo courtesy of the RSA)

References

Addison, J. 1714. *Spectator* 583, 20 August. In *The Spectator. Joseph Addison and Richard Steele* (1711–1714), ed. Y. D. Bond 1965. Oxford: Clarendon.

Bechmann, R. 1990. *Trees and Man – the Forest in the Middle Ages*. New York: Paragon House.

Bettey, J. 2000. Downlands. In *Rural England: An Illustrated History of the Landscape*, ed. J. Thirsk, 27–49. Oxford: Oxford University Press.

Brown, J. 1883. *The Forests of England and the Management of Them in Byegone Times*. Edinburgh: Oliver & Boyd.

Brown, N. 1997. Re-defining native woodland. *Forestry* 70 (3): 191–198.

Ciancio, O. and S. Nocentini. 2000. Forest management from positivism to the culture of complexity. In *Methods and Approaches in Forest History. IUFRO Research Series n°3*, eds. M. Agnoletti and S. Anderson, 47–58. New York: CABI, in association with IUFRO.

Clarke, G. 1985. Where did all the trees come from? An analysis of Bridgeman's planting at Stowe. *Journal of Garden History* 5 (1): 72.

Coates, P. 2005. *Nature: Western Attitudes Since Ancient Times*. Berkeley: University of California Press.

Cummins, J. 1988. *The Hound and the Hawk: The Art of Medieval Hunting*. London: Weidenfeld & Nicholson.

Daniels, S. 1988. The political iconography of woodland in later Georgian England. In *The Iconography of Landscape*, eds. D. Cosgrove and S. Daniels, 43–82. Cambridge: Cambridge University Press.

Darby, H. 1951. The changing English landscape. *The Geographical Journal* CXVII (4): 377–398.

Darby, W. 2000. *Landscape and Identity: Geographies of Nation and Class in England*. Oxford: Berg.

Delort, R. and F. Walter. 2001. *Histoire de l'environnement européen*. Paris: Presses Universitaires de France.

Dyer, C. 2000. Woodlands and wood-pasture in Western England. In *Rural England: An Illustrated History of the Landscape*, ed. J. Thirsk, 97–121. Oxford: Oxford University Press.

Evelyn, J. 1786. Silva. Or, a Discourse on Forest Trees. With notes by A. Hunter. York.

Finch, J. 2004. 'Grass, grass, grass': fox-hunting and the creation of the modern landscape. *Landscapes* 5 (2): 41–52.

Fitzgerald, B. 1966. *Portrait of the New Forest*. London: Hale.

Freeman, M. 1996. Plebs or predators ? Deer-stealing in Whichwood Forest, Oxfordshire in the eighteenth and nineteenth centuries. *Social History* 21 (1): 1–21.

Gandrillon-Purcell. 2003. *'Albion's Vital Tree': Arbre, société et identité dans la peinture de portrait britannique de 1709 à 1792*. Thèse de doctorat, Université de Paris VII – Denis Diderot. non publiée.

Gardiner. n.d. *Constitutional Documents of the Puritan Revolution: The Grand Remonstrance*. http://www.constitution.org/eng/conpur043.htm. Last accessed 6 August 2007.

Grant, R. 1991. *The Royal Forests of England*. Stroud: Alan Sutton.

Hinde, T. 1985. *Forests of Britain*. London: Victor Gollancz.

Hoskins, W.G. 1975. *The Making of the English Landscape*. London: Pelican Books.

James, N. 1981. *A History of English Forestry*. Oxford: Basil Blackwell.

Large, P. 2002. From swanimote to disafforestation: Feckenham Forest in the early seventeenth century. In *The Estates of the English Crown 1558–1640*, ed. R. Hoyle, 389–417. Cambridge: Cambridge University Press.

Manwood, J. 1598. *A Treatise of the Laws of the Forest; Wherein Is Declared Not Only Those Laws as They Are Now in Force, but Also the Original and Beginning of Forests and What a Forest Is in Its Own Proper Nature*. London: Reprinted by William S. Hein, Buffalo, 2002.

Miles, R. 1967. *Forestry in the English Landscape*. London: Faber & Faber.

Muir, R. 2005. *Ancient Trees, Living Landscapes*. Stroud: Tempus.

Nisbet, J. 1906. The history of the Forest of Dean, in Gloucestershire. *The English Historical Review* 21 (83): 445–459.

Paulson, R. 1989. *Breaking and Remaking: Aesthetic Practice in England, 1700–1820*. New Brunswick and London: Rutgers University Press.

Peterken, G. 2001. Ecological effects of introduced tree species in Britain. *Forest Ecology and Management* 141: 31–42.

Rackham, O. 1998. Trees and woodlands in a cultural landscape: the history of woods in England. *Forestry Sciences* 54: 139–147.

Rackham, O. 2000. Prospects for landscape history and historical ecology. *Landscapes* 1 (2): 3–15.

Rackham, O. 2001. *Trees and Woodlands in the British Landscape. The Complete History of Britain's Trees, Woods and Hedgerows*. London: Phoenix.

Rackham, O. 2004. Pre-existing trees and woods in country-house parks. *Landscapes* 5 (2): 1–16.

Readman, P. 2001. Landscape Preservation, 'advertising disfigurement' and the English National Identity c.1890–1914. *Rural History* 12 (1): 61–83.

Reed, M. 1990. *The Landscape of Britain from the Beginnings to 1914*. London: Routledge.

Schama, S. 1995. *Landscape and Memory*. London: Harper Collins.

Serovayskay, J. 2000. People's struggle against the institution of Royal forest reserves in England in the 11th–14th centuries. In *Methods and Approaches in Forest History. IUFRO Research Series n°3*, eds. M. Agnoletti and S. Anderson, 253–261. New York: CABI, in association with IUFRO.

Seymour, S. 1998. Landed estates, the 'spirit of planting' and woodland management in later Georgian Britain: a case study from the Dukeries, Nottinghamshire. In *European Woods and Forests: Studies in Cultural History*, ed. C. Watkins, 115–134. New York: CABI.

Short, B. 2000. Forests and wood-pasture in lowland England. In *Rural England: An Illustrated History of the Landscape*, ed. J. Thirsk, 127–149. Oxford: Oxford University Press.

Simmons, I. 2001. *An Environmental History of Great Britain*. Edinburgh: Edinburgh University Press.

Stamper, P. 2002. The tree that hid a king: the Royal Oak at Boscobel, Shropshire. *Landscapes* 3 (1): 19–34.

Steward, T., V. Pickett, T. Parker and P. Feidler. 1992. The new paradigm in ecology: implications for conservation biology above the species level. In *Conservation Biology. The Theory and Practice of Nature Conservation, Preservation and Management*, eds. P. Feidler and S. Jain, 65–88. London: Chapman & Hall.

Thomas, K. 1984. *Man and the Natural World. Changing Attitudes in England, 1500–1800*. Oxford: Oxford University Press.

Thompson, E.P. 1975. *Whigs and Hunters. The Origin of the Black Act*. New York: Pantheon.

Tsouvalis, J. 2000. Socialized nature: England's royal and plantation forests. In *Cultural Turns/ Geographical Turns: Perspectives on Cultural Geography*, eds. I. Cook, D. Crouch, S. Naylor and J. Ryan, 289–312. Harlow: Prentice Hall.

Whitelock, D. et al. 1965. *The Anglo-Saxon Chronicle. A revised Translation*. London: Eyre & Spottiswoode.

Williamson, T. 1995. *Polite Landscapes – Gardens and Society in Eighteenth-Century England*. Stroud: Sutton.

Williamson, T. 2000. The rural landscape 1500–1900: the neglected centuries. In *Landscape: The Richest Historical Record*, ed. D. Hooke, 109–117. Birmingham: Society for Landscape Studies, supplementary series 1, 2000.

Worster, D. 1993. *The Wealth of Nature. Environmental History and the Ecological Imagination*. Oxford: Oxford University Press.

Part I
The Institutionalisation of Forestry

Chapter 2
New Relationships with the Woodland

Communities are as much the result of their environment as the contrary

[Whitt and Slack 1994: 22]

Abstract To understand the present situation, both from the point of view of the management of wooded spaces and from that of their functions and meanings, one needs to take stock not only of the heritage mentioned in the previous chapter, but also of the dramatic changes that occurred during the late 19th and early 20th centuries. Thus, the second chapter revolves around the consequences of economic, scientific and social changes on English forests and forestry. First, the agricultural depression led to the replacement of the aristocratic landowners' power by that of the industrial capitalist. Then, the decline in shipbuilding led to the abandonment of the race towards broadleaf planting, while new scientific knowledge promoted a positivist vision of forestry. Lastly, the emergence of an overwhelmingly urban way of life modified the perception of the countryside in general, and of forests in particular, henceforth seen as sources of amenity and emblems of national identity. This led to the birth of modern forestry and provoked a visual and functional split between forests for production and forests for leisure, a trend which would inform 20th century British forestry.

Before the Industrial Revolution, woodlands had become a strong identity marker – firstly for the social elite, as landscapes of power, and secondly for the whole nation, especially after the Civil War, as emblems of belonging to the English people.

To understand the present situation, both from the point of view of the management of wooded spaces and from that of the functions and meanings of woodland, one needs to take stock not only of the heritage developed in the previous chapter, but also of the dramatic changes that occurred over the late 19th and early 20th centuries. What took place then modified the perception of woodland. First, the agricultural depression led to the replacement of the aristocratic landowners' power by that of the industrial capitalist. Then, the decline in shipbuilding led to the abandonment of the race towards broadleaf planting, while new scientific knowledge was promoting conifer plantations. Finally, the emergence of an overwhelmingly

S. Nail, *Forest Policies and Social Change in England.*

urban way of life resulted in new functions and representations of the countryside and its trees.

2.1 Agriculture in Crisis

The last quarter of the 19th century saw the beginning of agricultural decline in England, made worse by a series of bad harvests. Logically, the agricultural depression led to the conversion of arable land to woodland on poorer soils, so that the area of woodland began to increase from 1871. The movement was only interrupted during the two World Wars when the necessities of growing food at home prompted the ploughing-up of land that had been converted from arable to grassland. At the end of World War Two, arable land had more or less reverted to its 1879 level [Darby 1951: 388].

2.1.1 The End of the Landed Aristocracy

The crisis was not only economic: it entailed the disappearance of the power of the landed gentry whose flagship was the English countryside. With the gradual enfranchisement of the urban middle classes from 1832 onwards and the rapid growth of industrial capitalism, power moved towards the cities. The phenomenon was amplified by growing inheritance taxes from the beginning of the 20th century, which made it more costly for inheritors to enter into their own, and also by the First World War, which among its long list of casualties contained many heirs of the aristocracy.

Many estates were put on the market as a consequence of the changes in the circumstances of the landed classes: it is estimated that about 25% of England changed hands at the beginning of the 20th century, a figure unrivalled since the Norman Conquest. The number of demolished country houses is striking: 458 between 1921 and 1945, 595 between 1945 and 1980 (out of which 200 for the 1950s alone) [Thompson 1971: 332; Binney and Milne 1982: 12; Beard 1988: 109]. An exhibition at the Victoria and Albert Museum in 1974, entitled *The Destruction of the Country House 1875–1975*, displayed the photographs of over 1,000 country houses destroyed. It raised awareness as to the importance of preserving the heritage. A U-turn in public policies occurred in 1979, with ‘heritage’ entering the official vocabulary and preoccupations, both in Mrs Thatcher’s election campaign and in the Queen’s Speech. The safeguard of the old aristocratic emblems had become the concern and the responsibility of the government, and hence, of the nation.

2.1.2 Changing Needs in Woodland Products

With the countryside undergoing rapid decline, woodlands were experiencing changes. Not only did transfers of land have consequences on woodland, but needs were also changing. Firstly, firewood and charcoal were progressively replaced by coal in domestic and industrial uses. Secondly, wood products were replaced by metallic ones. Finally, the importation of cheap softwood timber made the light timber produced by coppicing unnecessary, apart from its use as cover for pheasants. This led to the disappearance of the skills which had been connected to the working practices of coppicing and pollarding since the Neolithic period.

As a consequence, and in spite of new markets opening for woodland products as the population grew, markets for coppice wood declined and whole areas of coppice fell into such disuse as to be classified as scrub in the Forestry Commission Census of 1947–1949. Old woodlands thus became less important for the economy, and associated with the past, while plantations of conifers signified social change, and 'came to signify Modernity *per se*' [Tsouvalis 2000: 306].

To add to the general decline of interest in hardwood planting, change in shipbuilding techniques put an end to the planting mania: in 1860, HMS *Warrior*, the first iron-clad ship, was launched and in March 1862, the battle of Hampton Roads witnessed the clash of two iron-hulled vessels during the American Civil War and announced the demise of wooden ships. This led to a climate of relative indifference with regards to woodland planting at the end of the 19th century, except among a few landowners who began to organise themselves.

2.2 The Emergence of Scientific Forestry

Plantations, as mentioned in the previous chapter, had started in the late 17th century, with acorns of oak and seedlings of ash, beech and sycamore being planted on previously ploughed land. The young trees were often sheltered by pioneer species such as holly, birch or thorn, which were thinned as the hardwoods grew and required less protection. The successor types were then thinned according to need, and felled when they reached full height. Yet, amateurism often prevailed in forest management as practised by land agents and foresters, even though the technical aspects of forestry had been the object of interest since the 18th century. Knowledge was far from systematic: even statistical knowledge of the area covered by woodlands was lacking.

2.2.1 The Disappearance of Amateurism

Changes occurred as a result of the development of scientific forestry following the principles of sustained yield. Leading the way to this new positivist conception of forestry was Germany, where fear of a 'wood famine' in relation to Prussian/German growth aspirations led to the organisation of rational silviculture, assisted by mathematics and geometry [McManus 1999: 190–191]. Germany was well organised to practise systematic forestry as early as the late 18th century, with schools, periodicals and even a foresters' association. From there, the concept of scientific forest management was exported to many parts of the world, including the British Empire. India in particular had a renowned Forestry Department, and it was there that Dietrich Brandis and William Schlich developed timber reserves according to German methods, before importing the knowledge of silviculture into Great Britain. Until the beginning of the 20th century, all the information available on high forest plantation management – as opposed to the more traditional coppice-with-standards management – came from Germany via the British Empire. The first forestry schools were created by Georg Ludwig Hartig in Hunden (Hesse) in 1789 and by Johann Heinrich Cotta in Tharandt (Saxony) in 1816 [Ciancio and Nocentini 2000: 47–49], while Britain had to wait until 1885 for a forestry department to be created at the Royal Indian Engineering College at Coopers Hill, Surrey. It was transferred to Oxford University in 1905, and in 1907 a readership in forestry opened at Cambridge University.

Conifers such as the native Scots pine were often used as pioneer species, but had progressively begun to be planted for their own sake in the previous century. The introduction of conifers from Western North America was to have dramatic consequences in terms of landscape. The Douglas fir (*Pseudotsuga menziesii*) was introduced in 1827, followed by the two species of conifers which are still the most commercial ones today, the Sitka spruce (*Picea sitchensis*) in 1831 and the Lodgepole pine (*Pinus contorta*) in 1853–1854 [James 1981: 184]. This led to conversion of coppice with standards into plantations of conifers in some areas, in particular in the cool and wet parts of the North of England. Forestry appeared for the first time in the Oxford Dictionary of 1859 as 'the science and the art of farming and cultivating forests' [Tsouvalis 2000: 300], and this formalised the treatment of forests as crops. The forests thus stopped being spaces of power to be perceived as spaces of production, to be enhanced according to scientific methods in the national interest.

2.2.2 Towards a National Forest Policy

Some enlightened landowners had started experimenting, essentially with conifers, and began to organise themselves in 1881 by creating the Royal Arboricultural Society, followed by the English Forestry Association. The purpose of these organisations was to lobby Parliament for the adoption of a national forest policy favouring the production and commercialization of national timber and wood products in the face of growing imports to satisfy the nation's needs. Indeed, a favourable balance of trade made softwood timber from America, Russia or Scandinavia cheap, all the more so as duties on timber imports had been removed in 1867. Timber prices declined and so did timber production. Imports, essentially of pine and fir, increased fivefold between 1850 and 1910, so that at the turn of the century, nine-tenths of all timber and wood products were imported into Britain [Miles 1967: 48; Tsouvalis and Watkins 2000: 380].

As a result, a Select Committee on Forestry was set up in 1885, which was the first step towards state intervention in the realm of forestry. In 1909, the final report of the Royal Commission on Coast Erosion, Reclamation of Tidal Lands and Afforestation advocated the reclamation of lands through two large scale planting schemes: even where they might not be economically profitable, they would benefit the community and provide employment. Under the *Development and Road Improvement Fund Act 1909*, a Development Commission was set up. Its duties included the promotion of forestry, for which it was empowered to make grants. In 1912, the Board of Agriculture created an Advisory Committee on Forestry, whose mission was to study the possible location of future plantations, to draw up plans, to experiment and to reflect on the training of future foresters. In its report in that same year, it took up the idea voiced by Prime Minister Lloyd George 3 years before, of an experimental forest on land acquired by the State.

So the conditions were ripe for the State to endorse responsibility for the forestry sector. It was however left to the First World War to bring together these strands and transform a national trauma into policy, as we shall see in the next chapter.

2.3 Urbanisation and the Rural Idyll

At the same time, the vision of the countryside was changing with the growth of urban life.

2.3.1 The Growth of Cities

Towns had started to grow in the Roman times, but most of the towns and cities that existed in 18th century England had grown from the 'fever of borough creation' in the 12th–14th centuries, and were the old administrative and market towns as well as cathedral cities. The population remained mostly rural and mostly in the South until late in the 18th century: urban dwellers only represented 8% of the population in 1600, 16.5% in 1730 and 34% in 1801, the date of the first official census. Most urban centres were small. At the beginning of the 18th century, the largest towns outside London were Norwich, York, Bristol, Newcastle and Exeter, their populations not exceeding 20,000. By 1801, they had been superseded by Manchester/Salford, Liverpool, Birmingham, Bristol and Leeds, whose populations however still did not reach the 100,000 inhabitants. Even at that time, only 19 towns in Britain exceeded 20,000 inhabitants, with the exception of London, which already numbered more than a million souls and 8% of the population in 1801 [Royle 1987: 21].

Things changed rapidly afterwards. The enclosure movement in the countryside pushed rural labourers towards industrial centres, while immigrants from Ireland swelled the ranks of unemployed manpower. At the beginning of the 19th century, improvements in the networks of communication and the generalised use of steampower allowed manufactures to settle further away from rivers. As a result of these two phenomena, cities began to grow at an unprecedented speed. Between 1821 and 1841, London saw its population grow by 20%, Birmingham, Leeds, Manchester and Sheffield by 40% and Bradford by 65% [Darby 2000: 114]. By 1851, for the first time in history, more than half of the British population lived in towns of more than 10,000 people, and the growth of industrial cities was dizzying, leading to six conurbations – Greater Manchester, Merseyside, West Yorkshire, Greater Birmingham, Tyneside and Greater London, and to many other big cities.

2.3.2 'Urban Hells' Versus Forest Heritage

The consequences of the Industrial Revolution and urban growth are well known and were described by Marx and Engels, among others: overcrowding, insalubrious living conditions, appalling sanitary conditions, rampant disease and demoralisation. As early as 1785, William Cowper concluded in *The Task* that 'God made the country, and man made the town' [Cowper in Parrinder 1990: 137], while 20 years later the Romantic poet Percy Bysshe Shelley summed up his feelings for the growing metropolis by saying that 'Hell is a city much like London' [Shelley in Darby 1951: 394]. As the growth of towns had been influenced by the construction of canals in the second half of the 18th century (Stourport is the perfect example), or was to be shaped by the railway mania in the second half of the 19th century (e.g. Crewe or Swindon), new industrial towns grew out of hamlets. Middlesborough, for instance, grew from a farmhouse in the 1830s to a town of over 80,000 persons in 1880. This led to cultural changes.

From the Middle Ages onwards, urbanity had been synonymous with refinement, culture and good manners, the place where humankind might fulfil its potential, while rusticity was associated with uncouthness, hence inferiority. At the end of the 18th century, as a result of the unfettered growth of cities and the rejection of urban life by the well-off, the two poles of nature and culture were suddenly inverted. Poets and artists praised the rural lifestyle of an idealised countryside. In the poems quoted above, as in the paintings by Helen Allingham, innocence, contentedness and social harmony supposedly reigned supreme in rural England. Rurality came to stand for superiority of character and holiness, while urbanity became another word for decadence. This association between an idealised vision of the pastoral, the 'religion of the fields', as John Clare described it [Clare in Williams 1986: 67], was the other side of the coin of urban sophistication, hence vice. This led to a different appreciation of rural landscapes.

One particular aspect of this vindication of rural scenery, only the beginning of which will be dealt with here, since it is an essentially recent phenomenon, is the development of forest history as an area of academic research. Even though the beginnings can be traced back much further, interest in forest history developed from the 18th century, in Germany and Italy first [Agnoletti 2000: 1], analysing the history of forests in relation with hunting history, botany, silviculture, agriculture or shipbuilding, among others. In the first decades of the 20th century, a specific focus on the relationship between forest history and human action appeared, inspired by new approaches (e.g. École des Annales in France in the 1930s, Carl Sauer and cultural geography in the United States around the same time) and the development of research groups: the Institute of Forest History at the University of Freiburg in 1943, the Forest History Society in the United States in 1946, the Forest and History Group within IUFRO in 1963, or the Groupe d'Histoire des Forêts Françaises in 1981. This new perception of forest landscapes as what later came to be known as 'cultural landscapes' marked the beginning of a rich area of research that has developed mostly since the last quarter of the 20th century and has been the cornerstone on which forest heritage has been interpreted and managed, as will be shown in a later chapter.

2.4 Woodland and Leisure for the Working-Class

The connection between urbanity and demoralisation was made obvious in the report of the Parliamentary Select Committee on Public Walks in 1833: according to its conclusions, it was for lack of open spaces that the 'humbler classes' resorted to 'low and debasing pleasures' [Darby 2000: 118]. Indeed, open spaces and trees, which had been present on the periphery of built-up areas, had no place in the cramped cities of the early 19th century: as cities grew haphazardly, they took in surrounding lands. W. Hoskins gives the example of Birmingham where 'the last remnant of Birmingham Heath was enclosed in 1799, and was built over forthwith with eight new streets' [Hoskins 1975: 226]. The idea grew that, as gardener Batty Langley had put it a century before, if more trees could be planted along the streets of London, 'then might we view a city in a wood' [Langley in Thomas 1984: 206].

2.4.1 The Public Park Movement

The 1833 report, together with the *Second Report on the State of Large Towns and Populous Districts* published in 1845 after the great epidemic of cholera, were the starting-points of the creation of public parks with free access to everyone. Royal parks in London had been open to the public in the 18th century or before, but such was not the case of pleasure gardens such as Vauxhall or Ranelagh. Hence, more open spaces like Victoria Park (1845) in the East End of London, were created, and areas like Hampstead Heath, Blackheath and commons at Clapham, Wandsworth, Tooting and Wimbledon, benefited from protective measures so as to serve as recreation areas for Londoners. Outside London, many municipal parks were created, such as Birkenhead Park on the Wirral peninsula, which was to inspire Frederic Law Olmsted for Central Park in New York, USA.

Fig. 2.1 Birkenhead Park, Wirral. Designed by Joseph Paxton in 1844, it was opened to the public in 1847. It was the first municipal public park created by a local authority on public funds (Photo courtesy of the Wirral Council)

Public parks were meant to provide recreation for the working-class families who could not afford to live away from the city centres where work could be found; they were also supposed to improve public health by the beneficial contact with nature. But they also had a hidden agenda. For the governing classes, they acted as safety valves and a derivative bulwark against revolutionary outbursts in a context of political unrest in Europe [Conway 1991: 54]; furthermore, the mixture of social classes in the parks was thought to provide a model for an improvement in the manners and morality of the working-class. On the other hand, for social reformers, public parks constituted a first step towards social reform. Unsurprisingly, open spaces such as the site of Victoria Park or Kennington Common in London, which were used as protest sites by Chartists in the 1840s, were turned into more civilised parks, thus neutralising their potential revolutionary powers under the well-groomed aesthetics of the rural landscape parks they were aping [Dennis 2000: 118]. Public parks became synonymous with civic pride.

The nostalgia for a mythical countryside thus fed on itself: on the one hand, country life was presented as a rural idyll; on the other hand, urban life could hopefully be improved if it only borrowed landscape elements from the idealised countryside. Humphry Repton used the rural model in his pattern-books for the suburban gardens of his middle-class clients, and municipalities tried to find inspiration from the parks around the country seats of the nobility for their urban public parks. Marylebone Park in London, created in 1811 by John Nash, was meant to provide Londoners with 'open Space, free air and the scenery of Nature', while Birkenhead Park was designed by Joseph Paxton 'in pastoral/Romantic style to copy 'natural' scenery, with lakes, undulating meadows, groves, rockeries and meandering paths' [Coates 2005: 160].

2.4.2 Woodlands for Recreation

Most provincial cities were still within walking distance of the countryside, and the commons became precious as recreational spaces for the working-class. But they were threatened by the closure of footpaths, the continuing enclosure of open fields and the enclosure of the commons, stimulated by the *General Enclosure Acts* of 1836, 1840 and 1845, which left ever less recreational space [Darby 2000: 28]. Awareness of the benefits of accessible countryside to urban dwellers led to the opposition to the enclosure of the commons and to the creation in 1865 of the Commons, Open Spaces and Footpaths Preservation Society to fight against the enclosure of the commons in the London area [Miles 1967: 54].

So, rather paradoxically, since the countryside was presented as a distinct 'other' from the urban world, the countryside began 'to be viewed as an urban amenity' [Darby 2000: 126], an extension of urban space for recreational purposes. This became even more so as the railways brought the countryside closer to cities and as the fashion for rambling developed from the mid-19th century onwards. Preservation for the sake of beauty and recreation became a credo while access became a bone of contention in the following decades. Although the term 'amenity' as an objective in planning only appeared in the *Housing and Planning Act* of 1909, the concept of amenity lay at the heart of the creation of public parks and the conservation of parts of some royal forests as unenclosed areas for recreation.

In the late 15th century, after the end of the War of the Roses, the term *Amoenitas*, conjuring up the pleasures of country living for the wealthy owners of a Roman *villa rustica* came back into fashion. As explained before, advocating the pleasures and virtues of rural living became even more fashionable in the second half of the 17th century, possibly as a 'compensatory myth' for the Royalists who had lost the Civil War and had taken to their rural retreats as a consequence. The application of the principle of amenity on a space devoted to recreation and beauty was first witnessed in the New Forest, with competition for land decreasing on account of diminished needs for shipbuilding. In 1875, a Select Committee pleaded for a balance in the missions of the New Forest between forestry and amenity, and put forward arguments that have since become classic in conservation policies: the fact that it was 'almost the only specimen left in England of a primeval forest, a priceless source of enjoyment to all classes of the nation', and thus should be 'managed with a view to the preservation of its natural aspect and condition as a Forest of surpassing beauty and unique character' [Grant 1991: 90; Reed 1990: 212]. After a lot of debates, parts of it were finally kept as sources of timber, while the rest was converted into an ornamental area for amenity purposes under the *New Forest Act* of 1877.

Legal battles and a rising awareness of the benefits of the peri-urban open spaces led to public authorities buying such areas for the benefit of the public. Burnham Beeches or Epping Forest provide early examples, the latter being all the more interesting as it was the remnant of the Royal Forest of Essex. In 1875 and 1876, the Corporation of the City of London bought 1,200 hectares of open waste lands and in 1878, through the *Epping Forest Act*, the Forest was disafforested, and almost 2,500 hectares of Epping Forest were turned into an open space for the recreation of the public.

Thus, at the end of the 19th century, the social construction of the reality of woodlands had operated in two distinct directions. On the one hand, forests were increasingly perceived as spaces of production, of conifers especially, through the introduction and practice of scientific forestry. On the other hand, broadleaved trees and woodlands, especially old ones around parks in the countryside, spelled status and identity and were increasingly used for amenity purposes. These two strands form the basis of present attitudes and perceptions to woodlands in England, with a dichotomy between use and beauty, production and values. The rest of this book aims to explore this duality, and what has become of it, in order to throw light on present attitudes and policies on trees and forests in view of cultural roots.

References

Agnoletti, M. 2000. Introduction: the development of forest history research. In *Forest History: International Studies on Socio-economic and Forest Ecosystem Change. IUFRO Research Series n°2*, eds. M. Agnoletti and S. Anderson, 1–20. New York: CABI, in association with IUFRO.

Beard, M. 1988. *English Landed Society in the Twentieth Century*. London: Routledge & Kegan Paul.

Binney, M. and E. Milne (eds.). 1982. *Vanishing Houses of England: A Pictorial Documentary of Lost Country Houses*. London: SAVE Britain's Heritage.

Ciancio, O. and S. Nocentini. 2000. Forest management from positivism to the culture of complexity. In *Methods and Approaches in Forest History. IUFRO Research Series n°3*, eds. M. Agnoletti and S. Anderson, 47–58. New York: CABI, in association with IUFRO.

Coates, P. 1998 (1st edition), 2005. *Nature: Western Attitudes since Ancient Times*. Berkeley: University of California Press.

Conway, H. 1991. *People's Parks: The Design and Development of Victorian Parks in Britain*. Cambridge: Cambridge University Press.

Darby, H.C. 1951. The changing English landscape. *The Geographical Journal* CXVII (4): 377–398.

Darby, W. 2000. *Landscape and Identity: Geographies of Nation and Class in England*. Oxford: Berg.

Dennis, R. 2000. Modern London. In *The Cambridge Urban History of Britain. Vol.3, 1840–1950*, ed. M. Daunton, 118. Cambridge: Cambridge University Press.

Grant, R. 1991. *The Royal Forests of England*. Stroud: Alan Sutton.

Hoskins, W. 1975. *The Making of the English Landscape*. London: Pelican Books.

James, N. 1981. *A History of English Forestry*. Oxford: Basil Blackwell.

McManus, P. 1999. Histories of forestry: ideas, networks and silences. *Environment and History* 5: 185–208.

Miles, R. 1967. *Forestry in the English Landscape*. London: Faber & Faber.

Parrinder, G. 1990. *The Routledge Dictionary of Religious and Spiritual Quotations*. London: Routledge.

Reed, M. 1990. *The Landscape of Britain from the Beginnings to 1914*. London: Routledge.

Royle, E. 1987. *Modern Britain, a Social History 1750–1985*. London: Arnold.

Thomas, K. 1984. *Man and the Natural World. Changing Attitudes in England, 1500–1800*. Oxford: Oxford University Press.

Thompson, F. 1971. *English Landed Society in the Nineteenth Century*. London: Routledge & Kegan Paul.

Tsouvalis, J. 2000. Socialized nature: England's royal and plantation forests. In *Cultural Turns/Geographical Turns: Perspectives on Cultural Geography*, eds. I. Cook, D. Crouch, S. Naylor and J. Ryan, 289–312. Harlow: Prentice Hall.

Tsouvalis, J. and C. Watkins. 2000. Imagining and creating forests in Britain, 1890–1939. In *Forest History: International Studies on Socio-economic and Forest Ecosystem Change*. Report N°2 of the IUFRO Task Force on Environmental Change, eds. M. Agnoletti and S. Anderson, 371–386. New York: CABI, in association with IUFRO.

Williams, M. and R. Williams. 1986. *John Clare, Selected Poetry and Prose*. London: Routledge.

Whitt, L. and J. Slack. 1994. Communities, environments and cultural studies. *Cultural Studies* 8 (1): 5–31.

Chapter 3
The Productivist Dream and Its Aftermath

Trees, woods and forests are among our most valuable natural assets although many of their attributes are difficult to price in cash terms

[Forestry Commission 1998: 4]

Abstract A lot has been written on the first 50 years of the Forestry Commission, and the third chapter does not aim to repeat previous publications. Rather, it aims at attracting attention to some characteristics which were dogma for several decades, not only in Britain but also on the international scene, and have shaped today's forestry, for example in terms of landscapes and economics. The chapter then explores the changes in emphasis following the economic failure of British forestry and presents the content and impacts of institutional changes, within the Forestry Commission and also in the UK at large, through devolution in the last years of the 20th century. It also presents the new focus of public policies through an analysis of the 1998 *England Forestry Strategy*.

A number of constraints have favoured or slowed down tree-planting throughout history, as has been seen in the previous chapters. Studying the history of forestry serves as an observatory of the economic, but also political and cultural contexts of society. The 20th century was no exception of course and the creation in 1919 of the Forestry Commission, the institution devoted to state forestry, partook of a series of factors – economic, political, sociological, scientific – unique to this period, which testify to a radical change: the history of forests became the history of state forestry. The context of the creation of the Forestry Commission conditioned its objectives and organisation. The institutionalisation of forestry had clear economic and strategic objectives. In this, the United Kingdom was not isolated from the main world trends, even though geographic isolation may have reinforced the feeling of urgency. It soon turned out, however, that the original purposes of the institution would not last.

S. Nail, *Forest Policies and Social Change in England.*

3.1 The Traumatic Context of World War I

Faced with the awesome efficiency of the German submarines and the naval blockade, Britain could no longer rely on imports of wood to provide for its needs, and now had to depend on its own wood production.

3.1.1 The Acland Committee

In 1915, the Board of Agriculture appointed a Home-Grown Timber Committee whose task was to organise timber supplies and to establish its own sawmills. This was replaced in 1917 by a Directorate of Timber Supplies, later Timber Supply Department of the Board of Trade. On top of controlling production, imports and timber stocks, it set up a system of rationing and setting maximum prices [Taylor 1945: 94–95]. Figures as to the area of woodlands felled during the Great War vary, but the woodland cover of Britain in 1919 is estimated at an all time low of 4% [Broadhurst and Harrop 2000: 185].

This state of urgency spurred the creation in 1916 of the Acland Committee, a subcommittee on forestry within the Reconstruction Committee. Its secretary was R.L. Robinson, a former pupil of Schlich, and who was to become the fourth chairman of the Forestry Commission. The report of the subcommittee in 1917 stressed a number of deficiencies and made recommendations to improve the situation. Not only was excessive reliance on imports dangerous in times of war, imports were also extremely costly in terms of the balance of payments in peace times as well. Furthermore, the country had at its disposal large areas of wasteland which would gain from being planted in order to increase their productivity and help rural employment. The Committee therefore recommended the afforestation (in the modern sense of establishing new forests on land not previously forested) of 700,000 hectares within the 80 years following the end of the war, as a strategic reserve of timber, to be compared with the estimated 800,000–1,000,000 hectares standing in 1914.

The report laid down the basis for the future national forest policy and formed the core of the *Forestry Act* 1919. The latter created the Forestry Commission, in charge of promoting afforestation and wood production. The first Commissioners were immediately appointed and the first tree-planting programmes, in Devon and in Scotland, implemented. The Commission was not placed under the responsibility of a minister, but one of the Commissioners, who was an MP, could speak for the Commission in Parliament.

The establishment of state forestry was completed with the *Forestry (Transfer of Woods) Act* 1923, mentioned above, which put Crown Forests under the Forestry Commission's responsibility. They had already been put under central administration to a certain extent in 1829, when the forest courts and other offices were vested in the First Commissioner of His Majesty's Woods, Forests and Land Revenues. These powers had passed to Commissioners of Woods, Forests, Land Revenues, Works and Buildings under the *Crown Lands Act* 1832. Finally, legislation passed in 1851 had transferred their control to Commissioners of Woods, Forests and Land Revenues, later Commissioners of Crown Lands and the Crown Estate Commissioners. As these disappeared, the remnants of the Royal Forests as landscapes of power disappeared, and productivity – soon accompanied by other missions – superseded status symbol.

3.1.2 The Forestry Commission at Its Beginnings

The implementation of principles imported from Germany such as the development of high forest plantations along scientific lines, including working plans, was twinned with institutional changes.

Powers were given to the Commission to buy or rent lands and woods suitable for planting, mostly on marginal land of little importance for agriculture (derelict woodlands, moorlands and heathlands, sand dunes, etc.). The Commission could give loans and grants to private owners as well as to local authorities. It provided advice on planting and managing privately-owned woodlands. Commissioners were also entitled to set up and manage industries to deal with wood, and to elaborate training and research programmes.

The prime purposes being strategic and economic, speed of growth was a crucial aspect, and the choice of single species, even-aged high forest plantations, mostly of those conifers imported over the 19th century, seemed to impose itself. It allowed the treatment of woods in a similar way to that of any other crop: planting seedlings on bare ground in regularly spaced rows, so as to make weeding and thinning easy, and clear-felling the mature crop for the cycle to be repeated. Work practices were also closely monitored so as to make manpower as efficient as possible. The priority of economic factors is clear in the principles set up by the first Commissioners. The last principle read

> Generally the elimination of all activities, however attractive, which do not conduce directly or in the long run to an increase of timber-production [Ryle 1969: 249].

Using these principles, about 162,000 hectares of the lands purchased by the Forestry Commission were planted in geometric patches between the two World Wars [Tsouvalis and Watkins 2000: 383]. Whereas in 1917, 97% of the wooded areas was in private hands, the remaining 3% being Crown lands [Coppock 1960: 127], by 1939, the Forestry Commission had become the largest landowner in Britain.

3.1.3 The First International Congresses

Parallel to domestic changes, forests and forestry became the object of international debates at the end of the 19th century. The International Union of Forestry Research Organizations (IUFRO) was founded in Germany in 1892 to promote international cooperation in forestry, forest operations and products research, through publications and workshops, as well as global conferences every 5 years. Membership gradually widened after the First World War to include members beyond Europe. In spite of the creation of the Food and Agriculture Organization (FAO) by the United Nations after the Second World War, IUFRO chose to remain a separate, non-governmental organisation based on voluntary action.

The first World Forestry Congress (WFC) took place in Rome in 1926, and since then congresses have taken place regularly, at first every decade, and then every 6 years. In 1947, on the occasion of its annual conference, FAO passed a resolution calling for a World Forestry Congress in 1949, and FAO has since then organised the WFC. The purpose of the Congresses is three-fold, in accordance with the guidelines agreed at FAO's 64th session in 1974: to assess the state of forests and forestry in the world, to provide a forum of discussion on trends and issues in the forest sector and to propose policies at all levels. They gather governments, the scientific community, international organisations, NGOs and the private sector. That is to say, they bring together knowledge, experience and stakeholders' view-points to provide guidance for policy-making, and try to elaborate world-wide agreements in relation to standards, terminology and research methods. However, their functions are advisory, not executive, although the conclusions of the congresses are communicated to the FAO which may – or may not – endorse the declarations made at the Congress through resolutions.

The trends in British forestry at the beginnings of the Forestry Commission were in-keeping with the themes of the first congresses: economic aims came foremost. The highly technical debates of the first congresses, in 1926 and 1936, focused on the preservation of the resource, which was linked to its protective (e.g. soil and water conservation) and productive (sustained yield, employment) values. Forest policy was clearly the resort of each country, yet the idea of international cooperation emerged at the Third WFC in Helsinki in 1949. This entailed, on top of clarifying forestry terminology in order to harmonise it, improving knowledge of the countries' forest domain and embarking on forest surveys in each country [*Unasylva* 1949]. At the same conference, proposals were made in order to improve the profitability of forestry by reducing wastage. The Fourth WFC, taking place in Dehra-Dun (India) in 1954, focused on the role of forests for economic development. It saw the specific problems of Europe and North America as revolving around the need to increase productivity against a backdrop of high fragmentation in forest ownership and competition for land [*Unasylva* 1954]. Similarly, the 6th WFC, held in Madrid in 1966, discussed the role of forestry in the world economy, and the 7th, in Buenos Aires in 1972, focused on forests and socioeconomic development. Productivism was clearly the order of the day.

3.2 Post-Second World War Priorities

By 1939, afforestation figures in Britain were below Acland's targets. They barely exceeded the losses brought about by the First World War when national reserves had to be drained again to provide for the needs of the Second World War. During the conflict, the state controlled prices, sales and felling of timber in a context of shortage, and an estimated third of the standing timber was consumed [Miles 1967: 82].

3.2.1 Post-War Forestry Acts

Logically, the two-fold objective of strategic reserve and improvement to the balance of payments was reiterated after the Second World War. The White Papers *Post-War Forest Policy* (1943), and *Post-War Forest Policy: Private Woodlands* (1944) expounded the means by which, it was hoped, one third of peace time requirements of wooden products would be produced at home. The *Forestry Act* 1945 effected a few administrative changes to the Forestry Commission. First of all, it placed it under the responsibility of the Minister of Agriculture and Fisheries (for England and Wales) and the Secretary of State in charge of agriculture for Scotland. These were responsible for the Forestry Commission's lands. Internally, a regional organisation by Divisions headed by Divisional Officers had been set up by the *Forestry Act* 1919, where District Officers were in charge of specific groups of forests. However, these Divisions were now replaced by eleven Conservancies with a Conservator at the head of each, assisted by District Officers.

The objectives of the *Forestry Act* 1951 had a lot in common with the *Acland Report*. It reaffirmed the Commission's duty of 'promoting the establishment and maintenance in Great Britain of adequate reserves of growing trees' [Winter 1996: 284]. The objective sought was 2,000,000 hectares of productive woodlands within 50 years of the end of the war. The powers granted to the Commission during World War II to license and control felling were made permanent. It is characteristic of the extent of the changes proposed that they envisaged the possible disappearance of deciduous woodlands to the benefit of conifer plantations, except in sites where broadleaves could produce the best quality timber and wood [Miles 1967: 89]. As a result, according to the 1947 *Census*, conifers accounted for 97% of Forestry Commission plantations since 1919. Although less was achieved than had been hoped in the decade after World War Two, a memorial celebrating the millionth acre planted by the Forestry Commission, most of it since the end of World War Two, was unveiled by Queen Elizabeth II in May 1956 at Eggesford Forest (Devon).

3.2.2 Economic Priorities

Progressively, the two-fold mission of the Forestry Commission gave way to the sole economic dimension. The Zuckerman *Report on Forestry, Agriculture and Marginal Land* of 1957 undermined the strategic function of forestry, partly because of the advent of the atom bomb which made the 3-year 'strategic reserve' useless in case of nuclear warfare. The report ended by urging woodland owners the Forestry Authority responsible for policy-making and grant-aiding, and a Policy and Resources Group for both Forestry Authority and Forest Enterprise. Forest Enterprise's functions were further divided in 1996, part of them related to forestry policies, and part managing the public forest estate through an Executive Agency. Forest Research, the Commission's research branch, also became an Executive Agency in 1997.

The Forestry Commission's annual report 2003–2004 described the times as 'challenging' [Forestry Commission 2005a: 11]. Indeed, between 1997 and 2002, the price of timber had collapsed by 60%, a phenomenon unheard of before; in 2003–2004 alone, timber income for the Forestry Commission fell by 8%. On top of historically low prices, forestry industries had to face the decline of traditional markets due to the use of recycled fibre in paper and woodpanel, plus the development of new materials and overseas competition.

3.4 New Labour and the *England Forestry Strategy*

To a large extent, forestry policies under the New Labour government have been a continuation of the previous Conservative governments' policies and have followed the widening scope of forestry. The resolutions voted in Helsinki in 1993 formed the backbone of *Sustainable Forestry: The UK Programme* (Cm 2429). In 1996, the Forestry Commission and the Countryside Commission published a discussion paper inviting comments on woodland expansion, which resulted in the publication the following year of *Woodland Creation: Needs and Opportunities in the English Countryside*. This document formed the essence of the new government's strategy published in 1998.

3.4.1 Changing the Emphasis of Forestry

Differences in policies may be defined in two ways. First of all, New Labour had committed itself to increasing the Forestry Commission's estate before being elected in 1997. The privatisation of the Forestry Commission estate, undertaken by the Conservatives, was indeed very unpopular and did not exceed 20% of its lands. As productivity stopped being a priority, the financial calculations which justified these sales were no longer the main means of assessing forest value.

Secondly, the New Labour government gave forestry, and the environment in general, more visibility by placing them much more at the centre of public policies, following international trends. By contrast, the 1990 White Paper, *This Common Inheritance*, gave two committees the responsibility of coordinating action on the environment within central government, but their action lacked visibility. Subsequently, the more important of the two was disbanded and replaced by a new one headed by the leader of the House of Lords. The New Labour government created a Cabinet Committee on the Environment, supported by a network of 'green' ministers. The Committee was chaired by the Deputy Prime Minister who also headed the Department of the Environment, Transport and the Regions (DETR), giving environmental questions a higher profile.

With devolution (transfer of powers from Westminster to a Scottish Parliament and a Welsh Assembly, both directly elected) on the horizon, each component of the United Kingdom began to prepare its own *Forestry Strategy*. In 1998, the policy framework for England, *The England Forestry Strategy. A New Focus for England's Woodlands*, was published. The text printed on the inside cover take up the four main axes of the policy:

> Forestry for rural development
> Forestry for economic regeneration
> Forestry for recreation, access and tourism
> Forestry for the environment and conservation

Four more words printed above and under the four objectives stress the principles underlying the delivery of the all-inclusive objectives of the new government in matters of forestry: 'quality, public support, partnership, integration'. After this, one hardly needs the words of Eliot Morley, the Forestry Minister for England, in the preface to understand that

> This Strategy (…) firmly closes the door on single-purpose plantations [Forestry Commission 1998: preface].

Indeed, it is clearly stated that a programme may come under more than one heading and fulfil various objectives.

'Forestry for rural development' sees wood production as a means to reinforce the economy and stimulate employment in rural areas, an objective reminiscent of the beginnings of British state forestry. 'Forestry for economic regeneration', while seemingly very close to the first objective, addresses the issue of the role of forestry in planning, and includes urban forestry as well as forestry on derelict industrial sites. 'Forestry for recreation, access and tourism' focuses on facilities for the public enjoyment of woodlands. Finally, 'forestry for the environment and conservation' places forestry firmly within the other government's policies in relation to conservation and biodiversity. The details and implications of each of these directions of policy will be analysed in later chapters. Suffice it to say here that the consequence of these changes for the Forestry Commission is that it now works much more in urban and peri-urban areas, environments which were not part of its traditional remit.

New Labour governments under Tony Blair (1997–2007) opted for the creation of larger woodlands, preferably in urban and peri-urban areas, where most of the population live, and on derelict land. Financing this programme has entailed the sale of lands of less than 100 hectares and/or which were unlikely to permit the realisation of multiple forestry objectives, for example because they were situated in inaccessible areas or were leasehold, and in both cases did not permit public access. The Commission has used the proceeds of the sales to purchase land in more strategic areas, closer to towns and cities. The latter trend goes against the whole history of the Forestry Commission, which never actively sought to buy land in peri-urban areas, and even sold the lands it did possess there in the 1980s in order to concentrate on producing timber.

Between 2000 and 2002, the Forestry Commission sold between 800 and 900 hectares and bought about 1,500 hectares, thus increasing the estate. This seems to have gone unnoticed, since most people questioned on the evolution of the Forestry Commission's estate for the *Public Opinion of Forestry* are not aware of the increase. In 2003, only 16% of English adults thought that the estate had increased (15% in 2004), whereas 61% thought it had actually decreased (60% in 2004) [Forestry Commission England 2003: 19; Forestry Commission England 2005: 20].

On top of buying land, the Forestry Commission has also leased lands from local authorities in order to plant trees. Such is the case in the East of London, in and around Manchester and Liverpool, and in the coalfields of Yorkshire.

3.4.2 Devolution and Decentralisation

As part of the programme of constitutional reform, the New Labour Government elected in 1997 rapidly implemented devolution to Scotland, Wales and Northern Ireland. The administration of the state forestry service was affected, except in Northern Ireland where forestry had been devolved since 1922.

In the new organisation, responsibility for Scottish forestry was transferred to the Forestry Department of the Scottish Executive; the Forestry Commission's activities in Wales were transferred from the Secretary of State for Wales to the National Assembly for Wales. Issues concerning the forest in England now rest with the UK government. In each individual nation a director was appointed and national committees deal with the Forestry Commission's policies. This entails among other things considering the respective positioning of the three nations in a competitive market where England consumes the most wood and produces the least.

The government followed the recommendations of Parliament, which advocated in a 1993 report the creation of a ministry which would deal with the environment, agriculture and rural affairs to centralise responsibility for rural affairs within a single department [House of Commons Environment Committee 1993]. In 2001, the Department for Environment, Food and Rural Affairs (DEFRA) was created, and the Forestry Commission has since been incorporated into DEFRA, in order to simplify implementation and control of forest policies. Thus, the Forestry Commission has become much more integrated with wider rural and environmental affairs and it is in a better position to coordinate the national implementations of forestry policies, within for instance the England Rural Development Programme set up in 2000.

Fig. 3.1 The new structure of the Forestry Commission following devolution (From Foresty Commission, 2004b: 63)

The Forestry Commission is answerable to DEFRA as far as England's forests are concerned. Since 2004, forestry policy in England has been transferred to DEFRA as part of the *Rural Strategy*, while the responsibility for Community Forests and the Farm Woodland Premium Scheme has come under the remit of the Forestry Commission. The Forestry Commission England is in charge of policy development, standards and regulations, the management of the State forests, grants and partnerships and research. Decision-making, delivery and funding have been devolved to the regional level. Funding of the Forestry Commission is now endorsed by the different countries, and the UK Parliament funds the English Forestry Commission. The separation of missions between the Forestry Authority and Forest Enterprise remains, but at country level. Forest Enterprise was replaced in March 2003 by three new agencies for the three countries, with a Chief Executive at the head of each one.

As a result of the *Forestry Devolution Review*, Forestry Commission England came into being on 1 April 2003. At the same date, an Executive Board (EB) for the Forestry Commission was appointed, comprising the Executive Commissioners, i.e. the Director General, and the three Country Directors, the Chief Executive of Forest Research and the Heads of Finance, Personnel Development and the Secretary to the Commissioners.

> The main aim of the EB is to assist the Director General and Country Directors in the effective management of the Forestry Commission in order to provide leadership for the Forestry Commission as an organisation, in particular setting policy on corporate matters [Forestry Commission 2005a: 25].

Indeed, a number of the past administrative characteristics remain, and questions concerning all of Great Britain are carried out by a Policy and Practice Division set up for that purpose in 1999. Ministers have expressed their desire to see questions related to international forest policy remain with the UK government, as was made clear in the *Forestry Devolution Strategy* published in August 2002. Similarly, forestry issues which affect the three countries, such as plant health, forest research, training and standards of sustainable forest management have remained at the British level of government. Policies concerning the trading of British wood products, and questions related to research, training and international agreements also come under cross-border topics.

As part of the decentralisation process, increased planning and delivery powers have been devolved to the regional level, with the creation of the Regional Development Agencies. After discussions between the Forestry Commission and other government partners, the Forestry Commission issued *Guidance Notes* on *Regional Forestry Frameworks*. Each region, under the responsibility of the Regional Assemblies, Regional Development Assemblies and Government Offices for the Regions, thus undertook 12-week public consultation exercises to test priorities. The corresponding *Regional Forestry Frameworks*, defining a set of rules and priorities according to each region's needs, were published between November 2003 and August 2005. They aim at reconciling the national objectives stated in the *England Forestry Strategy* with local and regional specificities and needs, e.g. access and recreation in the very urbanised South-East, rural development in the South-West, restoring the degraded landscapes of the Industrial Revolution in the North West and North East. The nine *Strategies* or *Forestry Frameworks* thus define their wide-ranging targets and the partnerships their implementation involves.

The targets will not be developed here since they will be the objects of the following chapters, except for one example, on account of its unique character: that of *A Trees and Woodlands Vision for the London 2012 Games*, part of the *London Trees and Woodland Framework*. Not only will tree-planting be part of the regeneration of East London that will be implemented on that occasion. Leaving a permanent green legacy of the Olympic Games in London, as was the case in Los Angeles in 1984, partakes of the traditional function of trees as markers of important events.

Concerning the partnerships, all *Strategies/Frameworks* list a wide array of partners who have participated in the elaboration of the document and/or who will be stakeholders in the projects. Typically, these include representatives from: the Government Office, the Regional Assemblies, the Forestry Commission's Regional

Advisory Committee, the Countryside Agency (now Natural England), the Department for Culture, Media and Sport, the Department for Environment, Food and Rural Affairs, English Heritage, Sport England, English Nature, the Environment Agency, the NHS Health Development Agency, as well as the Tourist Board, professional organisations like various environmental charities (Wildlife Trusts, Woodland Trust, Groundwork) and the Community Forests when there is one in the region. Delivery plans will follow, to translate the priorities into action in keeping with the priorities of the *England Forestry Strategy* [Forestry Commission 2005b: 14]. Regional progress in achieving national sustainability objectives is assessed against the measures outlined in the *Regional Sustainable Development Framework*. The ripple effect of the *England Forestry Strategy* also concerns local authorities, which are encouraged to set up their own tree/woodland strategies to contribute to the accomplishment of the regional and national objectives. These Strategies provide a vivid illustration of the wide remit of forestry and of the cross-government nature of the *England Forestry Strategy* today.

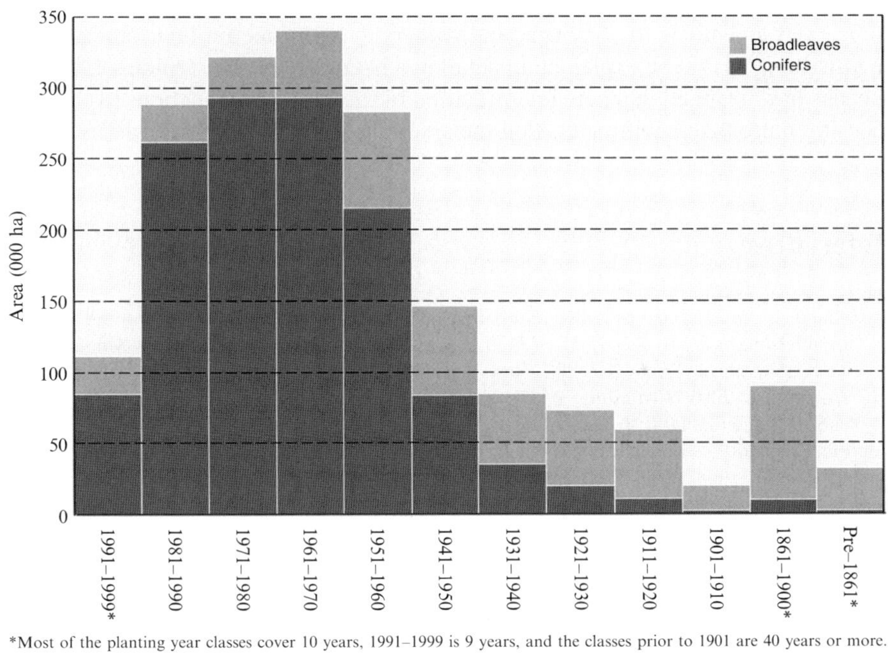

Fig. 3.2 The spectacular growth of plantations over 130 years (From Area by planting year class. Forestry Commission, 2003b: 21

Changes in the 20th century were dramatic. The traditional practices of coppicing and wood pasture were abandoned, ancient woodlands increasingly became isolated because of agricultural intensification, and conifers were widely introduced as even-aged plantations. Thanks to the creation of the Forestry Commission and state forestry policies throughout the 20th century, the proportion of woodland rose from 5% to 10%, fulfilling the objectives fixed at the foundation of the Commission.

Timber production was originally considered to be its prime function, which was in-keeping with international priorities. Yet, with time, it has appeared more and more clearly that the outlooks of the British timber industry were rather poor, and priorities have been modified to a great extent. This in turn has led to changes in the valuation of forest goods and services, to include non-market benefits such as environmental value, leisure and recreation and regeneration, which will be detailed in the next chapters.

The area of woodland in England went from 948,000 hectares in 1980 to 1,104,000 hectares in 2002, and to 1,115,000 hectares in 2004. Out of these, 805,000 hectares belong to the Forestry Commission, while the remaining 910,000 hectares are privately owned or belong to other public bodies [Forestry Commission 2004a: 1]. Most of the growth over the past 20 years has come from the private sector, as appears clearly from the evaluation of the Woodland Grant Scheme made in 2001 [CJC Consulting 2001: 10], and the role of the state has been to provide advice and incentives to convince land owners to plant tree on their lands. These and their results will be explained in detail in a later chapter. The policies of the last decades have resulted in a new interpretation of the economic functions of forestry, focusing more than anything else on the assessment of non-market benefits.

References

Broadhurst, R. and P. Harrop. 2000. Forest tourism. Putting policy into practice in the Forestry Commission. In *Forest Tourism and Recreation. Case Studies in Environmental Management*, eds. X. Font and J. Tribe, 183–200. New York: CABI.

CJC Consulting. 2001. *Evaluation of Woodland Creation in England Under the WGS and the WGS/FWPS*. Edinburgh: Forestry Commission.

Coppok, J. 1960. A decade of post-war forestry in Britain. *Economic Geography* 36 (2):127–138.

Forestry Commission. 1998. *England Forestry Strategy: A New Focus for England's Woodlands*. Cambridge: Forestry Commission.

Forestry Commission England. 2003. *Public Opinion of Forestry 2003: England*. Edinburgh: Forestry Commission.

Foresty Commission 2003. *National Inventory of Trees and Woodlands Great Britain. Edinburgh: Foresty Commission.*

Forestry Commission. 2004a. *Forestry Statistics 2004*. Edinburgh: Forestry Commission.

Forestry Commission. 2004b. *Great Britain and England Annual Report and Accounts 2003–2004. Edinburgh: Forestry Commission.*

Forestry Commission England. 2005. *Public Opinion of Forestry 2005: England*. Edinburgh: Forestry Commission.

Forestry Commission. 2005a. *Great Britain and England Annual Report and Accounts 2003–2004*. Edinburgh: Forestry Commission.

Forestry Commission. 2005b. *England Forestry Strategy Review. Implementation Progress Report 1998–2006*. http://www.forestry.gov.uk/pdf/efs-progress-report-2006.pdf/$FILE/efs-progress-report-2006.pdf. Accessed 14 April 2007.

Forestry Commission. 2006. *Forestry Statistics 2006*. http://www.forestry.gov.uk/website/ForestStats2006.nsf/byunique/imports.html. Accessed 24 December 2007.

HM Treasury. 1972. *Forestry in Great Britain: An Interdepartmental Cost/Benefit Study*. London: HMSO.

House of Commons Environment Committee. 1993. Paper 257. *Forestry and the Environment. First Report, Session 1992–93*. London: HMSO.

James, N. 1981. *A History of English Forestry*. Oxford: Basil Blackwell.

Miles, R. 1967. *Forestry in the English Landscape*. London: Faber & Faber.

Miller, H. 1995. *Forest Policy: The International and British Dimensions*. Aberdeen: University of Aberdeen.

Miller, R. 1981. *State Forestry for the Axe: A Study of the Forestry Commission and De-Nationalisation by the Market,* Hobart Paper 91. London: Institute of Economic Affairs.

Oosthoek, J-W. 2000. *The Logic of British Forest-Policy, 1919–1970*. Paper presented at the 3rd Conference of the European Society for Ecological Economics. *Transitions Towards a Sustainable Europe. Ecology - Economy – Policy*. Vienna, Austria, May 3 - May 6, 2000. http://www.eh-resources.org/documents/esee_paper.pdf. Accessed 1 November 2007.

Pringle, D. 1994. *The First 75 Years. A Brief Account of the History of the Forestry Commission*. Edinburgh: Forestry Commission.

Rickman, R. 1991. *What's Good for Woods: Promoting the Industry, Protecting the Environment*. Policy Study n°129, London: Centre for Policy Studies.

Ryle, G. 1969. *Forest Service: The First Forty-Five Years of the Forestry Commission of Great Britain*. Newton Abbot: David & Charles.

Salter, B. 1998. Urban forestry: UK experience in practice. *Arboricultural Journal* 22 (1), 1998: 11–23.

Taylor, W. 1945. *Forests and Forestry in Great Britain*. London: Lockwood.

Tsouvalis, J. and C. Watkins. 2000. Imagining and creating forests in Britain, 1890–1939. In *Forest History: International Studies on Socio-economic and Forest Ecosystem Change. Report n°2 of the IUFRO Task Force on Environmental Change*, eds. M. Agnoletti and S. Anderson, 371–386. New York: CABI, in association with IUFRO.

Unasylva, 3 (6), 1949, As a Memorial to the Third World Forestry Congress.

Unasylva, 8 (3), 1954, Fourth World Forestry Congress.

United Nations Forum on Forests. 2003. *National Report to the Third Session of the United Nations Forum on Forests. United Kingdom of Great Britain and Northern Ireland*. http://www.un.org/esa/forests/pdf/national_reports/unff3/united_kingdom.pdf. Accessed 17 November 2006.

Winter, M. 1996. *Rural Politics: Policies for Agriculture, Forestry and the Environment*. London: Routledge.

Zuckerman, S. 1957. *Enquiry into Forestry, Agriculture and Marginal Land*. London: HMSO.

Chapter 4
Widening the Scope

But the trees, woodland and green spaces that make up these new forests do not just provide a cosmetic fix.

[ODPM 2004: 51]

Abstract Focusing on the last 20–30 years, the fourth chapter demonstrates how today's new forestry paradigm stems directly from two main influences. The first one is the failure of the productivist ethos developed in the previous chapter, while the second one derives from a change of perceptions and priorities at international level. Three main levers thus articulate present policies: a response to negative reactions from the public to the productivist period, the growth of leisure and an international agenda tuned to sustainability. The tide of change which led to the inclusion of new objectives in forestry is made manifest in the renewed meaning of the word 'forest' itself. A whole new agenda has developed, for the benefit of the population, for which a new concept has developed, that of multiple-use forestry.

Multi-purpose forestry has become a major trend internationally over the last decades, in particular since the 5th World Forestry Congress in Seattle in 1960 made "Multiple Use of Forest and Associated Lands" its major theme.

Britain was a pioneer in a number of the aspects involved in multi-purpose forestry, as this chapter will analyse. In just a few decades, a purely economic approach became a thing of the past for the Forestry Commission and a new priority – or should we say a new range of priorities – emerged. Societal changes as well as economic results required a reassessment of the Forestry Commission's purposes. Thus, the inclusion of new demands, added to the lack of profitability of the forestry sector, have led to the Commission changing beyond recognition, with the 2003–2004 *Annual Report* stating clearly that the Report concluded that the main case for government intervention in forestry is to deliver public goods in the form of urban and peri-urban amenity, recreation and biodiversity [Forestry Commission 2005: 33].

S. Nail, *Forest Policies and Social Change in England.*

4.1 Amenity

Although the word was not used before the 20th century, amenity had always been part and parcel of the functions of the woods grown on country estates and Royal Forests. On top of timber, they provided aesthetic pleasures as well as leisure opportunities such as hunting. The concept of amenity as an objective for planning developed mostly in the first half of the 20th century. The term itself appeared in the *Housing, Town Planning Act* of 1909. It allowed local authorities to prepare schemes for town planning, and added amenity and convenience to sanitation as the objectives of such planning. The creation in 1929 of the Institute of Landscape Architects (ILA) reflected this new preoccupation of planning. Amenity lay at the basis of the *Town and Country Planning Act* of 1947, which made land use planning compulsory for the first time over the whole country and thus laid the foundation for a comprehensive system of control of development in Britain. The term amenity was defined as 'that element in the appearance and layout of town and country which makes for a comfortable and pleasant life rather than a mere existence' [Ministry of Local Government and Planning 1951: 8–11], hence it referred to quality, often visual quality, in the living environment. Section 28 of the Act referred specifically to trees and woods in the form of 'tree preservation orders' for trees deemed of special amenity value [*Town and Country Planning Act* Section 28 1947: 1385–1387].

4.1.1 Landscaping the Plantations

As early as the 1920s, the first voices were heard to warn against the most easily perceptible negative aspect of single species plantations: their unpleasant effect on landscape variety. The monotony of rectangles of dark conifers across the land was first stressed by John Perkins, a member of the Linnean Society in an article dated 1920 [Perkins 1920: 254–258]. Foresters themselves were aware of the fact that plantations detracted from the visual aspect of a landscape, and Lord Clinton, a former chairman of the Forestry Commission, suggested softening up the hard contours of the plantations and mixing species so as to obtain various colours. However, the economic priorities were such that it took a lot of protest for the discourse on aesthetic value to filter through. The recently founded Council for the Protection of Rural England (CPRE) acted as a sounding-board for many citizens when it complained of the disruption provoked by conifer plantations in the emblematic rural landscapes of Britain. This was all the more shocking as the First World War propaganda had played largely on the theme of the countryside landscapes (the 'rural idyll') to encourage Britons to defend their territory against the enemy. A well-known poster thus depicted a soldier pointing to the thatched cottages of his village, against a backdrop of hilly countryside. The text read 'Isn't this worth fighting For? Enlist now'.

Some particular sites loaded with cultural associations were the scenes of public protest. Such was the case of the New Forest, where conifer plantations in 1928 led to public indignation. Similar plantations in the Lake District, where the landscape sung by the Romantic poets was viewed as part of the national heritage, gave rise to opposition. Faced with public antagonism, the Forestry Commission began to include visual amenity in its objectives. In 1934, part of the fifteenth *Annual Report* was devoted to 'Hardwoods, Conifers and Amenity'. The latter referred essentially to visual pleasantness, as can be seen from the following passage:

> Planting hardwood belts along roadsides in flat country and other suitable positions;
> Avoiding straight outlines to plantations
> Laying out roads and rides with care for the country
> Introducing variety by changes of species within the plantations. [Miles 1967: 74]

These principles were irregularly implemented, and public protest continued, so that in the same decade, the Forestry Commission and the Council for the Protection of Rural England (CPRE) set up an informal committee to try and improve the aesthetics of the Commission's plantations. This resulted in 1936 in an agreement between the Forestry Commission and the CPRE over the limitation of forestry in the Lake District, which precluded the central zone of the Lake District from any land acquisition by the Forestry Commission. In 1934, the Commissioners and the CPRE formed an informal joint committee to provide advice on how to improve the appearance of plantations. But it took until the late 1950s' for the Forestry Commission to take into account the arguments in favour of a better integration of conifer plantations into the surrounding landscapes. After a declaration by the Director-General in 1962, the Commission appointed their first landscape consultant, Sylvia Crowe, the famous landscape architect and former President of the Institute of Landscape Architects, who kept the position until she retired in 1976. In 1966 she wrote a booklet which gave foresters basic principles of composition for plantations that were sympathetic with the character of the landscape: 'the age of square-sided plantations on the hillside' progressively became 'a thing of the past' [Broadhurst and Harrop 2000: 186].

Amenity as synonymous with pleasant appearance slowly penetrated the official discourse, and its definition, as well as its beneficial effects, began to be the object of a consensus and finally taken for granted. To give just one instance, just after the Second World War, the Department of Health in Scotland advocated planting more trees in and around urban areas in order to bring the 'pleasant and beneficial effects' of the colour green to people [Lambert 2005: 22]. This probably also had to do with research into the positive psychological effect of the colour green, which began before the Second World War. The simple, quiet appreciation of the beauties of a landscape seemed to conduce to unquestionable benefits for anyone and everyone, and this had consequences both in planning and in conservation policies.

4.1.2 Providing for Leisure Needs

Even though it opened the door to intensive conifer planting for the next 2 decades, the 1943 White Paper also stipulated that the leisure function of woodlands should be taken into account. It was indeed becoming increasingly difficult to ignore public demands as leisure and tourism developed. An agreement having been reached between the Youth Hostels Association and large estates, a successful attempt had been made to test the leisure potential of forests in the inter-war years. Following the report of the Committee on National Parks in 1931, the Forestry Commissioners decided to launch the first instance of a Forest Park in Argyll Forest (Scotland) in 1936. By 1964, there were eight Forest Parks in Britain. For the first time, the *Forestry Act* 1951 mentioned leisure as part of the Forestry Commission's functions. On 24 July 1963, a declaration on forestry was made in both the House of Commons and the House of Lords, stating that, on top of planting 180,000 more hectares of forests between 1964 and 1973, the Commission should henceforth pay more attention to such issues as leisure, access and amenity. The *Forestry Act* 1967, which repealed and updated the Acts of 1919, 1945 and 1951, went further, by making it an obligation for the Forestry Commission to provide for leisure needs in its forests, and by also stressing its role in conservation. The Forestry Commission was henceforth entitled to

> Manage, plant and use land placed at its disposal,
> Manage and supervise woods and forests and give assistance and advice in regard to forest planting and management
> Purchase and dispose of standing timber and promote the supply, use and conversion of timber
> Establish, carry on and promote woodland industries
> Regulate private felling and planting
> Make orders for the control of timber pests and diseases
> Collect, prepare and publish forestry statistics, promote training in forestry and conduct forestry research
> Provide tourist, recreational and sporting facilities [NDAD 2003].

In the same decade, landscape degradation through 'ribbon development' and its accompanying infrastructure of by-pass roads and motorways was leading to a sense of loss of the English countryside. The *Countryside Act* of 1968 created the Countryside Commission to cater for recreation in the countryside by providing country parks and picnic sites. This Act reinforced the 1967 one as far as the Forestry Commission was concerned, since it gave the latter permission to plant and manage its lands according to amenity purposes and to provide recreational facilities. The *Wildlife and Countryside (Amendment) Act* 1985 took a step further, urging the achievement of a *balance* between, on the one hand, the management of woodlands and the production of wood and, on the other hand, the enhancement of natural beauty and the conservation of special sites. The growth of tourism and conservation preoccupations in woodlands went hand in hand.

4.2 Conservation

As a reaction to the excesses of scientific forestry, voices had begun to praise a return to 'naturalistic silviculture' [Ciancio and Nocentini 2000: 49] as early as the end of the 19th century (Karl Gayer, Bernard Lorentz, and Adolphe Parade, for instance).

The tables had begun to turn in relation to the perception of the ecological and conservation value of woodlands.

4.2.1 The Ecological Value of Forests

In Britain, Arthur Tansley's ecosystem concept (1935) paved the way for the recognition of forests as complex ecosystems harbouring a wide range of wildlife, topics developed in the 1950s by Mark Anderson, Professor of Forestry in Edinburgh and Research Officer of the Forestry Commission, and Frank Fraser Darling, the human ecologist. The 1949 Act which set up National Parks also established the Nature Conservancy to advise government on conservation issues. It set up regional offices and 'started the immense tasks of selecting, demarcating and securing by purchase, lease or agreement, National Nature Reserves throughout the country' [Miles 1967: 114]. By 1965, there were in England 54 National Reserves, and 16 in the Forest or Local categories. Nature Conservancy became the Nature Conservancy Council in 1973. It was split into three organisations corresponding to the three countries of Great Britain in the early 1990s, with English Nature in charge of conservation in England, and a Joint Conservancy Council overseeing the three agencies. Pressure grew from such charities as the CPRE, the Royal Society for the Protection of Birds (RSPB) and, in the 1970s, emphasis was officially placed on the environmental role of the Forestry Commission and a Wild Life Officer was appointed. Under the *Wildlife and Countryside Act* 1981, amended in 1985, greater protection was given to special sites, Sites of Special Scientific Interest (SSSI): woodlands and forests could be declared SSSIs.

Tension between conservationists and foresters led to a conference jointly organised by the Forestry Commission and the Institute of Chartered Foresters in 1982, which resulted in the formulation of a Broadleaves Policy. *Broadleaves in Britain*, as the conference was entitled, aimed at investigating the future of broadleaves and improving their management, and it proved to be a turning-point in the Forestry Commission's thinking and policy. In 1985 (incidentally declared 'International Year of the Forest' by the FAO Council), only 1 month before the vote of the *Wildlife and Countryside (Amendment) Act* 1985, the Commission announced its new policy in favour of broadleaves. Ancient semi-natural woodland would benefit from special measures and grants to ensure that their area did not diminish further; more deciduous trees would henceforth been planted on the Commission's lands, and grants of a higher amount than for conifers would be available for planting broadleaves on private lands, even if timber production was not the main objective of the plantings. Areas of broadleaves felled by the Commission should be replanted with broadleaves [Forestry Commission 1985a, b]. This was a far cry from the *Forestry Act* 1951 which had contemplated the disappearance of deciduous woodlands to the benefit of conifer plantations, but it is less surprising if we put this U-turn in relation with changes of perception of forestry internationally.

Indeed, the widening scope of the Forestry Commission's remit derives, not only from public pressure, but also from the evolution of the perception of forestry issues by foresters themselves, as can be witnessed in the topics chosen for the World Forestry Congresses. In 1978, the 8th WFC held in Jakarta (Indonesia) selected the theme of 'Forests for People' and dealt with the various demands people put on forests, thus opening forestry to more missions than productivity. The topic of the following Congress, held in Mexico in 1985, was 'Forestry in the Integral Development of Society', and it underlined the importance of the conservation of forest ecosystems. The 10th WFC held in Paris in 1991 focused on 'Forests, a Heritage for the Future', and its conclusions put forward the need to think of forest management as a long-term pursuit. The shift in foresters' perspective shows that the seeds sown at the Seattle WFC in 1960, the first one to deal with recreation and wildlife as part of the multi-purpose forestry, had had time to grow and become integrated into a wider vision of the benefits of forestry.

The results of the Broadleaves policy have been tangible in Britain. Whereas in the mid-1980s, 25,000–30,000 hectares of trees (95% of which conifers) were planted yearly by the Forestry Commission, this figure had dropped to 12,000–13,000 hectares at the beginning of the 1990s, and the plantation of conifers had all but stopped [Clegg & Co 2002]. In the late 1990s, the Forestry Commission was planting annually 5,000 hectares of trees in England, 80% of which were broadleaves. The proportion of broadleaves increased by 36% between 1980 and 1998 in England, while the relative proportion of broadleaves in relation to conifers passed from 56% to 65% [Forestry Commission 2001: 2].

4.2.2 Protecting Ancient Woodlands

Not only were broadleaves to be planted: they were to be protected, especially if they stood in old woodlands. In order to do this, they had to be recognised, then classified.

As was alluded to in chapter two, forest history as an area of academic research grew in the last 25 years of the 20th century, often in relation to the development of historical ecology, historical geography and cultural geography. Thanks to a wealth of academic research (see bibliography) and especially, in connection with trees in the British landscape, to those of Oliver Rackham, Richard Muir and Chris Smout, landscapes have increasingly been perceived as settings where layer upon layer of interactions between humans and nature has transformed over time 'natural' landscapes into palimpsests of meanings related to identity. Bearing witness to this development, a 'cultural landscapes' category was added to the World Heritage List of UNESCO in 1992. Another significant testimony of this growing concern was the choice of 'Landscape and Garden History: Issues, Approaches, Methods' as the topic for the 1989 symposium of the Garden and Landscape Studies programme at Dumbarton Oaks (Washington D.C.).

Once recognised as cultural landscapes, woodlands had to be classified. According to the classification elaborated by Oliver Rackham and George Peterken, 'ancient' woodland refers to those areas which have never been used for other purposes than woodland since at least 1600, while 'recent' woodland refers to those woodlands which have grown on unwooded land since 1600, with cartographic evidence being used to draw the line between the two categories. As a consequence of this categorization, an *Ancient Woodland Inventory* was started by the Conservancy Council in 1981, and management grants were introduced in 1992 for woodlands of high environmental quality.

An amendment in the 1985 *Wildlife and Countryside (Amendment) Act* formalised the multiple objectives that the Forestry Commission must aim for:

> In discharging their functions under the Forestry Acts 1967 to 1979 the Commissioners shall, so far as may be consistent with the proper discharge of their functions, endeavour to achieve a reasonable balance between:
> the development of afforestation, the management of forests and the production and supply of timber, and
> the conservation and enhancement of natural beauty and the conservation of flora, fauna and geological or physiographic features of special interest [Miller 1995: 68].

This was taken up in 1991 by the Commission itself in a policy statement entitled *Forestry Policy for Great Britain*, which endorsed the principle of multiple use forestry and defined the main functions of the Forestry Commission as being

> The sustainable management of our existing woods and forests
> A steady expansion of tree cover to increase the many, diverse benefits that forests provide [Miller 1995: 45, 70; Winter 1996: 297].

4.3 Sustainability: A New Preoccupation in Forestry

The emphasis put on cultural landscapes, ecology and multi-purpose forestry went hand in hand with the emergence of a 'new scientific paradigm', 'holistic' or 'systemic', based on the concept of 'the intersubjectivity of science' [Ciancio and Nocentini 2000: 51], i.e. dependent both on a network or relationships involving the observer as well as the environment. O. Ciancio and S Nocentini argue that this has greatly contributed to forests acquiring a new cultural dimension.

At the same time as these changes were taking place, the UK Government was coming under increasing scrutiny from various NGOs, at the head of which was the WWF, for imports of wood products resulting from unsustainable sources and illegal logging. As one of the main importers of wood in the world, the UK was challenged to prove that it respected its international engagements. This reflects a wider trend which has become all-important in the last decade of the 20th century, that of sustainability.

4.3.1 International Policies

The concept of sustainable development gained growing recognition on the international scene, with important landmarks: the Club of Rome report *The Limits to*

Growth (1972), the first 'Earth Summit' in Stockholm in 1972, the Brundtland report in 1987 and the United Nations Conference on Environment and Development (UNCED) in Rio in 1992. At the latter, forest-related matters proved controversial, yet intense negotiation led to the *Statement of Principles for a Global Consensus on the Management, Conservation and Sustainable Development of all Types of Forests*, also known as *Statement of Forest Principles*, a non-legally binding set of principles which was one of the main documents issued in Rio de Janeiro.

Although it had predecessors, e.g. in the form of 'progressive forestry' in the 1949 WFC resolution in Helsinki, the *Statement of Forest Principles* was widely advertised as the first global agreement on the sustainable management of the forests of the world. It included reforestation and forest conservation objectives. Figuring among the important breakthroughs of the 'Earth Summit' was *Agenda 21*, hailed as a turning-point in that it integrated sustainable development objectives at all levels of government. Chapter 11 and chapter 28 were particularly relevant, since they were devoted respectively to the fight against deforestation and to local authorities' initiatives. The motto 'think globally, act locally' was to serve as a guideline and the stress was put on cross-sectoral participation (the participatory principle had already been enunciated in *The World Charter of Nature* (clauses 16 and 23) in 1982), which had implications for the forest sector, as we shall see.

Since Rio, the stress has been put on the development of coherent policies embracing the management, the conservation and sustainable development on all forests. To achieve these objectives, the Intergovernmental Panel on Forests (IPF) was appointed in 1995 under the auspices of the United Nations Commission on Sustainable Development. It reported in 1997 to the UN, urging governments to turn guidelines into National Forest Programmes as national frameworks [Font and Tribe 2000: 7]. This gave birth to the adoption of Forest Protocols and Forest Standards. The IPF was followed by the Intergovernmental Forum on forests (1997–2000), and between them, they issued 270 proposals for action towards sustainable forest management, the *IPF/IFF Proposals for Action*, which the countries are expected to enforce. To follow up on the work of IPF and IFF, the United Nations Forum on Forests (UNFF) was established (2000–2003), promoting multi-stakeholder dialogues to develop its participatory agenda with a wide range of stakeholders (minorities, NGOs, trade unions, businesses, local authorities, among others).

The topics of the World Forestry Congresses reflect this holistic vision. The 1997 WFC, held in Antalya (Turkey), was entitled 'Forestry for Sustainable Development: Toward the 21st Century' and aimed at responding to the challenges of the IPF. The *Antalya Declaration* called for increased political will, progress in forestry education, the promotion of participatory approaches to forest planning, methodologies concerning the valuation of forest goods and services, and the elaboration and adoption of criteria and indicators of Sustainable Forest Management [*XII World Forestry Congress Bulletin* 2003: 1]. The latest (12th) WFC, took place in Quebec City (Canada) in September 2003, and its topic was 'Forests, Source or Life'. It took up the various trends of forestry over the previous congresses, i.e. sustainable development, social needs and cultural values, as can be seen in its final declaration, which put forward a vision for a future

> with social justice, economic benefits from sustainable forest management, participatory governance, and responsible use of forest resources [Environment News Service 2003].

In parallel, moves have been made to harmonise standards. An international network, Forest Stewardship Council (FSC), was set up in 1993 to promote 'environmentally responsible, socially beneficial and economically viable management of the world's forests, by establishing a worldwide standard of recognised and respected Principles of Forest Stewardship' [FSC 2004]. It coordinates one of the most widely known forest certification systems.

By the time the World Summit on Sustainable Development was held in Johannesburg in 2002, it was clear that the Earth Summit of 1992 had led to more words than action, both in environmental and in societal terms. This is probably due to the strength of the positivist culture which makes it difficult to replace the old scientific paradigm, based on economic management for sustained production, by a new, holistic, one, taking into account ecological and cultural values. However, the vision of forestry put forward at the Rio Summit probably provided an impetus for action on forests at the European level, with clear implications for British policies.

4.3.2 The European Dimension

Forests were the great absent of the European construction to begin with, both on environmental and economic fronts. The Treaty of Rome neither included environmental protection in its objectives, nor did wood produce figure among the products included in the common policy of the Common Market.

The year when Great Britain joined Europe (1973) saw the First Environmental Action Programme (1973–1976) being formulated at the European level. It focused on the improvement of the quality of life of the European peoples and therefore established principles which have since been refined in their definition and applications: the precautionary principle and the 'polluter pays' principle. A special policy sector around environmental questions at the European level was born with the creation of the Environment Directorate General (DG XI) within the European Commission in 1981. DG XI has been instrumental in driving forward environmental legislation at the European level. Legal recognition of the inclusion of environmental issues in the European agenda came with the 1987 Single European Act. Forests came into their own in the 1980s', as the analysis of the series of reports on *European Timber Trends and Prospects Study*, published by the CEE-ONU/FAO from 1952, reveals. A clear change occurred between the 1976 report (*European Timber Trends and Prospects Study III*) and the 1986 report (*European Timber Trends and Prospects Study IV*). In the latter, a chapter was devoted for the first time to products and services other than wood [Parlement Européen 1994: IV.2.2 §2.2.2]: the economic, environmental and social dimensions were being woven together at the European level. Figuring prominently among the main landmarks on the road to recognition of the forest sector at the European level is the International Conference on Trees and Forests (*SILVA Conference*) held in Paris in 1986. It acted as a forum of discussion for European heads of state and a landmark in the awareness of the assets and problems of forests.

Ministerial Conferences on the Protection of the Forests of Europe (MCPFE) were initiated in Strasbourg in 1990, and the declarations and resolutions passed at each of these show a growing emphasis on social and environmental aspects, on top of biological and environmental ones. The six resolutions passed during the Strasbourg conference centred on forest ecosystems and tree physiology, genetic resources, forest fires and mountain forests. The second conference, held in Helsinki in 1993, was geared in its four resolutions towards questions of biodiversity, cooperation with developing countries and adaptation to climate change. It provided a precise definition of sustainable forest management, since the Brundtland definition of 1987, namely development 'that meets the needs of the present without compromising the ability of future generations to meet their own needs' [WCED 1987: 8], was not easily applicable to forestry and open to too many interpretations. The following has become the reference definition of sustainable forest management:

> the stewardship and use of forest lands in a way, and at a rate, that maintains their biodiversity, productivity, regeneration capacity, vitality and their potential to fulfil, now and in the future, relevant ecological, economic and social functions, at local, national and global levels, and that does not cause damage to other ecosystems [MCPFE 1993: 1].

The research dimension, which was lacking, was added with the foundation of the European Forest Institute (EFI) in 1993. It is a non-governmental organisation which, in 2005, totalled 130 members (research institutes, universities, NGOs and industries) from 39 countries. The EFI undertakes research on forest policy, including its environmental aspects, on the ecology, multiple use, resources and health of European forests and on the supply of and demand for timber and other forest products and services in order to promote the conservation and sustainable management of forests in Europe.

In 1994, a key report was published on the forests of Europe by the European Parliament [Parlement Européen 1994]. It took stock of the recent evolution of forests and forestry and noted that until recently, publications on the functions of forests had remained vague: interest in forests and forestry at the European level was first of all subjected to policies in other areas, a sideline of agricultural, environmental and rural development policies. Moreover, forestry was only considered in its own right in cases of short-term predictable difficulties. The authors explained this by the low share that wood produce represented in the economy of the countries (2–3% of the GDP). It may also be linked to the fact that, contrary to the Third World, forests of Europe had seen their surface area grow in the 20th century, especially since the 1970s', so that forestry was not perceived as a problem-area.

The *Third Ministerial Conference on the Protection of Forests in Europe* took place in Lisbon in 1998, a year after the Antalya congress, and reflects recent changes. In the *General Declaration* and the two resolutions, the signatories asserted the responsibility of Europe's governments to strengthen the links between forests and society and expressed the need to integrate 'all forest functions', including 'wood and non-wood forest products and services' [MCPFE 1998: 1]. The social aspect formed the backbone of Resolution L1, which advocated partnership to raise the level of awareness, through education and participation, of the role of forests in sustainable development. The general declaration and the five resolutions of the *Fourth Ministerial Conference on the Protection of Forests in Europe* (Vienna 2003), read like a synthesis of the evolution of the vision and objectives of forests in Europe over the last 15 years, and therefore reflects the integration of global changes, apparent in the Québec WFC for instance. Far from being a sideline of agriculture, forestry has come to embrace objectives related to sustainable development and social and cultural welfare both in the countryside and in cities, as well as enhancement of biodiversity and instrument against climate change, not forgetting economic viability. The tools to fulfil these aims include cross-sectoral co-operation, and participation is seen as an essential component of 'National Forest Programmes'. The very title of the general declaration – 'Vienna Living Forest Summit Declaration'– testifies to the growing role that forests have been invested with. The first sentence of the Vienna declaration boldly asserts: 'Living forests are the basis of life on Earth. By sustaining forests, we sustain life' [MCPFE 2003: 1].

4.3.3 Sustainability in British Forests

The progressive development of the concept of sustainability applied to forestry led to the publication in 1998 of the *UK Forestry Standard*. It established a comprehensive benchmark for sustainable forestry according to the Helsinki agreements. It was a milestone in that it identified forestry practices which were, or were not, appropriate, and thus established sustainable forestry practices [Forestry Commission 1998], a commitment which was reiterated in the *UK National Forest Programme* (2003). The British government, as one of the main importers of timber and wooden products in the world, has adopted a policy of purchasing wood products that, eventually, will come only from legally and sustainably managed sources. Following the Asia Forest Law Enforcement and Governance in 2001, for instance, the UK entered into a bilateral agreement with Indonesia in order to tackle illegal logging and trading. At home, the whole UK public forest estate managed by the Forestry Commission and the Northern Ireland Forest Service was assessed in 1999–2000 against the Standard. The certification process is a complex one; timber and wood products are assessed against the strict standards enunciated in the UK Woodland Assurance Standard (UKWAS) to preserve biodiversity and sustainability. If the standards are achieved, wood produced from an UKWAS accredited source can display the Forest Stewardship Council (FSC) logo to advertise the sustainability of the product. Other certification schemes exist, such as PEFC, SFI and CSA, but the FSC label is probably best known, hence the choice of the Forestry Commission for its woodland. A second edition of the *UK Forestry Standard* was published in 2004 to take into account recent changes, including devolution.

The *UK Woodland Assurance Scheme*, launched in 1999, was based on the principles defined in the *UK Forestry Standard*. It came as an answer to the queries of NGOs about the origins of timber used in the UK, and was the result of an original tripartite agreement signed by the Government, NGOs, and the processing sector (i.e. timber producers ad woodland owners) to implement a certification standard to ensure buyers and users that the wood products come from sustainably managed woodlands. As a consequence of the *UK Forestry Standard*, a set of indicators of Sustainable Forestry was produced and published in 2002, along the same lines as the national set of 15 'headline' indicators of sustainability which had been developed in 1999 to help monitor progress at the national level. The 40 indicators cover six themes: 'woodland, biodiversity, condition of forest and environment, timber and other forest products, people and forests, and economic aspects' and aim at measuring the contribution of forests to the sustainability agenda [United Nations Forum on Forests 2003: 5].

In order to monitor progress, factual data were necessary. The range of statistical information has thus been extended. *British Timber Statistics* provide figures on production and consumption of wood products by the British industry. *Forest Statistics* have been added to the existing *Forestry Facts and Figures*, and include some information about environment and recreation in forests. Finally, a *National Inventory of Woodlands and Trees* was embarked on; it was completed in 2002 and provides factual information about British forests (size, distribution, type, general condition and treatment) as well as a benchmark of the evolution of these data.

4.4 Integrating All Forestry Missions

As a consequence of the widening scope of forestry, forestry policies have recently been drawn into a web of other governmental policies, which requires a close co-operation, both horizontally between the different departments and agencies of the government, and vertically with an integrated approach from the local to the UK level, through to regional and national, and even international.

4.4.1 Governmental Policies

This appears clearly in numerous official publications, including the Forestry Commission's *Corporate Plan for England and Great Britain Activities 2005–2006*. It summarises the various governmental policies that have affected forestry. They include, first of all, policies on sustainable development, expressed most recently in the governmental Strategy, *Securing The Future* published in March 2005. The implications of this for forestry are the need to connect the economic, environmental and social values of forests [Forestry Commission 2006: 9]. Other influences include the *England Rural Development Programme*, which redirects some support for agriculture and forestry, as well as the *Countryside and Rights of Way Act 2000*, with its provisions for access to the countryside, the *UK Energy White Paper* [DTI 2003] in its focus on renewable sources of energy, and the *England Biodiversity Strategy* (2002) and its stress on ancient and native woodland. Forestry activity in the private sector now comes under the Country Rural Development Programme, according to the principles laid down by the Rural Development Commission in Europe.

Co-operation extends to the business sector, with the development of the English Forest Industries Partnership. According to Paul Hill-Tout, Director of the Forestry Commission England,

> since the Rural White Paper was published, significant progress has been made across all four programmes of the England Forestry Strategy. A more integrated role for forestry and woodlands has developed in both rural and urban areas. (...) The opportunities woodlands provide for employment, education, health and recreation need to be exploited, along with efforts to diversify the range of economic goods they provide. There is also a need to find new ways to ensure ancient woodlands - those which have continuity back to the original wildwood, and are most valuable for wildlife and heritage - are managed and used sustainably [Forestry Commission 2005: 34].

The complexity of state forestry at the beginning of the 21st century is clearly reflected in the list of missions that Forest Enterprise England is meant to fulfil, according to the Forestry Commission's *Annual Report 2003–2004*:

> To manage the estate sustainably in all respects (Sustainable forest management);
> To increase the contribution of the estate to the rural economy (Economic);
> To conserve and enhance the biodiversity, landscape and cultural heritage of the estate (Environmental);
> To promote public access to the estate and develop diverse recreational opportunities (Social);
> To expand the role of the estate in improving the environment of disadvantaged communities and bringing brownfield land into beneficial public use (Social) [Forestry Commission 2005: 141–142].

The new Director General of the Forestry Commission presented the order of priorities of the Forestry Commission in 2004 in a way unthinkable 3 decades before:

> As the biggest single provider of countryside recreation, the guardian of a huge wildlife resource, and the largest producer of wood, we work with a very wide range of interests, whose needs are continually changing [Forestry Commission 2005: 4].

4.4.2 From the Global to the Local

European policies have a marked effect on British ones. First of all, as was mentioned before, the entry into the European Union of Finland, Sweden and the Baltic countries, with their huge reserves of timber, contributed to changes in forest policy in the United Kingdom. In addition, finances for the *England Rural Development Programme*, which supports agriculture and forestry, partly come from the European Union. This has also been the case for the Woodland Grant Schemes and Farm Woodland Premium Schemes since the beginning of 2000. The *2004 Mid Term Review of the Common Agricultural Policy European Agricultural Fund for Rural Development*, which officially broke the link between production and subsidies, could indeed have repercussions on land use in the countryside and possibly on woodlands in England.

As has been shown, the quantitative objective is no longer of primary interest for the government. Quite the reverse, the management and increase in forest area are perceived as a means to an end – the delivery of other public benefits. An emblematic example of policy changes can be seen in the New Forest. Protests in relation to conifer plantations in 1928 were mentioned above. Following the tide of change in the forestry objectives, the whole of the New Forest was classified as a Site of Special Scientific Interest in 1971 and a consultative commission was finally created to take into account public opinion in the management of the New Forest. On 1 March 2005, it became a National Park to preserve its environment and its character.

The tide of change which led to the inclusion of new objectives in forestry is made manifest in the renewed meaning of the word 'forest' itself. After being used with a legal meaning associated with absolutism under the Normans, and with an economic one in the 19th and 20th centuries, the word 'forest' in Britain today has come to cover a range of meanings which have nothing to do with the previous definitions of the term. Forests in the 21st century have little to do with economy, less so with exclusion, two senses of the word which made forestry unpopular. A whole new agenda has developed, for the benefit of the population, for which a new concept has developed, that of multiple-use forestry.

One of the main institutions implementing multi-purpose forestry is the new National Forest in the Midlands, which encompasses parts of Leicestershire, Derbyshire and Staffordshire. The 510 square kilometre area concerned includes, on top of towns, agricultural lands, mineral land, as well as remnants of two ancient forests and transport corridors. Without going into the details of the project in this chapter, it seems important to stress two things here. First, the discourse around the concept of forest that his project has given rise to. The decision was taken in the early 1990s to transform this area into a 'forest', supposedly in the medieval sense of the word. What is meant by this is before anything else a mixed use landscape, the idea of a 'mosaic' often developed in official literature, made up of settlements, agricultural lands and up to 30% woodland cover. Secondly, the multifarious aspects entailed in his regeneration project are worth noting. On top of landscape and environmental improvement in an area damaged by coal extraction, what is sought is economic regeneration, leisure and tourism development, timber resources and the diversification of rural businesses. Quite a heavy agenda to develop in 20 years' time.

The various forms and functions of multi-purpose forestry will be the object of the next two parts of this study. But beforehand, one must stop and analyse the development of a brand new branch of forestry, typically multi-purpose in its remit, that of urban forestry.

References

XII World Forestry Congress Bulletin 10 (12), 2003.

Broadhurst, R. and P. Harrop. 2000. Forest tourism. Putting policy into practice in the Forestry Commission. In *Forest Tourism and Recreation. Case Studies in Environmental Management*, eds. X. Font and J. Tribe, 183–200. New York: CABI.

Ciancio, O. and S. Nocentini. 2000. Forest management from positivism to the culture of complexity. In *Methods and Approaches in Forest History. IUFRO Research Series N°3*, eds. M. Agnoletti and S. Anderson, 47–58. New York: CABI, in association with IUFRO.

Clegg & Co. 2002. *Evaluation of Woodland Creation in England under the Woodland Grant Scheme and the Farm Woodland Premium Scheme*. London: Department for Environment, Food and Rural Affairs.

Environment News Service. 2003. World Forest Congress Harmony Broken by Canadian Logging. http://www.ens-newswire.com/ens/sep2003/2003–09–30–04.asp. Accessed 17 November 2006.

Font, X. and J. Tribe. 2000. Recreation, conservation and timber production: a sustainable relationship? In *Forest Tourism and Recreation. Case Studies in Environmental Management*, eds. X. Font and J. Tribe, 1–22. New York: CABI.

Forestry Commission. 1985a. *The Policy for Broadleaved Woodlands. Policy Paper N°5*. Edinburgh: Forestry Commission.

Forestry Commission. 1985b. *Guidelines for the Management of Broadleaved Woodlands*. Edinburgh: Forestry Commission.

Forestry Commission. 1998. *The UK Forestry Standard*. Edinburgh: Forestry Commission.

Forestry Commission. 2001. *National Inventory of Woodlands and Trees – England*. Edinburgh: Forestry Commission.

Forestry Commission. 2005. *Great Britain and England Annual Report and Accounts 2003–2004*. Edinburgh: Forestry Commission.

Forestry Commission. 2006. *Corporate Plan for England and Great Britain Activities 2005–2006*. Edinburgh: Forestry Commission.

FSC (Forest Stewardship Council). 2004. *FSC Principles and Criteria for Forest Stewardship*. http://www.fsc.org/plantations/docs/FSC-STD-01–001%20FSC%20Principles%20and%20Criteria%20for%20Forest%20Stewardship%202004–04.pdf. Accessed 11 November 2007.

Lambert, R. 2005. *Trees and People: Tracing the Early Provision of Forest and Woodland Recreation in Twentieth Century Britain*. http://www.nottingham.ac.uk/ttri/pdf/2005_3.pdf. Accessed 5 March 2006.

MCPFE (Ministerial Conference on the Protection of the Forests of Europe). 1993. *Resolution H1. General Guidelines for the Sustainable Management of Forests in Europe*. Helsinki. http://www.mcpfe.org/system/files/u1/helsinki_resolution_h1.pdf. Accessed 12 November 2007.

MCPFE (Ministerial Conference on the Protection of the Forests of Europe). 1998. *General Declaration*. Lisbon. http://www.mcpfe.org/mcpfe/resolutions/lisbon/general_declaration.pdf. Accessed 12 November 2007.

MCPFE (Ministerial Conference on the Protection of the Forests of Europe). 2003. *General Declaration*. Vienna. http://www.mcpfe.org/mcpfe/resolutions/vienna/Vienna_Declaration.pdf. Accessed 12 November 2007.

Miles, R. 1967. *Forestry in the English Landscape*. London: Faber & Faber.

Miller, H. 1995. *Forest Policy: The International and British Dimensions*. Aberdeen: University of Aberdeen.

Ministry of Local Government and Planning. 1951. *Town and Country Planning 1943–1951: Progress Report*. HMSO: Cmnd 8204.

NDAD (National Digital Archive of Datasets). 2003. *Forestry Commission*. AH/3. http://www.ndad.nationalarchives.gov.uk/AH/3/detail.html. Accessed 1 March 2006.

ODPM (Office of The Deputy Prime Minister). 2004. *Making it Happen – The Northern Way*. London: ODPM.

Parlement Européen. 1994. *L'Europe et la forêt – EUROFOR*. Coordonné par l' Office National des Forêts (France). 2 vols. Complete text available on http://www.europarl.eu.int/workingpapers/agri/default_fr.htm. Accessed 16 September 2007.

Perkins, J. 1920. A plea for the consideration of the aesthetic side in restocking our warfelled woods. *Quarterly Journal of Forestry* 13: 254–258.

Town and Country Planning Act. 1947. London: Hansard, Geo.6, 10&11.

United Nations Forum on Forests. 2003. *National Report to the Third Session of the United Nations Forum on Forests. United Kingdom of Great Britain and Northern Ireland*. http://www.un.org/esa/forests/pdf/national_reports/unff3/united_kingdom.pdf. Accessed 17 November 2006.

WCED (World Commission on Environment and Development). 1987. *Our Common Future*. Oxford: Oxford University Press.

Winter, M. 1996. *Rural Politics: Policies for Agriculture, Forestry and the Environment*. London: Routledge.

Chapter 5
Forestry Comes to Town

If you are not sure what urban forestry is, then neither are we

[Johnston 1999: 234]

Abstract One of the most dramatic changes of the last 3 decades, in England as in many other countries, has been the rise of urban forestry. It is important to put this trend into an international context, in order to determine the English characteristics of the movement and, conversely, to stress the part that England has played in the urban forestry movement. The fifth chapter thus analyses the definition, remit and tools devised to deliver the objectives of urban forestry. It also stresses elements of resistance to this new concept, linked with the cultural background. Among these are to be found the historic connection of forests with social status, the association between forestry and the Forestry Commission's work and the dominantly private ownership of land. However, at the beginning of the 21st century, urban forests seem to have become widely accepted, as the figures of their frequentation by the public testify.

In a country like England where 90% of the population live in urban areas, compared to a European average of 70%, developing multi-purpose forestry has meant both managing peri-urban forestry for new purposes as well as bringing woodlands closer to towns and cities. This chapter explores the significant milestones in the development of urban forestry, from its genesis to its adaptation in England, with a view to setting the scene for the qualitative analysis of the following chapters.

5.1 The Origin of the Concept

Although the term 'urban forestry' made this new way of practising forestry official and visible in the second half of the 20th century, the practice of urban forestry in England had begun a century before, in the aftermath of the 1833 report of the Parliamentary Select Committee on Public Walks, as was shown in chapter two. Likewise, although the concept is often associated with North America, there were precursors elsewhere.

S. Nail, *Forest Policies and Social Change in England.*

5.1.1 Genesis and Definitions

Indeed, the term 'urban forestry' appeared for the first time in 1965 at the University of Toronto, in the title of Willem Morsink's thesis on the evaluation of tree planting projects in a borough of Metropolitan Toronto. However, the concept had antecedents in Europe. A particular case in point is the Amsterdamse Bos in the Netherlands, 'the first modernist landscape in Europe based on urban forestry principles' in Alan Simson's terms [Simson 2005: 481]. Started in 1934 and finished in 1967, this 10-square kilometres recreational park was conceived following ecological principles rather than as an English style landscape park. Furthermore, its informal aesthetics, based upon a natural mix of woodland, water and meadows, and its provisions for various activities, reflected its democratic agenda. Finally, it introduced the notion of resource management into the urban landscape by viewing urban forestry as an agent of inward investment.

The term 'urban forestry' itself was coined by Willem Morsink's tutor, Eric Jorgensen, professor at the Faculty of Forestry at Toronto University, who had started working in this field in the early 1960s [Van Wassenaer 2004: 17]. In Jorgensen's words, urban forestry is

> a specialised branch of forestry (which) has, as its objective, the cultivation and management of trees and forests for their present and potential contributions to the psychological, sociological and economic well-being of urban society. These contributions include the overall ameliorating effect of trees on their environment as well as their recreational and general amenity value [Jorgensen 1970: 43–51].

This went against the traditional approach to forestry in several respects. First of all, the very objectives expounded in the above quotation are a far cry from the management of forests aimed essentially at producing economic revenue. It also testified to a new approach, dubbed 'ecosystem approach' [Duvigneaud 1974], which to a certain extent superseded the traditional urban-rural distinction. The sectoral management of forests was ill adapted to this holistic vision, which united under one umbrella term what had so far been separated into 'arboriculture', which tended to concentrate on individual trees, 'municipal forestry', which was concerned with public tracts of wooded land in urban areas, and traditional forestry as such. The management of trees inside urban areas itself was so far the realm of arborists, horticulturists or landscape architects, while forestry outside cities was in the hands of foresters, with their own approaches and methods [Konijnendijk 1999: 15].

As appears clearly from Jorgensen's quotation, urban and peri-urban forestry entails specific problems and benefits, which implies changing the definition of the word 'forest' itself. In the first textbook on urban forestry, published in the United States in the late 1970s, the authors define the urban forest as including 'all vegetation within the environs of all populated places, from the tiniest villages to the largest cities' [Grey and Deneke 1978: 12]. According to a later publication, it encompasses 'all the vegetation in an urbanised area' [Rowntree 1995: 43], that is to say

> street and residential trees, urban woodlands, wildlife habitats, open spaces, windbreaks, green belts, roadside screens, kerb areas, parks and other areas within the urban development capable of supporting vegetation (…) [Nobles 1980: 53–56].

In other words, urban forestry can be perceived 'as a continuum, ranging from small parks to peri-urban woodlands'[Konijnendijk 1999: 132]. This multifaceted conception of urban forestry led J. Paris, Professor of Environmental Planning in Montreal, to coin in 1972 the term 'citification' of the forest [Paris 1972: 119–122], which enhances the symbiotic nature of the forest-urban settlement relationship. Urban forestry, by broadening the scope and the scale of arboriculture in urban areas, requires the participation of more experts than traditional forestry, more particularly planners, social scientists and economists to fulfill the objectives of amenity, recreation and environmental conservation [Konijnendijk 2004: 3–5].

5.1.2 Coming of Age

Although the concept met with some opposition to begin with, it progressively took root, first of all in the United States. The reason for this was that the term filled a gap in foresters' terminology with regard to the integrated treatment of trees in urban areas. This came at a time when American foresters were facing growing demand from the urban public, and were officially urged to respond to this [Johnston 1996: 258]. The International Society of Arboriculture (ISA) played a crucial role in promoting urban forestry among professionals. The Society of American Foresters (SAF), for its part, acknowledged urban forestry as a specialist branch of forestry and formed an Urban Forestry Working Group in 1972. The American Forestry Association (AFA) was instrumental in founding in 1981 what was to become the National Urban Forest Council, which led to the creation of Councils in various states of the country. Spreading the good word on urban forestry among professionals was helped by the publications of these groups. The ISA devoted articles to urban forestry from 1978 onwards in the *Journal of Arboriculture*, then in *Arborists News* from 1992. The *Journal of Arboriculture* changed names in 2005 to become *Arboriculture and Urban Forestry*, thus showing signs of both the unquestioned acceptance of the distinction between the two terms and the validity of dealing with them jointly. *The Journal of Forestry*, published by the SAF, also included papers on the subject, while the AFA began publishing a newsletter in 1981, *Urban and Community Forestry Forum*, later to become *Urban Forests* and incorporated in 1995 into *American Forests*.

Urban forestry initiatives were progressively incorporated into federal (e.g. 1990 *Farm Bill*) and state law, as well as into academic syllabi. The first urban forestry conference took place in Washington, D.C., in November 1978, and was a success in that it brought together researchers and practitioners, and resulted in a surge of interest. In that same year, the USDA Forest Service established in its Northeastern Research Station in Syracuse an Urban Forest Research Unit, which focuses its investigations on the human and environmental effects of urban forests as well as the development of tools for so doing.

Some well-publicised community actions such as the 'Million Tree Campaign' during the Olympic Games in Los Angeles in 1984, or the national campaign 'Global ReLeaf' launched in 1989, focused on environmental objectives: the one million trees in Los Angeles, once mature, were meant to absorb 200 tonnes of particulate smog every day, and the slogan of ReLeaf was "Plant a Tree - Cool the Globe".

By that time, the USA had launched an international network, ensuring that in spite of political and economic contingencies, urban forestry was there to stay, albeit in different forms according to the country.

5.2 Importing the Concept

Urban forests like the Amsterdamse Bos paved the way for much of what urban forestry has developed in Europe over the past decades.

5.2.1 Urban Living in Europe

In Europe, urban forests display specific characteristics: 'the focus of urban forestry is more on forests than on urban green space at large' [Konijnendijk 1997: 31]. Ecosystems in which trees dominate, they differ from other forests by displaying different species composition as well as class age distribution. They are situated in or near densely-built up areas, and as a result, they are often fragmented. Because they are subject to high urban pressure, their conservation often involves protective regulations. By the same token, they often shelter a lot of recreational facilities, and may also be subject to conflicting uses. Because of their past history, the municipalities are often involved in their management. Finally, the 'dominance of local, urban dwellers in decision-making processes affecting urban forests is a crucial aspect', so that communication and participation are essential tools, which is not the case in traditional forestry [Konijnendijk 1999: 38–40].

At the end of the 20th century, not unlike at the beginning of the 19th, it was gradually realised that improving environmental conditions in towns and cities could help make urban living more attractive, while maintaining their role as dynamos of the economy. A pilot study commissioned by the EEC in 1984, to study the potential of urban forestry for job creation in three European towns, showed that the development of urban forestry could significantly improve the environment *and* promote direct and indirect employment, in other words that environmental protection and economic development need not be conflicting interests, provided urban forestry management practices were used [Stirrat et al. 1984]. It recommended an education and training programme to promote urban forestry among professionals and the wider public.

At the same time as it was realised that urban forestry followed specific modalities in Europe, it was felt that European countries could benefit from comparison and cooperation between them: more and more European citizens lived in urban areas and at times suffered for it. Research on urban forestry at the European level was lacking, so research networks got underway in the mid-1990s.

5.2.2 Research and Education

In 1995, the research project 'Urban Forestry: overview and analysis of European urban forest policies' was launched by the European Forest Institute to compare urban forest policy-making in 16 large European cities. In 1997, the Tree-ROUTE (Research On Urban Trees in Europe) Network, a database of European researchers via the internet, was set up by the Danish Forest and Landscape Research Institute, in order to improve coordination between researchers so as to promote knowledge, recognition and good practice. So as to have better knowledge of urban forests and urban trees and coordinate research at a European level, COST (European Cooperation in the field of Scientific and Technical Research) Action E12 'Urban Forests and Trees', funded by the European Commission, was set up and operated from 1997 to 2002. It gathered over 80 urban forestry experts from 22 countries and its main objective was

> to improve the knowledge base needed for better planning, design, establishment and management of urban forests and urban trees in Europe, and by doing this, to establish urban forests and urban trees as a scientific domain in Europe [Randrup and Nilsson 1997].

Four working groups for Tree ROUTE and COST Action E12 were set up. They concerned respectively the objectives and functions of urban forests and urban trees, the identification and selection of species for urban uses, the establishment of trees for urban uses and the management of urban forests and urban trees. Another recent European programme (2001–2004), entitled NeighbourWoods, aimed at producing a comprehensive toolbox for socially-inclusive planning and design of urban woodlands. Its report, *NeighbourWoods for Better Cities* [Konijnendijk and Schipperijn 2004] included a case study of Telford in England, a New Town which boasts twice the national average wood cover, at 18%, and which 'illustrates how urban forestry can successfully be used to structure large areas of new development' [Simson 2005: 485]. These European programmes, on top of the information they have delivered, have permitted to delineate urban forestry as a distinct scientific domain at EU level.

Cooperation on urban forestry also takes place via the International Society of Arboriculture, which has extended its member-base to Europe, and via the activities and annual meetings of the European Forum on Urban Forestry (EFUF), a network of urban foresters set up in 1998 under the auspices of the Working Party on Urban Forestry of the International Union of Forestry Research Organisation (IUFRO). One of the aims of EFUF is to provide a place for researchers as well as practitioners to exchange. Finally, the European Forestry Institute, based in Finland, created in 2001 the European Forestry Research and Information Centre (EUFORIC), also to promote interaction and exchange between the worlds of research and practice.

Fig. 5.1 Ironbridge Gorge (Telford), the birth place of the Industrial Revolution. Benthall Edge, in the background, is an ancient natural woodland in which a coppice regime has recently been reintroduced for habitat and access (Photo courtesy of Alan Simson)

All these initiatives, on top of improving the knowledge of urban forestry, aim at better communication between researchers and professionals for better practice, taking into account both technical aspects (tree species resistant to pollution, etc.) and societal aspects (trees and urban stress, noise, vandalism, etc.), so as to foster improved training, higher efficiency and cost-effective policies. *Urban Forests and Trees. A Reference Book*, the first reference book dealing with urban forestry throughout Europe, edited in 2005 by experts who actively contributed to the development of urban forestry, shows the uneven development of urban forestry in the various European countries presented [Konijnendijk et al. 2005].

Over the past 2 decades, urban forestry has gained wide recognition and become integrated, both into institutions and into research and practice in Europe. It is now common to find urban forest management plans in European cities, essentially based on recreation and environmental protection. Knowledge has made headway, too, and the number of journals which include articles on urban forestry on either side of the Atlantic is considerable. The Forestry Library of the University of Minnesota indexes no less than 87 journals in English on the topic, ranging from *Acta Horticulturae* to *Women in Natural Resources*, and covering the full range of preoccupations related to urban forestry, from law to economy and from psychology to forest management and uses.

Although most of the urban forestry educational programmes are to be found – logically enough – in the United States, they have recently begun to develop in Europe, too. The 6th European Forum on Urban Forestry, held in Arnhem (Netherlands) in 2003, was devoted to the education of foresters. Before that, within the framework of the EU-funded COST Action E12, the Danish Forest and Landscape Research Institute had led a European-wide study of the offer and status of higher education in urban forestry (1999–2001). It pointed out the need to transform the education to include mobility (of staff and students), multi-disciplinary and trans-disciplinary approaches. It also stressed the need to educate the new profession of urban foresters. Indeed, the need to integrate the natural sciences and the social sciences, communication skills, forest design and management for leisure pursuits, as well as increasing exchanges among European foresters, demanded a new, multi-disciplinary approach to higher education curricula. Bearing in mind that the inclusion of urban forestry in higher education was still in its infancy and that the number of students concerned was still modest, the report – the first one of its kind – noted the growth of urban forestry as an educational discipline, albeit as a specialised niche. In quantitative terms, a core of six disciplines emerged from the multidisciplinary approach involved: arboriculture, biology, forestry, horticulture, landscape architecture and (landscape) ecology. These disciplines also happen to lie at the heart of multi-disciplinary research on urban forestry according to the authors. They concluded that on the whole, higher education had placed little emphasis so far on disciplines related to social sciences and aesthetics [Andersen and Konijnendijk 2002: 507]. The same authors pointed out at the 2003 Forum that, more than a separate subject,

> urban forestry is primarily taught as a part of educational programmes in related disciplines rather than at a more integrative level, dealing with all elements of urban green structures [Randrup and Konijnendijk 2004: 23].

However, professional education has taken urban forestry on board, with the European Arboricultural Council (EAC) harmonising the certification of tree care professionals through its European Tree Worker Examination.

5.3 Urban Forestry in Britain

The strengths of the concept of urban forestry thus lie in three directions. First, it is aimed at multiple benefits, and involves multiple partners. Secondly, reflecting this first characteristic, it involves a wide range of approaches and disciplines. Finally, it relies on long-term strategic planning of *all* trees in an urban area. In Britain as in the rest of Europe, the definition of urban forestry was subject to a lot of discussion and doubts among professionals, and tree management practices took time in integrating the width and breadth of urban forestry.

5.3.1 Building Up a Toolkit

The first paper given on urban forestry in Britain was delivered, unsurprisingly, by Eric Jorgensen, at the 10th Commonwealth Forestry Conference held in Oxford in 1974, and the first contribution on urban forestry published in Britain appeared in

1978 in *Arboricultural Journal*, later to be subtitled *International Journal of Urban Forestry*.

While planners and landscape architects welcomed the concept of urban forestry, a number of professionals from the Arboricultural Association perceived this new approach as a threat to arboriculture [Johnston 1997: 111] and many examples show this debate around questions of definition and remit. A few of them will enhance the issues at stake. In a seminar organized in November 1988 by the northern branches of the Royal Town Planning Institute and the Landscape Institute, the publicity leaflet bluntly announced 'If you are not sure what urban forestry is, neither are we!' [Johnston 1999: 235]. The few urban forestry projects achieved by then (mostly Motherwell in Scotland and Tower Hamlets in Inner London) had indeed a limited impact and failed to achieve their initial purpose, hence to convince.

An important British book on trees in urban areas, *Trees in the Urban Landscape: Principles and Practice*, published in 1995 by Anthony Bradshaw, Ben Hunt and Tim Walmsley, shows the hesitation in terminology that was still going on 30 years after the term 'urban forestry' had first been introduced in Canada. After devoting fifteen chapters to technical questions of site, planting, choice of species, diseases, etc., the authors ended with two chapters entitled respectively 'Urban forestry' and 'The integrated approach'. In the former, the passage related to the definition of urban forestry merits being quoted at some length.

> Recently, some people have taken urban forestry to cover all trees grown in, and close to, urban areas. Most of these trees have previously been thought of as the remit of arboriculture, or horticulture. There is therefore confusion. It would seem best to keep the term forest to apply to 'an extensive tract of land covered with trees and undergrowth…', and forestry to apply to 'the science and art of forming and cultivating forests, management of growing timber' (Oxford English Dictionary, 1971). It is perhaps most sensible then to think that what can be called an urban forest should have a minimum size of 0.5 acre (0.25 ha). Trees on such an area will not only form a recognizable and practicable small woodland, but it is also the minimum size for which a Forestry Commission grant is available [Bradshaw et al. 1995: 232].

The authors go on to say that, because of constraints on urban land uses which make it unlikely for a forest to have enough time to mature, 'urban forestry is unlikely to be conventional forestry as we have come to know it'. They suggest short-term tree crops 'for commercial wood production' in urban contexts and define as 'land available for urban forestry', undeveloped land, 'recent farming land, sand and gravel pits, old railway lines, old factory areas' [Bradshaw et al. 1995: 232–234]. The mention of the minimum area (0.25 hectare) to qualify for forestry status and to permit to apply for grants is telling of the importance of definitions and institutional criteria in the construction of the forest reality. It also stresses the fact that forestry is still thought of by the authors as first and foremost an economic pursuit.

Objections to a wide-ranging definition of the term 'urban forestry' can still be found in the arguments put forward by the Chief Executive of Milton Keynes Parks Trust Ltd in his address to the National Arboriculture Conference in 1997. He argues that what he calls the 'pot pourri' of urban trees can hardly be seen as an entity, and that forests, so as to be planned and managed, have to be made up of one entity and under one management. He therefore excludes private lands and isolated trees from the definition. Moreover, he wonders whether Epping Forest or Burnham Beeches should be considered as urban forests, and concludes that, since their very existence 'has been a function of resisting urban development rather than embracing it' [Salter 1998: 16], the answer should be no. Historical antagonism prevails over present realities in his analysis. Considering the fact that Epping Forest had purposefully been kept for the leisure pursuits of Londoners, this judgement may seem extreme, but the point of the author is that 'the urban forest must have a permanent love affair with its town and city to qualify'. This very restrictive definition excludes all old forests and limits urban forestry to the municipal parks of the 19th century set up as 'green lungs' for the asphyxiated cities of the Industrial Revolution, and the woodlands which form an integral part of the New Towns established in the *New Towns Acts*, from 1946 onwards, among them Milton Keynes and Telford.

5.3.2 From Resistance to Acceptance

From these examples, several arguments seem to stand out as obstacles to the acceptance of the concept of urban forestry as described above.

The term 'forest', in English history as in many others, is a rural one, and it has cultural connotations which may have made the concept of urban forestry difficult to come to terms with in the beginning. It was associated with Royal Forests and aristocratic estates and their history of exclusive leisure and social conflict. Then, in the 20th century, forestry was mostly perceived as the – contested – domain of the Forestry Commission's work and very much linked to its conifer plantings for timber production, in Scotland in particular. In both cases, the term conjured up visions of tracts of planted land which had very little to do with what the majority of the population was experiencing on a daily basis. Superimposed on this first layer of interpretation lie two others, linked to the specific context of the end of the 20th century.

Forests in Britain tend to be associated in the public mind with public ownership, which allows local and central governments to implement policies easily and on a large scale. In urban areas, this is far from being the case if all green spaces are taken into account. The privately owned part of the 'forest' of London, for instance, was found in 1992–3 to represent 69% of the total number of trees in the capital [Cobham Resource Consultants 1993]. Furthermore, forests are thought of as blocks, as the above quotation from Bradshaw, Hunt and Walmsley shows. In urban areas, this is seldom the case, apart from public parks. The forestry characteristics (ecosystem, etc.) are therefore less obvious. Counting all the individual plots where trees grow of course entails a change in the way policies can be implemented.

Finally, resistance may have to do with something more intangible, which is the cultural status of forests in relation to urban settlements. The history of mankind has been in good part effected against the forests, which emblematise pristine nature in European history while cities represent a host of environmental problems. To come to terms with this dichotomy, environmental writer David Nicholson-Lord recommends to 'unmake cities and bring forests back into them' [Nicholson-Lord 1991: 315], thus adopting the opposite posture to J. Paris and advocating what we might call the 'forestification' of cities rather than the 'citification' of forests. This symbiotic vision of forests and cities is however not at all shared by everybody. For some, the return of the forest to the human settlement possibly feels too much like a Horse of Troy, wild nature being reintroduced into civilisation, threatening the very basis on which societies have been able to develop.

It is interesting to note that the Institute of Chartered Foresters, the only professional body for foresters and arborists in the UK, has used since 1982 a much wider definition of forestry which embraces urban forestry:

> the practice of all aspects of tree management, including forest and woodland management, arboriculture, urban forestry and environmental forestry and research, education and training in these fields [ICF 2006].

In spite of the difficulty of agreeing to a wide definition of urban forestry, the first British National Urban Forestry Conference, entitled 'A Seed in Time', took place in March 1988 (10 years after its American counterpart), in the Black Country, where a major urban forestry programme had begun to develop. It was successful in attracting over 300 participants, and helped advertise the concept. It also gave birth to *Urban Forests*, the only publication dedicated to urban forestry in Britain at the time (though *Arboricultural Journal* had begun to integrate contributions on urban forestry as early as 1978). Finally, it helped fund the creation of the Black Country Urban Forestry Unit.

Putting urban forestry into practice has thus been an incremental process over the last 30 years, and similarly it has taken a long time to become integrated in education and training. The COST ACTION E12 revealed that at the end of the 1990s, the United Kingdom had 'the longest track record in education in urban forestry' [Konijnendijk et al. 2005: 471]. In British higher education, five urban forestry related degree programmes and 19 urban forestry related courses were on offer, while 20 departments offered higher education related to urban forestry. However, more often than not, the courses are included as modules in degrees in forestry, amenity horticulture or landscape architecture/management. Only one MSc course in urban forestry is presently available: created in April 2007, it is run jointly by the National School of Forestry and Myerscough College, and is taught at Myerscough College in Preston (Lancashire). Unsurprisingly, urban forestry professionals often feel insufficiently prepared, in particular for the human dimension of their tasks, covering such skills as communication, public involvement, conflict management or marketing [De Vreese et al. 2004: 7]. Professional training qualifications can also be obtained through the Professional Diploma in Arboriculture of the Royal Forestry Society [Dowson 2001: 353–359]. There is plenty of scope to develop part-time or correspondence training courses for professionals wishing to update their knowledge in the context of urban/community forestry, and to improve foresters' training to include skills relating to communication, not only to production.

5.4 Governmental Involvement

Public policies have been slow to integrate the multiple benefits of urban forestry, mostly because, as was stressed before, the Forestry Commission started with a remit to increase productivity and provide timber for the national good, essentially in rural areas. From the ivory tower where it operated, it was hard to see the justification for urban forestry, all the more so as its economic advantages were difficult to perceive. Moreover, trees had not yet received visible public support and the adage was that 'trees [did] not win votes' [Chambers 1987: 4], therefore leading to poor political attention. Thus, in the White Paper *This Common Inheritance*, published in 1990, urban forestry was not tackled as such [DoE 1990].

5.4.1 Raising Awareness

Yet the tables had begun to turn, helped by circumstances, as is often the case. The storm that affected the South East of England on the 15th and 16th of October 1987, reportedly the worst one since 1703, resulted in 15 million uprooted trees and considerable public outcry, and it spurred the government into action: £25 million were put aside to replant the trees in the devastated areas. Task Force Trees was set up by the Countryside Commission to help with the replanting by grant-aiding local authorities in southern England, including urban areas. It was at the initiative of Task Force Trees that Cobham Resource Consultants was commissioned in 1992 to undertake a study of the trees of London, the first of its kind [Cobham Resource Consultants 1993]. The results, if only in numerical terms, exceeded the expectations: an estimated 4.5 million trees were recorded in the capital alone [Johnston 1999: 249]. In 2005, another study recorded around seven million trees in Greater London [Mayor of London 2005: 3].

Another essential study was undertaken on behalf of the Department of the Environment by Land Use Consultants between 1991 and 1993, entitled *Trees in Towns* [DoE 1993]. Based on a new approach combining aerial photography with ground survey, it aimed at assessing the state of trees in English towns by using a sample of 66 towns and cities representing a total of 3% of the English urban area. These towns and cities of more than 2,500 inhabitants were distributed throughout the country and stood for different types of settlement: industrial cities, market towns, etc. The study made a connection between housing density and proportion of trees, stressing the fact that most trees were to be found in low density residential areas. Of the land in urban areas, 'open space' accounted for 30%, including schools and hospitals, churchyards, derelict land and areas of remnant countryside, playing fields and golf courses, informal open space as well as parks and gardens and allotments, and of course woodland.

Lobbying had promoted urban forestry among professionals and authorities in the United States; likewise, the National Urban Forestry Unit (NUFU) sought to raise awareness of the positive contribution that trees could make in British urban areas, as well as to demonstrate best practice in urban forestry, thus standing at the interface between science and practice. It was created thanks to a grant from the Millenium Commission obtained in 1995, and through its numerous publications, NUFU has acted as a forum of information on the strategic development of urban forestry. On the other hand, its practical implementation of urban forests in the Midlands, as well as its support to regional forestry initiatives (East Lancashire, West Yorkshire, Thames Gateway) and to three city forests (Stoke-on-Trent, Hull and Newcastle-upon-Tyne), which stressed cheap and easy-to-implement solutions, placed it in a position to provide technical expertise to local and regional urban forestry initiatives. First supported by government funds, it became a fully independent charity in 2001, until its existence stopped, exactly 10 years after it was created, in July 2005, for lack of the sufficient resources to continue its wide-ranging activities, and maybe also because urban forestry had by then become part and parcel of forestry programmes, and made a champion less essential.

The *England Forestry Strategy* published in 1998 asserted the importance of trees and woods both in rural and urban environments for their contribution to amenity, environment and quality of life. It envisaged action in two ways, directly through Forestry Commission policies and indirectly by supporting initiatives that would demonstrate partnership approaches, especially those aiming to reclaim derelict industrial land in deprived areas, working jointly with regional authorities, private owners and charitable organisations. Since then, urban forestry has become a buzz word in a host of policy areas. According to the *England Forestry Strategy Implementation Progress Report*, published in October 2006, it can contribute to 'neighbourhood renewal, healthy living, social inclusion, education and lifelong learning, environment sustainability, and culture and heritage' [Forestry Commission 2006: 12]. This wide remit is probably the result of the publication of numerous studies between 1998 and 2006 which stress the importance of green spaces and woodlands in people's lives. Among these feature the 2002 Urban Green Spaces Task Force's final report [DTLGR 2002], the Office of the Deputy Prime Minister's 2003 *Sustainable Communities' Plan* [ODPM 2003], the Forestry Commission's *Putting Greenspace on the Map* in 2004 and the Countryside Agency and Groundwork's 2005 report on the countryside in and around towns [Countryside Agency 2005].

Judging from the uses of the term 'urban forest' in recent publications, it can be argued that the long road to acceptance in public policies has come to an end and that urban forestry, including all types of open space, plus the 'green' areas around industrial sites, commercial centres and transport corridors, has come of age and become an integral part of the urban agenda. It is now an integral tool to counter the lack of permeability provoked by fragmentation in towns and cities.

5.4.2 Local Government

Within local authorities, until the 1990s, questions relating to trees were usually split between various departments: Recreation and Leisure, technical services, etc., which went against coherent policies and the visibility of Tree Officers. It also meant that there was no budget for trees as such. Things worsened still in the 1980s with drastic cuts in local authorities budgets and the introduction of Compulsory Competitive Tendering in 1988 which required local authorities to put a number of their services out to tender. This often resulted in the contracting out of services such as the maintenance of green spaces to outside companies which did not always provide the most competent service.

However, a number of experiments of urban forestry were launched by local authorities, which made sense since they had a long tradition regarding environmental protection and the prevention of pollution, as was stressed in the Audit Commission's *It's A Small World – Local Government's Role as a Steward of the Environment* (1997). Local authorities were also in the best position to implement urban forestry projects since their Tree Officers were in charge of planting and maintaining trees in urban areas, as well as of recreation. It was local authorities, for instance, who computerised inventories and management systems in the 1990s. Experiments such as that of the London Borough of Tower Hamlets in 1985 proved positive in spite of the difficult political context (abolition of the Greater London Council). The 'Forest of London' (FoL) urban forestry project contributed a lot to advertising urban forestry. Launched in September 1987, it was given a high profile and was the first one to attract high amounts of sponsorship. Its aim was three-fold:

> to get Londoners working together to plant trees where they were most needed, to increase Londoners' awareness, appreciation and sense of responsibility fro the capital's trees, and

Fig. 5.2 Publicising the Forest of London: Richard Branson, the famous entrepreneur, poses for the press after climbing the tallest tree in central London (Photo courtesy of Mark Johnston)

to encourage community involvement in the planting and care of trees on publicly owned land [Johnston 1997: 119].

Although the FoL project did not develop as was originally planned, it was responsible for planting and distributing 50,000 trees in 18 months and it did a lot to pioneer urban forestry in Britain.

A major project involving local as well as central government was the Black Country Urban Forest (BCUF) in the Midlands, already mentioned in relation to the first British National Conference on Urban Forestry. The BCUF was launched in 1990 in order to regenerate an area of 360 square kilometres gathering a population of over one million. The area, previously the seat of primary industries such as coalmining and metallurgy, was by the 1980s the victim of economic depression and physical dereliction. Through planting programmes, meant to demonstrate that urban forestry could be achieved at a low cost, 500 hectares of new woodland were planted over 10 years and access policies implemented [Jones 2001: 60–62]. The Black Country Urban Forest contributed greatly to the credibility of urban forestry. Gathering numerous partners (central government agencies, local authorities, private companies, NGOs, and charitable organisations), it served as a precursor to many other projects.

Among these projects are the two major forestry programmes of the last decade of the 20th century: the Community Forests and the National Forest. The latter will be developed in the chapter devoted to regeneration, let us focus here on the Community Forests.

5.4.3 The Creation of the Community Forests

At the end of the 1980s, in a context of conflicting viewpoints between the Forestry Commission and the Countryside Commission, a publication by the latter, entitled *Forests in the Countryside*, launched the idea of Community Forests as multi-purpose forests linking the countryside to the cities, to be used on a daily basis by urban dwellers. The Forestry Commission came down the hills where it produced timber to join the Countryside Commission. The term 'community forestry' emerged in the rural areas of developing countries, as a tool to address deforestation, promote economic sustainability and fight against democratic deficit, yet it was first used in North America to make explicit the idea that the forest was managed as a public resource. While urban forestry, as was explained before, focuses more on the holistic management of urban green spaces including trees, both urban and community forestry aim at contributing to sustainability as defined by Gro Harlem Brundtland by combining economic, environmental and socio-cultural objectives, and both terms have become more or less interchangeable in England. The advantage of the term Community Forest for the government's initiative lies in the fact that it bypasses the traditional opposition rural-urban to focus on public benefits rather than on the location of woodlands. The objectives of Community Forests are thus to contribute to environmental regeneration, to economic development by providing employment and rural diversification, and to social welfare by providing recreation opportunities to the 26 million British citizens living within 10 miles of the 400,000 hectares concerned. New objectives have since been added, such as health, and the challenge is now to regenerate derelict land, improve public access, protect areas of high environmental, conservation or archaeological value, provide sites for educating children to nature and outlets for rural diversification, encourage inward investment and social cohesion, not to forget provide timber. In order to achieve this heavy agenda, the forest teams concentrate on seven priorities, summarised by Clive Davies, Director of the North East Community Forest:

1. Creating well-designed woodlands of more than 20 ha, or less than 20 ha but meeting certain location requirements (e.g. being immediately adjacent to development)
2. Securing access to, and good management of, existing woodlands which meet the criteria of Priority 1
3. Creating or re-opening good quality walking/cycling/riding networks for leisure, recreation and/or commuting
4. Securing involvement in the Community Forest by a wide range of local communities, especially those presently socially excluded from their environment.
5. Securing financial and in-kind investment from local businesses and external sources to support major restoration and recreation projects leading to environmental regeneration, and to develop a woodland economy
6. Creating new woodlands on sites other than those in Priority 1, where there are significant landscape, heritage and/or biodiversity gains
7. Securing landscape, heritage and/or biodiversity benefits on the non-wooded parts of the Community Forest, where opportunities are presented by new developments or on agricultural or reclaimed land, including hedgerow restoration [Davies 2001: 26–30].

The Community Forests programme, the largest national urban forestry programme to date, was launched by the Prime Minister John Major in July 1988. Three Community Forests were announced, before the programme was extended to 12: Mersey Forest (1,101 square kilometres), Red Rose Forest (756 square kilometres), Forest of Avon (572 square kilometres), Tees Forest (567 square kilometres), South Yorkshire Forest (518 square kilometres), Greenwood Forest (417 square kilometres), Great Western (363 square kilometres), Great North Forest (249 square kilometres), Forest of Mercia (238 square kilometres), Watling Chase (187 square kilometres), Forest of Marston Vale (158 square kilometres) and Thames Chase (104 square kilometres).

The Community Forests are situated on the edge of the major English cities and, although they display very different characteristics in terms of surface and setting, they share a strategic vision, laid out in the individual Plan of each one of them, which underlies landscape changes over 30 years, and indicators against which to monitor progress. These Forests are meant to revitalise the urban fringe and contribute to the quality of life of urban dwellers by providing accessible and sustainable landscapes.

Participation in the programme is voluntary and the involvement of a wide range of partners, from local authorities, charitable organisations, businesses as well as the local population, lies at the core of the concept.

Fig. 5.3 Map of the 12 English Community Forests (Map courtesy of the Tees Community Forest)

Total planting in Community Forests between their inception in 1990 and 2004 amounts to over 13,000 hectares, not to mention the existing woodland brought under management, the planting or restoring of hedgerows and the adaptation and opening of woodlands to recreational use. Despite its success, however, the programme is not beyond criticism. It is getting harder and harder for the teams to implement the objectives, as the goalposts keep moving (now 450,000 hectares) and as the easiest part of the programmes was achieved first (planting on public grounds), leaving the most challenging parts for later. This means in practice that even though on the whole the achievements are on target, when one looks in detail, some parts are well behind schedule. Since the Community Forests represent one of the main forms in which forestry is being implemented today in England, the analysis of their achievements will be presented in the next part.

The rapid progress of urban forestry in England since the late 1980s can be measured in several ways. One of them is the rising use of the apparent oxymoron 'urban countryside'. Although one has to be wary of terms, as they may reflect political wishful thinking rather than realities, this term translates a progression inside urban areas of the tenets of rural natural landscapes. In the case of the Parks Strategy in Manchester, for example, the term stresses more than anything else the establishment of links: corridors between various open spaces within the city and between the city and its surroundings, links between urban sites and naturalistic management reminiscent of rural countryside. Another, more tangible, indicator of the rising popularity of urban forests is the increase in their frequentation by the public. The preliminary report of the *Public Opinion of Forestry 2007* on England notes an increase, not just in the proportion of respondents who had visited forests and woodlands for recreational purposes in the last few years (76% against 65% in 2005 and 66% in 2003), but also in the share of people who had gone to urban woodlands: out of the 2,554 people questioned, if 82% had visited rural woodlands, no less than 61% had gone to urban woodlands [Forestry Commission 2007: 18].

The report *Trees in Towns II*, a study commissioned by the Office of the Deputy Prime Minister, the publication of which has been postponed several times since the end of 2006, will no doubt provide very valuable information on the real impact of urban forestry in terms of landscapes and uses of the English urban forests.

References

Andersen, F., C. Konijnendijk and T. Randrup. 2002. Higher education on urban forestry in Europe: an overview. *Forestry* 75 (5): 501–511.

Bradshaw, A., B. Hunt and T. Walmsley. 1995. *Trees in the Urban Landscape: Principles and Practice*. London: E&FN Spon.

Chambers, K. 1987. Urban forestry in the London borough of Tower Hamlets – an account of the European urban forestry project sponsored by the EEC. *Arboricultural Journal* 11 (1): 1–14.

Cobham Resource Consultants. 1993. *London Tree Survey*. Abingdon, Oxfordshire: Prepared for Task Force Trees, Countryside Commission.

Countryside Agency. 2005. *The Countryside in and Around Towns*. Wetherby: Countryside Agency. http://www.countryside.gov.uk/Images/CAT_tcm2–22089.pdf. Accessed 11 February 2007.

Davies, C. 2001. Planning for community forests, 'the green lungs of European cities'. In *Partnership-Led Regeneration. Papers of the 4th European Forum on Urban Forestry Held in Durham 15–19 May 2001*: 25–32. http://www.vbv.be/efuf/_docs/EFUF4_Papers.PDF. Accessed 10 November 2007.

De Vreese, R., K. Bonsen, G. Borgman, B. van Alfen, J. van Rooijen and M. Borgman. 2004. *Educating the Urban Foresters. Proceedings of the 6th European Forum on Urban Forestry*, Arnhem (The Netherlands). Velp: University of Professional Education.

DoE (Department of the Environment). 1990. *This Common Inheritance: A Summary of the White Paper on the Environment*. Cm 1200, London, HMSO.

DoE (Department of the Environment). 1993. *Trees in Towns: A Survey of Trees in 66 Towns and Villages in England*. Prepared by Land Use Consultants. London: HMSO.

Dowson, D. 2001. Arboricultural education and qualification opportunities. *Arboricultural Journal* 25: 353–359.

DTLGR (Department of Transport, Local Government and the Regions). 2002. *Green Spaces, Better Places*. Final Report of the Urban Green Spaces Task Force. London: DTLGR.

Duvigneaud, P. 1974. *La synthèse écologique*. Paris: Doin.

Forestry Commission. 2006. *England Forestry Strategy. Implementation Progress Report. 1998–2006*. Cambridge: Forestry Commission. http://www.forestry.gov.uk/pdf/efs-progress-report-2006.pdf/$FILE/efs-progress-report-2006.pdf. Accessed 12 February 2007.

Forestry Commission. 2007. *Public Opinion of Forestry 2007: England (preliminary report)*. Edinburgh: Economics and Statistics Unit, Forestry Commission.

Grey, G. and F. Deneke. 1978. *Urban Forestry*. New York: Wiley.

ICF (Institute of Chartered Foresters). Home page. http://www.charteredforesters.org, Accessed 26 February 2006.

Johnston, M. 1996. A brief history of urban forestry in the United States. *Journal of Arboriculture* 20 (3): 257–278.

Johnston, M. 1997. The early development of urban forestry in Britain, Part I. *Arboricultural Journal* 21: 107–126.

Johnston, M. 1999. The springtime of urban forestry in Britain - developments between the 1st and 3rd Conferences, 1988–1993, Part I. *Arboricultural Journal* 23: 233–260.

Jones, N. 2001. Political partnerships for regeneration. The Black Country Urban Forest: a UK case study. In *Partnership-Led Regeneration. Papers of the 4th European Forum on Urban Forestry Held in Durham 15–19 May 2001*: 60–64. http://www.vbv.be/efuf/_docs/EFUF4_Papers.PDF.

Jorgensen, E. 1970. Urban forestry in Canada. In *Proceedings of the 46th International Shade Tree Conference*, 43a–51a.

Konijnendijk, C. 1997. A short history of urban forestry in Europe. *Journal of Arboriculture* 23 (1): 31–39.

Konijnendijk, C. 1999. *Urban Forestry in Europe: A Comparative Study of Concepts, Policies and Planning for Forest Conservation, Management and Development in and Around Major European Cities*. Doctoral dissertation. Research Notes No. 90. Faculty of Forestry, University of Joensuu.

Konijnendijk, C. and J. Schipperijn (eds.). 2004. *NeighbourWoods for Better Cities. Tools for Developing Multifunctional Community Woodlands in Europe*. Frederiksberg: Danish Center for Forest, landscape and Planning. http://www.vbv.be/neighbourwoods_leaflet.pdf. Accessed 2 November 2007.

Konijnendijk, C., K. Nilsson, T. Randrup and J. Schipperijn (eds.). 2005. *Urban Forests and Trees. A Reference Book*. Berlin: Springer.

Land Use Consultants. 1993. *Trees in Towns: A Survey of Trees in 66 Towns and Villages in England*. London: HMSO.

Mayor of London. 2005. *Tree and Woodland Framework for London*. London: Greater London Authority.

Nicholson-Lord, D. 1991. Conference paper. In *A Seed in Time. The Second UK/international Conference in Urban Forestry*, eds. C. Field, and M. Johnston, 311–315. Edinburgh: Seed in Time.

Nobles, R. 1980. Urban forestry/arboriculture programme. *Journal of Arboriculture* 6: 53–56.

ODPM (Office of the Deputy Prime Minister). 2003. *Sustainable Communities. Building for the Future*. London: Her Majesty's Stationery Office.

Paris, J. 1972. The citification of the forest. *Canadian Pulp and Paper Magazine* 9: 119–122.

Randrup, T. and C. Konijnendijk. 2004. Higher education of urban foresters in Europe: status and prospects. In *Educating the Urban Foresters. Proceedings of the 6th European Forum on Urban Forestry*, eds. De Vreese, R., K. Bonsen, G. Borgman, B. van Alfen, J. van Rooijen and M. Borgman, 23–32. Velp: University of Professional Education.

Randrup, T. and K. Nilsson. 1997. *Research on Urban Trees in Europe*. Third European Arboricultural Congress, Merano (Italy). http://www.gemeinde.meran.bz.it/giardinerie/congresso/atti/research.htm. Accessed 10 November 2007.

Rowntree, R. 1995. Toward ecosystem management: shifts in the core and the context of urban forest ecology. In *Urban Forest Landscapes: Integrating Multidisciplinary Perspectives*, ed. G. Bradley, 43–59. Seattle: University of Washington Press.

Salter, B. 1998. Urban forestry: UK experience in practice. *Arboricultural Journal* 22 (1): 11–23.

Simson, A. 2005. Urban Forestry in Europe: innovative solutions and future potential. In *Urban Forests and Trees. A Reference Book*, eds. K. Cecil, K. Nilsson, T. Randrup and J. Schipperijn, 479–504. Berlin: Springer.

Stirrat, R., R. Scruton and N. Sargeant. 1984. *Urban Forestry and Job Creation*. Edinburgh: PEFS.

Van Wassenaer, P. 2004. Educating urban foresters in Canada: past, present and future. In *Educating the Urban Foresters. Proceedings of the 6th European Forum on Urban Forestry*, eds. De Vreese, R., K. Bonsen, G. Borgman, B. van Alfen, J. van Rooijen and M. Borgman, 17–22. Velp: University of Professional Education.

Part II
Multi-Purpose Forestry: Another Name for Utopia?

Chapter 6
The Economy of Postproductivist Forestry, the Impossible Challenge?

Forest and recreation managers will have to balance the needs for timber against the populistic views of woodlands which could sometimes
be described as 'Bambi and Robin Hood's home'

[Font and Tribe 2000: 3]

Abstract The second part of the book focuses on the main tenets of English forestry today. Although economic returns are given a low priority, and consumption looms larger than production in multi-purpose forestry, the calculations made to assess the benefits of forestry nevertheless rely more often than not on economic types of valuation. This leads to a certain ambiguity: what does 'profitability' mean, then, how is it calculated and with what results? The new challenges of economic profitability of forestry in England are the topic of the sixth chapter. The traditional categories of market and non-market benefits are defined, with their underlying expert methods, and their new interpretations and outputs are explored. It clearly appears that the 'halo' effect of forestry on the surrounding economy often constitutes a valid reason for sustaining the post-industrial forest resource from an economic perspective.

With the reduced emphasis on production and the advent of multi-purpose and urban forestry, production has often given way to consumption in woodlands and forests.

A clear translation of how much the postproductivist ethos has permeated the social construction of forests can be seen in the results of the *GB Public Opinion of Forestry Statistics*, which are biennial surveys of public attitudes to forestry-related matters conducted systematically since 1995 [Bellringer and Gillam 1998]. Although finances are not the main object of the surveys, public opinion throws an interesting light on what the public thinks public money should finance, with or without returns on investment. When asked whether they believed that 'public benefits are good reasons to support forestry with public money', this is how respondents reacted (in percentages) in the six surveys undertaken between 1995 and 2005:

S. Nail, *Forest Policies and Social Change in England.*

Table 6.1 Why support forestry with public money (From Forestry Commission 2005a: 9)

	2005	2003	2001	1999	1997	1995
To provide good places for wildlife to live	67	72	70	66	75	72
To provide good places to visit and walk in	57	62	58	49	47	47
To help prevent the 'greenhouse effect' and global warming	55	57	58	57	60	61
To improve the countryside landscape	53	58	55	55	48	48
To support the economy in rural areas	41	46	52	52	48	52
To create pleasant settings for developments around towns	37	41	40	37	–	–
To help rural tourism	36	42	35	–	–	–
To make woods more accessible to all in the community	36	40	41	–	–	–
To provide places to cycle or ride horses	35	40	–	–	–	–
To restore former industrial land	30	35	37	34	–	–
So Britain can buy less wood products from abroad	30	33	33	29	34	32
To provide wood as a renewable fuel	28	32	31	20	–	–
To provide timber for sawmills and wood processing	23	28	29	21	24	20
None/don't know	11	10	8	11	4	4

The categories used, it is to be noticed, match up with the objectives of multipurpose forestry and sustainable forestry: economy, landscape amenity, recreation and tourism, rural and urban regeneration, biodiversity, inclusion, health. The results call for several comments. First of all, an overwhelming majority can think of at least one public benefit to support forestry with public money. Within the benefits, it is striking to note that the purely economic arguments (i.e. reduction of the balance of payments, renewable fuel and feeding the wood processing industry) come last in the interviewees' preoccupations. It seems that there is a disconnection between investment and wood production; qualitative arguments (as in quality of life) are paramount in the vision of the forest's functions. This would lead one to think that, should these wishes be translated into policies, profitability as an objective of forestry would be given up, and forestry would become for the State an area of expenditure rather than one of income. Yet this goes against the grain of forestry. As was shown in the first part of this book, the history of forestry displays two fundamental characteristics. First, forestry only exists when there is no better use for the land. Secondly, forestry must be profitable. How do these truths translate, if they do, in the postindustrial, postproductivist English forests, as A. Mather calls them [Mather 2001: 249–268]? The numerous economic studies produced over the last decades tend to show that there still is a strong interest in demonstrating the profitability of the forestry sector, not least to the government. The oil shocks of 1973 and 1979 have probably had their share of responsibility in this, as they vividly demonstrated the urgency of resource conservation and efficient use of resources other than oil [Lowe and Flynn 1989: 258]. When Forest Enterprise was set up as an agency in 1994 to manage the state forests, its remit included producing an annual corporate plan including 'costed options for recreational, environmental, conservation and heritage outcomes' [Hill-Tout 1999: 165].

What does 'profitability' mean, then, how is it calculated and with what results? The new challenges of economic profitability of woodlands in England will be the topic of this chapter.

6.1 The Meanings of 'Profitability'

Most of the numerous studies undertaken recently on the topic of the economic costs and benefits of forestry, particularly urban forestry, separate market and non-market benefits. Let us define these categories before analysing what new profitable uses of forests have recently been devised.

6.1.1 Market Benefits

The marketable outputs of forestry are the easiest to ascertain. They include income from timber sales and other forest products (meat, fruit, nuts, charcoal, mulch, composting materials, firewood, fencing, furniture, building and garden products, as well as nursery plants), and employment. The price of timber (softwood especially) has been falling since the 1970s and has collapsed over the last 10 years. Private landowners as well as the Forestry Commission have seen their timber income decline (by 8% for 2003–2004 for example [Rollinson 2005: 5]). Similarly, the number of sawmills, mostly utilising imported timber, has kept declining over the last few years (from 482 in 1994 to 242 in 2002 for Great Britain, 321–139 respectively for England, and down to 120 in 2005 [Forestry Commission 2002: 12; Forestry Commission 2006a: table 2.5]) with inevitable consequences on employment in the forestry sector.

A study was commissioned by the Forestry Commission in 2000 to find out the contribution of forestry to the national economy and to the rural one in terms of output and employment. It concluded that

> total **gross output** generated by the forestry and processing industries in England amounts to £2,939 million, 37% of which is attributable directly to the forestry and processing activities (£1,085 million). The rest is indirect and induced [...].
>
> The total number of **jobs** (FTEs) supported by the forestry and processing activities in England is estimated to be 34,100, 84% more than the number of jobs attributable directly to the forestry and processing sectors (18,500) [...].
>
> Amongst forestry activities, **harvesting** accounts for a large proportion of direct jobs, 5,200 compared with 3,600 in maintenance and 2,200 in establishment [PACEC 2000: vii. Original emphasis].

It also assessed five different scenarios for the future of the forestry sector. It concluded that the most favourable of these for the English economy, in terms of outputs and employment, would be a 50% increase in English timber harvesting, while the worst one would be the removal of all forestry activities, with dramatic impact on income and employment. This tends to show that, however unprofitable the forestry sector might appear, it is still worth maintaining with public funds, which is what the study concluded.

The first attempt to measure the health and viability of the forest industry businesses in Britain was commissioned by the Forestry Commission in 2000, to find out information on the private businesses on which the sustainable management of woods and forests partly depends, as well as to identify potential innovations to support forest industries [Firn Crichton Roberts 2000]. The study listed 3,200 such businesses, over half of which were sole traders. It concluded that they were important to the rural economy as well as for the delivery of national and international policy requirements for sustainable forest management, but that their performances depended on the development of a market for home-grown timber. It therefore emphasised the need to maintain information on these industries in the future.

This suggests that in order to maintain and increase the market benefits of forestry, mechanisms have to be implemented in order to find new outlets for wood products, attract funding and business through regeneration programmes, develop tourism and finally create employment [Crabtree 2000].

6.1.2 Non-market Benefits

Non-market benefits (NMBs) derive, not from the exploitation of the wood resource itself, but from the added value created by the presence of woods. Some make an additional distinction between use value and non-use value.

The main NMBs of forestry are recreation, landscape amenity, biodiversity and carbon sequestration, while water quality and stormwater control, pollution absorption, health and the preservation of archaeological artefacts are sometimes included [Willis et al. 2003: i–vii]. In spite of their importance, without adequate monetary quantification of the NMBs of forestry, it is difficult to capture politicians' attention; furthermore, under-valuation of the social and environmental benefits 'impedes the efficient allocation of resources to achieve sustainable forest management' [United Nations Forum on Forests 2003: 7–9]. Thus, recent studies [National Audit Office 1986; Roper and Park 1999; Willis et al. 2003; CJC Consulting 2003] have concentrated on assessment methods likely to translate into figures to substantiate their claim that non-market benefits represent a more and more relevant part of an overall appraisal of forestry.

To a certain extent, the host of publications based on economic calculations since the 1970s remind one of those made by improvers in the 17th century to try and convince the public (especially landowners and the sovereign at the time) of the economic validity of plantations against all odds. According to Colin Price,

Professor of forestry economics, the main types of calculations used include: indirect economic benefits, in the form of reduced medical costs; savings made by the survival of trees that do not therefore need replacement; reduction in the costs induced by anti-social behaviour; willingness to pay for landscape quality or right of access (contingent valuation); expert valuation according to which the monetary value of trees is estimated by multiplying several factors (often used in settling legal claims); and finally the price of houses in relation to the aesthetics of the surroundings (hedonic pricing) [Price 2003: 124–125].

From the appointment of Professor David Pearce[1] as a Special Adviser to Chris Patten, the Secretary of State for the Environment, in the late 1980s, neo-classical economics as a way of assessing environmental value has gained momentum in government agencies and academic research. According to these 'expert methods', what people are willing to pay is interpreted as an indicator of the value they place upon the environment, which in turn informs decision-making. From the benefits and costs derived either from real market prices or from hypothetical situations constructed through social survey techniques (Contingent Valuation), economists derive the monetary value of the non-market benefits of woodlands. Cost-benefit analysis (CBA) is then applied to decide what course of action is to be taken. Contingent Valuation is the favoured method used in relation to green spaces and woodland, and that favoured in political circles, essentially through the Hedonic Pricing Method and the Travel Cost Method. Hedonic Pricing assesses the economic ripple effect of green areas and woodland on the value of nearby residential property and/or rental rates of office properties or commerce [Laverne and Winson-Geideman 2003: 281–290]. A typical conclusion can be found in the following passage:

> NUFU sought to quantify the influence that the presence of trees/greenspace has upon domestic property values in England. In-depth interviews with estate agents and other property professionals were undertaken supported by detailed analysis of aerial photography to quantify the tree and woodland cover in three selected Community Forests representing the North/Midland and South of England. The study indicated that property values could be enhanced by up to 18% [Forestry Commission 2005b: 16–17].

The Travel Cost Method infers the value of a site from the cost incurred by users to reach it. Kathleen Wolf, from the University of Washington, has repeatedly used these methods to calculate the 'externality effect' of forestry in relation to the economic impact of urban nature on consumer behaviour. She concludes that consumers perceive their environment more positively in business districts with trees than in districts without trees; their Willingness To Pay (WTP) increases between 11% and 50% for parking and for consumer goods in a landscaped business district [Wolf 1998a, b, c, 1999, 2003]. This can be related to the results of another American study, which found out that one of the main reasons given by respondents for having trees in cities is that 'trees in shopping areas make people think the stores care about the environment' [Lohr et al. 2004: 30]. The perceived

[1] He was at the time a 'leading protagonist of cost-benefit/surrogate valuation methodologies in the environmental domain' [Grove-White 1997: 21–22].

care may be part of the reason why customers are willing to pay more. Similarly, consumer perceptions are recorded as being correlated with the quality of roadside landscaping, with trees in particular [Wolf 2000]. In England, a study in the Marston Vale Community Forest (Bedfordshire) using the Contingent Valuation Method, obtained a WTP of £6 per person per year (for people living within 6 kilometres of the Forest). Once extrapolated, this translates into £768,000 for the people living in the area, which makes the investment required by the planting and maintaining of the Forest financially viable.

These methods, which calculate the expenses 'consumers' are prepared to make in order to use or be close to urban nature, are not without shortcomings, the first of which being that they do not always address the same commodity with the same criteria. The methods have been widely discussed, and we shall come back to them in a later chapter.

The Forestry Commission is keen to find out how to assess the market and non-market benefits of forestry; it therefore funds two kinds of research. The first kind aims at assessing the role of forests in tourism, in terms of expenditure and employment induced. The second type of research is meant to calculate values for a range of non-market, social and environmental benefits. The Forestry Commission uses two principal NMB values: a value for recreational visitors, assessed at £1.42 per visit in 1999, and a value for biodiversity, but only in remote coniferous forests, assessed since 1996. The latter being mostly situated in Scotland, this assessment will not be included in the discussion.

In order to achieve the objectives of multi-purpose forestry, new mechanisms have been implemented.

6.1.3 New Mechanisms

Until the late 1980s, private landowners were encouraged to plant trees by tax relief, and business people as well as famous TV and show business personalities made the most of this opportunity, in Scotland in particular. In 1988, the fiscal environment changed. The new budget removed the possibility to offset forestry management costs against income tax liability and introduced the first Woodland Grant Scheme in 1989. The change from tax relief to grants meant passing from a system with no strings attached to one in which grants are conditional and include requirements on access, conservation, species, and so on. Grants therefore permit to focus investments more precisely according to targets, and they give more flexibility, which is all the more adequate as targets can change [Miller 1995: 45, 70]. The Forestry Commission has thus focused on attracting and advising private landowners, and on funding initiatives implemented mostly by the private sector. This means that in the Community Forests for example, a lot of the teams' energy is devoted both to bringing together public and private funding partners, which can be quite complex and time-consuming, and to looking for new partners (landowners) to reach the objectives of the Forests in terms of new planting and management.

The Woodland Grant Scheme (WGS) was meant to support the creation and management of woodland. It allowed private woodland owners to claim planting grants ranging from £700 (for conifer planting) to £1,350 (for broadleaf planting in woods less then 10 hectares) [MAFF/Forestry Commission 1998: 11]. To complement it, following the vote of the *Farm Land and Rural Development Act* in 1988, the first Farm Woodland Scheme was introduced for a trial period. Accordingly, if woodland was being established on agricultural land (a minimum of 3 hectares and a maximum of 40 hectares per farm), participants could receive, on top of WGS initial establishment grants, annual payments for up to 15 years to compensate for agricultural income foregone as a result of conversion to woodland. Farmers could thus claim £600 per hectare for converting arable land, improved grassland or cropped land into woodland (Better Land Supplement), as well as a one-off payment of £950 per hectare for planting close to urban areas and giving public access (Community Woodland Supplement), £600 per hectare for woodland planting in specially targeted areas (Locational Supplements) and grants ranging from £400 to £600 per hectare for short rotation coppice (SRC Grant).

This diversity allowed to cater for the specific needs of each area, and hopefully to attract new participants to the schemes. Grants for existing woodlands included restocking grants, via direct seeding (£325 per hectare for conifers, £525 per hectare for broadleaves), and/or natural regeneration (discretionary payment on top of restocking grant); an Annual Management Grant, given to enhance environmental value or give public access (£35 per hectare per year); and a Woodland Improvement Grant to undertake special work in existing woodland (single discretionary payment).

With devolution, the implementation of the schemes became differentiated in England, Scotland, Wales and Northern Ireland. The revised scheme for England, the English Woodland Grant Scheme, was launched by the Forestry Commission in June 2005. It is part of the England Rural Development Programme, which sets out government support for agriculture, forestry and the rural economy. It comprises six grants – or, as the Forestry Commission puts it, 'a suite of environmental support measures' [Forestry Commission 2006b: 3] – meant to promote the public benefits of forestry. The new scheme includes five different grants for maintaining existing woodland and one for creating new woodland. Woodland creation grants are now on a competitive and regional basis; they are selected according to their best fit to public benefit priorities, and the grant rates have been increased (£1,800 for broadleaf planting per hectare, £1,200 for conifer planting). In the case of regeneration for felled areas, the rates of grants reflect the priorities of the scheme. The highest grants go to conifer and/or broadleaf plantations on Ancient Woodland Sites being regenerated with native broadleaves (£1,760 per hectare), and to conifer plantations/broadleaf plantations/ancient and other semi-natural woodland being regenerated as broadleaf plantations with native trees (£1,100 per hectare).

The previous distinction between woodland on rural land (Farm Woodland Premium Scheme) and elsewhere (Woodland Grant Scheme), deemed too complex, has been abandoned in favour of selection according to woodland benefits. However, within the Woodland Creation Grant, there is a special Farm Woodland Payment to compensate for agricultural revenue foregone on arable land.

The fact that the new scheme has been simplified and the rate of payment increased can be seen as a response to the criticisms made both by Forestry Commission officials and by the potential users: one interviewee who had planted more than 50,000 trees with his father on his farm in the Wirral, commented that he had never applied for a Forestry Commission grant as they were far too complicated and the constraints too stringent.

6.2 New Outlets for Forest Products

The English Woodland Grant Scheme focuses more on purchasing public benefits than timber, but it can also encourage production, hence the need to find outlets for wood products, whether as direct fossil fuel or material substitution. Woodland products fall into three main categories. Nationally, products include paper, packaging and chipboard products. Regionally, they consist of sawn products (construction timber, pallet boards), turned wood products and fencing. Fencing, crafts and firewood represent the local production [Crawford and Cooper 2004: 20]. Demand on big markets like paper and packaging or pallet is constant but in the case of the latter two, cheaper production in the Baltic countries makes British competition difficult. Similarly, furniture-making is a sector in relatively good health, but most of the hardwood is imported. The need to innovate and add value to wood products, especially timber, has led to research and initiatives, including from the professional sector. A conference involving professionals was organised in 1994 at the University of Surrey to tackle the demise of many traditional markets utilising home grown hardwood and try to renew interest in this mostly wasted resource [Betts and Claridge 1994]. It is all the more important to find outlets as the softwood plantations from the 1960s–1970s, essentially in Scotland, will soon be reaching maturity and need processing [Smith et al. 2001].

6.2.1 Certification

One of the avenues explored over the recent decades to add value to wood products concerns certification, presented in a previous chapter. The Forestry Commission's woodlands in England, Scotland, Wales and Northern Ireland now carry the FSC stamp of approval, thus making the Forestry Commission the largest supplier of certified wood in the UK. The market for certified forest products has developed slowly, however, and the British timber industry has shown concern about certification not being visible enough. If the visibility and attractiveness of certified wood products is not sufficient, producers of certified wood can easily suffer from competition from other, cheaper wood. The challenge is therefore to increase the profile of certified UK timber and wood products. Taking this context into account, the Forestry Commission and the Countryside Agency commissioned a study aimed at finding means to promote the sustainable woodland products industry in England in order to make the public more aware of their existence [Crawford and Cooper 2004]. The study, entitled *Sustainable Woodland Products*, was published in 2004; it confirms that even though a substantial minority of consumers (15%) are aware of sustainable woodland issues and products, this fails to translate into a corresponding volume of sales for these products. Difficulties are also identified from within the forestry sector. Traditionally conservative, this sector is not necessarily responsive to the interest of 'branding' British timber products originating from sustainable management.

6.2.2 Niche Markets

Another avenue explored to add value to wood products concerns niche markets, especially in smaller woodlands in private ownership. This is to be related to the renewed practice of woodland management. Until the last decades of the 20th century, a lot of coppice woodlands in particular had been left derelict, partly because it was believed that this was best for conservation purposes. Awareness that productive and well managed woodlands were not necessarily bad for conservation, but could even add to their conservation value, brought about a change of attitudes and led to the reintroduction of coppice management, hence to the need to find markets for these local products. Medium rotation coppice crops (12–15 years) permit the production of roundwood poles (for fencing and hurdles, for instance) as well as local certified softwood timber for housing construction (e.g. in the Forest of Dean). One of the first implementations of Added Value Projects (AVPs) took place between January 1999 and September 2001 in the Bowland Forest, an Area of Outstanding Natural Beauty in North Lancashire. Although it was not a complete success according to the project manager, it raised a lot of interest. The AVP aimed at encouraging the sustainable management of small woodlands by looking into new sources of income for the owners. In practice, this meant replacing harvesting and timber-selling by the roadside, an unprofitable practice for many small woodland owners, by suppressing the middleman and dealing with all the aspects of timber processing, 'from wood to shop' [Jacovelli 2001: 5]. This entailed minimising production costs as well as encouraging the use of local timber locally. Awareness-raising was therefore an important part of the programme. A Woodland Skills Training Scheme (WSTS) was set up to raise the skills' level of contractors and owners alike in such varied topics as chainsaw use, tree-planting, hedge-laying, charcoal production or wood sculpture. Awareness also concerned customers' needs, as well as marketing. The final report on the programme concluded that niche markets, 'producing high value, low volume products - clearly offer the best opportunity for developing a successful local timber industry' [Jacovelli 2001: 5]. Manufactured products that proved successful included floor boards, interior and garden furniture. Other outlets include firewood and charcoal. The Bowland Forest AVP led to the Lancashire Woodlands Project, the first phase of which ran between

Fig. 6.1 Creating a sense of ownership while using the timber resource: magnets, kitchen spatulas and bread boards made from cherry, sycamore and beech trees from the Mersey Forest (Photo courtesy of the Mersey Forest)

April 2005 and March 2007. It embraced objectives of social forestry (improved access, promotion of woodland walks, community involvement, woodland training sessions for children at risk of exclusion from school) and Added Value objectives strictly speaking (business support, Woodland Improvement Grant Workshops) [Bullimore 2001]. The outstanding success of the seminars on Risk Management and Wood Fuel Systems testify to the current interest in the management and production of small woodlands.

Economic development, although it does not loom as large as non-market benefits in their agenda, is one of the purposes of the Community Forests. Innovative market outlets have been found there, in particular to use residues or timber without commercial value. 'Green Gluing' uses poor quality hard and softwood from the Greenwood Community Forest to make flooring and joinery. 'Timber Stations' in the Mersey and the Red Rose Forests make a wide range of products [Mersey Forest and Red Rose Forest 2004], while a marketing co-operative sells wood products from the Forest of Avon. A recent trend, named 'heed to the view', consists in selling objects (chopping boards, table mats, etc.) made of the wood taken from the Community Forests themselves, with the Forest logo on them, thus establishing a link between the wooded landscape people enjoy visiting and objects made locally. All these initiatives aim at promoting the local woodlands to the local population and at conveying the message that through their purchase, customers are contributing to the conservation of the woodlands.

It may seem odd to classify under 'new' markets what is one of the oldest uses for wood: woodfuel. More than new, then, 'revisited' to fit 21st century needs might be a more appropriate term. In the Forest of Mercia, it was estimated in 1999 that 6% of the electricity needs for the area could be covered from wood within 6 years, with half of the wood coming from waste currently being sent to landfill sites [Countryside Agency/Groundwork 2005: 12]. The mission of the Biomass Task Force, appointed by the government in October 2004, was indeed to discuss the prospects for the development of biomass energy in Britain, biomass being defined as any biodegradable fraction derived from plant or animal matter, including material from forests, crop-derived biomass including timber crops, short rotation forestry, straw, chicken litter and waste material [Biomass Task Force 2005]. This is in-keeping with the demands for renewable energy enunciated in the 2003 *Energy White Paper* and the Renewables Obligation (electricity providers having to deliver 10% of the energy from waste), on the one hand, and the government's dedication to sustainable development through the Kyoto Protocol (1997) and its own *Climate Change Programme*, on the other hand. The report of the Biomass Task Force published in October 2005 clearly asserts that even though 'biomass is unique as the only widespread source of high-grade renewable heat' (including landfill gas and waste combustion, biomass contributes 84% of the total UK renewable energy supply), renewables only account for 1.5% of the electricity and 1% of the heat markets [Biomass Task Force 2005: 1, 4, 8]. It proposes to raise this to 3% by 2010 and 7% by 2015. The government's response was favorable [DTI/DEFRA 2006]: it introduced a 5-year grant scheme for biomass boilers, accepted to support energy crops under the new Rural Development Programme starting in 2007. It also created the

Biomass Energy Centre within the Forestry Commission's Research Agency, with a remit to provide information, advice and best practice guidance on the whole range of biomass. This will complement the work of the Carbon Trust – focusing on energy efficiency, low carbon building design and a range of low-carbon technologies in the non domestic sector – and that of the Energy Saving Trust, which promotes energy efficiency and renewable energy at the domestic, small business and community level. In 2006, the Forestry Commission published a brochure entitled *Woodfuel Meets the Challenge*, to explain the advantages of 'green' woodfuel in comparison with 'brown' woodfuel (coming from fossil fuel), and promote energy crops, from short rotation coppice (3–5 years) for example.

Fig. 6.2 Bags of kindling wood and small firelogs for sale at a petrol station (Photo courtesy of Mark Johnston)

Charcoal provides another example of ancestral use for coppice, recently revamped as a result of the fashion for *al fresco* eating in England. Some local woodland owners have set up a lucrative, albeit seasonal, business by reviving the almost extinct practice of charcoal burning. The charcoal is then sold to the local petrol stations and even in some cases to national supermarkets (this is the case of the BioRegional Charcoal Company). Artists' charcoal is another outlet. Yet it can prove difficult to market British charcoal and 95% of the charcoal sold in Britain is imported.

Managing trees can also generate employment, for example through the 'Modern Apprenticeship Schemes', experimented in various parts in England (e.g. the New Forest, Lancashire, South Yorkshire and London). They are meant to give young people, or long-term unemployed people, training in the practical skills necessary to manage woodland. According to the National Urban Forestry Unit, around 80% of the long-term unemployed trainees have found employment as a result [NUFU n.d].

In order to organise the wood supply chain, numerous partnerships, supported by the Forestry Commission, the Regional Development Agencies and various other organisations, have been created, such as the England Forest Industries Partnership. They aim to strengthen business networks, develop marketing strategies and improve competitiveness, as well as access advice, support and training. Such is the case of the project 'Heartwoods' in the West Midlands. Another example is to be found in the Woodworks Project II, delivered by the Silvanus Trust in Devon, Cornwall and Somerset. On top of creating jobs in the woodland management and wood-using sector, it aims to establish business clusters to bring all the related businesses in a particular area to network in order to solve their problems. Community woodlands and charitable organisations have paved the way in marketing and commercialising the products of the forest locally, in the form of firewood, fencing material and woodchip: the Hampshire Wildlife Trust, for instance, has produced a guide entitled *Selling your Wood* to that effect. Similarly, the National Trust runs two sawmills and, on top of selling timber and other forest products, use as far as possible their own timber in estate fencing and building maintenance. All these initiatives seem to be bearing fruit, since the *UK Timber Statistics 2006* show a significant increase in the use of small British roundwood for fencing, fuelwood, shavings, poles and woodwool.

The revival of old practices like coppicing and wooden crafts is not only an indicator of new economic outlets and a new understanding of conservation. These practices also fulfill a nostalgic streak in the target audience which has sometimes nothing to do with economic viability. Such is the case of some craft activities like broom-making, or reviving extinct practices like hedge laying or pole-lathing or using horses to remove trees from woods, and these will be analysed in the last chapter.

Even though the latest trends seem to indicate that demand for sawn softwood has increased [United Kingdom Forest Products Association 2007], forestry as a market-driven proposition for timber and wood production, as well as employment, is rather weak and the returns of conventional forestry-based activities are often not enough for landowners to make ends meet.

6.3 The Markets of Recreation and Tourism

As the authors of a contribution on the role of tourism in the Forestry Commission put it,

> tourism is about much more than forestry. Likewise, forestry is about much more than tourism. Forest tourism is, nevertheless, of immense value [Broadhurst and Harrop 1999: 183].

Tourism provides an interesting output, especially close to urban areas. Forest sites can generate income through entry charges, parking charges, and various other services provided on site (tea-rooms or shops), but they also generate costs in maintenance and the provision of facilities and staffing:

> the majority of research regarding the value of forests for recreation show a great difference between the value visitors allocate to the forest landscape and the income made by the owner [Font and Tribe 2000: 5–6].

6.3.1 The 'Leisure Explosion'

The amount of free time available to people for leisure pursuits (including tourism) has steadily increased since World War Two. In parallel, the decline of the manufacturing sector has given way to the rise of the service sector (of which leisure and tourism are a part). Tourism has reached a macro-economic weight and grown to be a major contributor to the national balance of payments [Clarke et al. 1994: 22]. When *The Fourth Wave: the Challenge of Leisure* was published in 1965 [Dower 1965], a new wave of rural visitors was beginning to make its effects felt on the countryside, and better knowledge of these visitors was necessary to provide adequate planning and management. Since then, leisure has come into its own as a field of study, not only for landscape architects and planners, but also for social geographers, sociologists, anthropologists, environmental historians, economists and psychologists. Numerous studies from the perspectives of these various disciplines have been devoted to the changes in leisure patterns in Britain and their correlation with the evolution of society at large. In 1965, the British Tourist Authority and the University of Keele published the first *Pilot National Recreation Survey*. It identified that hiking, cycling and camping were less important than they had been between the two World Wars. On the other hand, driving to the countryside still was one of the most popular weekend activities. The attraction of driving as part of leisure has continued unabated: Countryside Commission figures showed that in 1985–1986, 42% of mileage travelled in Britain was for social entertainment and leisure purposes, a figure which had risen to 46% by 1992 [Clarke et al. 1994: 40]. Traffic on rural roads, according to a study by the Oxford University Transport Unit, was expected to keep increasing by between 127% and 267% before 2025.

The countryside, and in particular some beauty spots which have been tourist destinations since the 19th century, have tended to become victims of the development of leisure and of this attraction for driving to the countryside. Such is the case of the most famous of these 'loved to death' sites, the Lake District, where congestion has reached dramatic proportions, leading to campaigns to promote public transport, and to traffic being prohibited in some parts: the *Ambleside Relief Road Campaign*, the *Give the Driver a Break* itineraries published by the Lake District National Park Authority, or the *Empty Roads* initiative promoted by the Youth Hostel Association are examples of this. The need to protect beautiful landscapes from over-use (erosion of footpaths for instance) has played an important part in the commodification of parts of the countryside. The creation of the Countryside Commission under the *Countryside Act 1968* was an early response to the influx of day-trippers and holiday-makers to the countryside. Creating 'honeypots' to channel visitors to places which are more robust and less susceptible to trampling for example, has been one way of tackling this sort of problem. In order to deflect pressure from vulnerable natural sites and from national parks, the Commission has created country parks providing opportunity for recreational activities (picnics, etc.), and helped other organisations finance the purchase of open space for similar purposes (e.g. the National Trust, Wildlife Trusts, the Woodland Trust, the Royal Society for the Protection of Birds). Around 250 of these country parks have come into existence over the last 35 years in England and Wales, attracting some 57 million visitors a year, according to DEFRA.

Over the last 2 decades of the 20th century, the multi-faceted forms that the 'leisure explosion' has taken have led to an increased diversification of leisure provision. Even though the trebling of demand in recreation forecast in *The Fourth Wave* has not occurred, demand for recreation has kept growing, while the resources to access leisure facilities have decreased for some. The growth of individualism in British society at large has led to leisure becoming more individualised, and less focused on group tourism, with quality becoming an increasing concern. The Center Parcs holiday villages, the first of which was created in Sherwood Forest in 1987, followed by Thetford and Longleat, are emblematic of this new trend. Their association of pristine environments and high-class leisure seems to be the key to their success and the reason why they are recognised 'as a sector champion for biodiversity in the leisure industry' [Font and Tribe 2000: 19]: between them, these three sites attracted 650,000 visitors in 1998 [Lambert 2005: 20].

The countryside is now under increasing pressure both from traditional low-impact leisure pursuits (walking, bird-watching, camping) and also from new, at times 'high-tech' uses. The growth in the number of golf courses, which in 1994 represented no less than 0.5% of land use in England, is but one aspect of these developments. Theme parks, holiday cabins, war games, paintball, mountain biking, dog-sled racing, rally-driving, music concerts, wood sculpture trails, are among the numerous forms that recreation has taken during recent decades, which bring in substantial income but also require adequate provision.

Successive governments have recognised the economic attraction of 'green' tourism, while they have also perceived its challenges in terms of environmental protection. The first official attempt to describe and promote the concept of 'sustainable tourism' can be found in *Tourism and the Environment. Maintaining the Balance*, published in 1991 by the Department of Employment and the English Tourist Board's Tourism and Environment Task Force [Department of Employment/ English Tourist Board 1991]. *Policy Planning Guidance* (PPG) *Note 21* on tourism, published in November 1992, united the two objectives of tourism and environmental protection:

> This PPG outlines the economic significance of tourism and its environmental impact, and therefore its importance in land-use planning [DoE/Welsh Office 1992: 1].

The New Labour Government put together a strategy for sustainable tourism in 1999, *Tomorrow's Tourism*, and the *Sustainable Tourism Indicators*, first published in 2001, identify three core actions:

> Protect and enhance the natural environment and destinations;
> Support local communities and culture;
> Benefit the local economies of destinations [DCMS 2005: 5].

The need to reconcile the view that the countryside should fulfill the leisure needs of the population with the protection of the same countryside from the wear and tear engendered by tourists, without resorting to access restrictions smacking of exclusion, is a crucial issue. One way of responding both to sociological trends and to the threats caused to the countryside has been the creation of accessible outdoor leisure spaces close to urban areas, where most people in Britain (and in England in particular) live. The Government's White Paper *Sport and Leisure* (1975) and the Countryside Review Committee (1977) recommended the creation of facilities for recreation in the urban fringe, in particular in the green belts around cities. If the countryside is in the eye of the beholder, then giving urban dwellers a piece of nature close at hand should detract them from using their cars at the weekend to go and visit (and risk damaging) vulnerable spots of countryside. Hence, since the 1960s, provision of land for recreation has gone from being resource-based – where quality of resources dominated, irrespective of location – to user-orientated, where location primes over inherent quality [Patmore 1970]. Another important factor underpinning the provision of open spaces close to urban areas is that, in spite of the increase in car ownership over the last 3 decades, 26% of the population still did not own a vehicle in 2003 [ONS 2006: 179], which in a lot of cases is a factor of exclusion from outdoor recreation.

Britain is not isolated in the recent provision of green spaces for recreation in the fringes of urban areas: as the table of contents of the journal *Urban Forestry and Urban Greening*, from its inception in 2002–2007, reveals, numerous initiatives have been taken to green cities for recreation purposes all over Europe.

6.3.2 Woodland and Tourism

Woodlands have become favourite venues for recreational purposes because they are resilient and give a relatively high number of people the opportunity to enjoy different types of activities without getting in each other's way. As the *England Forestry Strategy* points out,

> woodlands are robust and absorb people better than open countryside [Forestry Commission 1998: 18].

The association of woodlands with public recreation is not new, as was said previously; it started in the second half of the 19th century, with Burnham Beeches and Epping Forest. The *New Forest Act 1877*, nicknamed *the Commoners' Charter*, was the first act to make provisions for the maintenance of the amenity and picturesque character of a forest. The Forestry Commission itself started providing for recreation in its forests long before the appointment of Recreational Planning Officers to most of its regional offices in 1974, or the publication of its first information booklet on *Recreation* in 1980. Faced with the increase in car ownership and outdoor recreation such as rambling and camping, the Commission started opening some of its less productive lands for access in the early 1930s, giving birth to the National Forest Parks in the 1930s. After 1947, National Parks were created; in England there are twelve National Parks to date, and interestingly enough, the latest of them is the New Forest, classified in 2005 after almost 5 decades without a new National Park.

What is novel, however, is the stress put on the economic benefits of forest-related tourism. As a result of its commitment to multi-purpose forestry in the

1980s, and in order to increase its income from non-market benefits, the Forestry Commission has provided more recreational facilities for day-trippers and tourists at its sites. The Forest Holidays Division of the Forestry Commission offers 166 forest cabins across four sites, plus caravan and campsites spread across Britain [Broadhurst and Harrop 1999: 188]. In the North York Moors National Parks, the 60 cabins were rebuilt in 2003 and are expected to provide 50,000 bed nights per year. Another recent example of adaptation to the market is the launch in 2004 of a project to create 'world class' visitor facilities (craft units, office space, bike hire outlet) in Dalby Forest (North Yorkshire). Yet another case is to be found in North Kent, where work began in 2005 to convert a 147 hectare arable farm into a community woodland within the Thames Gateway Growth Area.

In a cost-benefit study of 1972, it was estimated that 15 million visits to Forestry Commission forests were taken annually and that this figure would increase by 10% per year. The recreation function of forests would, it was suggested, take precedence over timber production [Price and Chambers 2000: 55]. Indeed, in 1998, the *UK Day Visit Survey* assessed visits taken to woodlands every year at no less than 350 million [Broadhurst and Harrop 1999: 183]. This reality goes beyond the British borders, since it was estimated in 2001 that 'recreation and tourism are the second largest use of forest resources in Europe, after timber production' [Font et al. 2001: 154]. An attempt was made in 1992 by the Forestry Commission to place a value on the non-market benefits of recreation in its woodlands, and it concluded that the aggregate value of informal recreation exceeded that of timber sales [Benson and Willis 1992]. In the mid-1990s, the potential income from the recreation activities of the Forestry Commission was reckoned at £100 million, in comparison with only £88 million for timber products [Hermeline and Rey 1994: section iv, 2, 1.3]. No wonder in these conditions that the Forestry Commission began to be associated in the mid-1990s with the provision of recreational facilities. A two-stage study was commissioned by the Forestry Commission in 2000 in order to quantify the economic benefits of forest-related day visits, the share of forests in rural tourism and the link between attitudes and behaviour towards forests. It used a Trip Generating Functions Method based on a modelling exercise considering a wide range of predictor variables. The results have to be taken with caution, because the approach requires costly primary data and statistical analysis which are not available in a lot of sites. The authors themselves acknowledge that their results represent '"best guess" estimates of the economic significance of forest-related day visit tourism' [Hill et al. 2003: 95]. Bearing this in mind, they reckon that forest-related expenditure amounts to 3.4% of total tourism spending in England (3.3% in Great Britain).

A look at the recent evolution in the number of facilities and activities provided in the English woodlands of the Forestry Commission testifies to the efforts of the Commission to follow the new leisure/tourism trends.

The increase in the number of categories testifies to the desire to provide for an increasing range of leisure pursuits, with some clearly on an upward trend (e.g. arts, cycling), while others stagnate (e.g. camping, picnics), which reflects their relative decline in popularity. On the other hand, more income was generated in 2005 from fee-paying services (car parks, visitor centres, shops, refreshments) than ever before.

Table 6.2 Facilities and activities provided at Forestry Commission sites (adapted from *Forestry Commission Economic and Statistics Unit* 2001–2005: Table 11)

	2001	2002[a]	2003	2004	2005
Facilities					
Parking: free	356[b]	363	155	165	167
Parking: paid			33	34	31
Information	n/a	n/a	99	101	106
Easy access	n/a	n/a	78	82	87
Toilets	57	55	62	66	66
Refreshments	n/a	n/a	45	47	47
Visitor centre	20	19	22	24	25
Shop	n/a	n/a	18	18	19
Activities					
Walking	233	249	157	172	181
Cycling	84	81	87	98	98
Picnic	209	214	89	94	94
Horse riding	32	45	60	61	62
Education	23	23	48	51	46
Wildlife activities	12[c]	14	51	47	54
View point	0	0	12	16	17
Heritage	n/a	28	19	19	19
Play area	27	n/a	28	28	27
Orienteering	35	n/a	28	28	27
Fishing	n/a	n/a	11	12	14
Barbecue	n/a	n/a	22	24	24
Camping	n/a	n/a	10	10	10
Arts	n/a	n/a	11	14	15
Forest drive	6	6	13	13	13
Arboretum	n/a	n/a	8	7	7
Water sports	n/a	n/a	3	2	2

[a] The categories changed between 2002 and 2003, making comparison difficult for some items.
[b] In the 2001 and 2002 editions, no distinction is made between free and paying car parks.
[c] In the 2001 and 2002 editions, the category is 'wildlife hides'.

The New Forest provides an interesting example of how the protection of the environment and the promotion of the heritage can go hand in hand with economic returns. Research which aimed at providing background information for the economic analysis undertaken for the *Forestry Review* as part of the *Spending Review 2002*, established that the annual cost of the management of its 26,000 hectares was about £5 million, but £3.5 million were generated by timber and recreation, leaving a net cost of £1.5 million to Forest Enterprise. Furthermore, it is estimated that the 30 million visitors who go there every year bring in an additional £60–70 million to the local economy.

Similarly, though on a lesser scale, many local businesses benefit directly from visitors coming to watch birds in various Forest Parks. Visitors coming to view breeding ospreys in Whinlatter Forest Park in the Lake District are an added bonus to the local economy, and funding has been secured from the European Regional Development Fund to re-develop the Whinlatter visitor centre using just that argument. In the Forest of Dean, tourism employs 1,500 people and the 50,000 visitors who come every year to the Peregrine Observation Point at Symonds Vat contribute significantly to the local economy, too.

Another trend which makes good economic sense consists in providing mountain bike trails in various forests. Demand is high, due to the current popularity of mountain biking in many countries (4.5 million bicycles were sold in the UK in 2004, according to the Bicycle Association of Great Britain). Bikers on the whole have a high purchasing power, and they spend more money locally than visitors in a car do, as they have little space to carry food, for instance. On top of that, providing trails for mountain bikes can give new lease of life to a number of woodlands, especially softwood plantations which are no longer needed for timber. The Forestry Commission boasts more than 2,600 kilometres of waymarked mountain bike trails, out of which the 7 Stanes in Scotland and five centres in Wales, and there are numerous examples in private woodlands (Coed Llandegla Forest in Wales is one of the largest). A market study for the development of mountain bike trails in Whinlatter Forest Park in 2004 estimated the possible benefits at £1.3 million in additional net annual value to the local economy as well as 10 new jobs [Rural Innovation 2004: 17].

Thus, tourism and leisure clearly demonstrate the 'halo' or 'shadow' effect [Slee 2006: 545–549] of forestry on the surrounding economy, and the need to take it into account when calculating the economic benefits of forestry. Tourism and recreation alone often constitute a valid reason for sustaining the post-industrial forest resource from an economic perspective. Oliver Rackham concluded the second edition of his *Trees and Woodland in the British Landscape* thus:

> The future of woodland is now less often presented in economic terms. [...] If forestry has to be justified, there are much better grounds on which to do so. The c onfidence with which landowners planted trees in the 1960s has dwindled into a sad plea that 'the sale of the timber will pay for the replanting'. This bubble of fashion has burst [Rackham 1990: 199].

Seventeen years later, with the wider meaning of economic profitability, nothing seems to justify this comment any more, especially if tourism and regeneration are taken into account. Or, maybe we should say that the holistic approach of multiple use forestry permits to give credibility to the economic approach of forestry. Will this be the next bubble to burst?

References

Bellringer, A. and S. Gillam. 1998. *Forest Visitor Surveys. 1994 and Earlier Years*. Edinburgh: Forestry Commission HQ Statistics.

Benson, J. and K. Willis. 1992. Valuing informal recreation on the Forestry Commission Estate. *Forestry Commission Bulletin* 104. London: HMSO.

Betts, A. and J. Claridge. 1994. *New Markets for Old Woods – Proceedings of a Conference at the University of Surrey*. Cambridge: Forestry Authority for England.

Biomass Task Force. 2005. *Report to Government*. London: DEFRA.

Broadhurst, R. and P. Harrop. 1999. Forest tourism. Putting policy into practice in the Forestry Commission. In *Forest Tourism and Recreation. Case Studies in Environmental Management*, eds. X. Font and J. Tribe, 183–200. New York: CABI.

Bullimore, P. 2001. *The Lancashire Woodlands Project. Project Report April 2005 - March 2007*. Edinburgh: Forestry Commission/Lancashire County Council. http://www.lancashire.gov.uk/environment/forestry/pdfs_and_publications/Lwp%20Project%20Report%2005%20–07s.pdf. Accessed 3 August 2007.

CJC Consulting et al. 2003. *Economic Analysis of Forestry Policy in England, Final Report*. London: DEFRA and the Treasury.

Clarke, G., J. Darrall, R. Grove-White, P. MacNaghten and J. Urry. 1994. *Leisure Landscapes. Leisure, Culture and the English Countryside: Challenges and Conflicts*. Lancaster: Council for the Protection of Rural England.

Countryside Agency/Groundwork. 2005. *Countryside in and Around Towns*. Wetherby: Countryside Agency.

Crabtree B. et al. 2000. *Economic Benefits of Community Forestry – Phase 1 Study*. Research contract for the Countryside Agency and Forestry Commission. Oxford: CJC Consulting with MLURI.

Crawford, C. and T. Cooper. 2004. *Sustainable Woodland Products. Review of Consumer Awareness and Best Marketing Practice*. Prepared for the Forestry Commission and the Countryside Agency. Guilford: Arkenford.

DCMS (Department For Culture, Media and Sport) Tourism Division. 2005. *National Sustainable Tourism Indicators*. London: Department for Culture, Media and Sport.

Department of Employment/English Tourist Board's Tourism And Environment Task Force. 1991. *Tourism and the Environment. Maintaining the Balance*.

Department of the Environment/Welsh Office. 1992. *Planning Policy Guidance 21. Tourism*.

Dower, M. 1965. *The Fourth Wave: The challenge of Leisure*. London: The Civic Trust.

DTI (Department of Trade and Industry). DEFRA. 2003. *Energy White Paper. Our Energy Future – Creating a Low Carbon Economy*.Cmnd 5761. London: Stationery Office.

DTI (Department of Trade and Industry). 2006. *Renewable Energy: 2005/6 Review of the Renewables Obligation Statutory Consultation Document*. London: DTI.

Firn Crichton Roberts Ltd. and John Clegg & CO. 2000. *Monitoring the Health and Financial Viability of Forest Industry Businesses: Business Survey Report to the Forestry Commission & Forest Industry Associations*. Edinburgh: Forestry Commission.

Font, X. and J. Tribe. 2000. Recreation, conservation and timber production: a sustainable relationship? In *Forest Tourism and Recreation. Case Studies in Environmental Management*, eds. X. Font and J. Tribe, 1–22. New York: CABI.

Font, X., K. Yale and J. Tribe. 2001. Introducing environmental management systems in forest recreation: results from a consultation exercise. *Managing Leisure* 6: 154–167.

Forestry Commission. 2003. *British Timber Statistics 2002*. Edinburgh: Forestry Commission.

Forestry Commission. 2005a. *GB Public Opinion of Forestry Statistics 2005*. Edinburgh: Economics and Satistics Unit, Forestry Commission.

Forestry Commission. 2005b. *England Forestry Strategy Implementation Progress Report 1998–2006*. http://www.forestry.gov.uk/pdf/efs-progress-report-2006.pdf/$FILE/efs-progress-report-2006.pdf. Accessed 11 April 2007.

Forestry Commission. 2006a. *Forestry Statistics 2006*. http://www.forestry.gov.uk/website/ForestStats2006.nsf/byunique/ukgrown.html. Accessed 27 February 2007.

Forestry Commission. 1998. *England Forestry Strategy. A New Focus for England's Woodlands*. Cambridge: Forestry Commission.

Forestry Commission Economics and Statistics Unit. 2001. *Forestry Facts and Figures 2001*. Edinburgh: Forestry Commission.

Forestry Commission Economics and Statistics Unit. 2002. *Forestry Facts and Figures 2002*. Edinburgh: Forestry Commission.

Forestry Commission Economics and Statistics Unit. 2003. *Forestry Facts and Figures 2003*. Edinburgh: Forestry Commission.

Forestry Commission Economic and Statistics Unit 2004. *Forestry Facts and Figures 2004. Edinburgh*: Forestry Commission.

Forestry Commission Economic and Statistics Unit 2005. *Forestry Facts and Figures 2005. Edinburgh*: Forestry Commission.

Forestry Commission England. 2006b. *General Guide to EWGS*. Cambridge: Forestry Commission England.

Hermeline, M. and G. Rey. 1994. *L'Europe et la forêt*. Paris: Eurofor; Luxembourg: Parlement européen/Office des publications des Communautés européennes.

Hill, G., P. Courtney, R. Burton et al. 2003. *Forests' Role in Tourism : Phase 2. Final Report*. Aberdeen: Macaulay Institute.

Hill-Tout, P. 1999. Costed options for managing state forests in Britain. In *Non Market Benefits of Forestry. International Symposium Held in Edinburgh*, ed. C. Roper and A. Park, 165–170. London: Stationery Office.

Jacovelli, P. 2001. *The Bowland Initiative. Added Value Project 1999–2001. Final Report*. Edinburgh: Forestry Commission/Lancashire County Council.

Lambert, R. 2005. *Trees and People: Tracing the Early Provision of Forest and Woodland Recreation in Twentieth Century Britain*. http://www.nottingham.ac.uk/ttri/pdf/2005_3.pdf. Accessed 5 March 2006.

Laverne, R. and K. Winson-Geideman. 2003. The influence of trees and landscaping on rental rates at office buildings'. *Journal of Arboriculture* 29 (5): 281–290.

Lohr, V., C. Pearson-Mims, J. Tarnai and D. Dillman. 2004. How urban residents rate and rank the benefits and problems associated with trees in cities. *Journal of Arboriculture* 30 (1): 28–35.

Lowe, P. and A. Flynn. 1989. Environmental politics and policy in the 1980s. In *The Political Geography of Contemporary Britain*, ed. J. Mohan, 255–279. London: Macmillan.

Maff/Forestry Commission. 1998. *A Guide to the Woodland Grant Scheme*. Edinburgh: Forestry Commission Grants and Licences Division.

Mather, A. 2001. Forests of consumption: postproductivism, postmaterialism and the postindustrial forest. *Environment and Planning C: Government and Policy* 19: 249–268.

Mersey Forest. Red Rose Forest. 2004. *The Mersey and Red Rose Forests Timber Stations Project. September 2003-March 2004. Report to the Countryside Agency and Forestry Commission North West England Conservancy*. Forestry Commission England/Countryside Agency, 2004. http://www.merseyforest.org.uk/files/TSReport_v1.pdf. Accessed 30 October 2007.

Miller, H. 1995. *Forest Policy: The International and British Dimensions*. Aberdeen: University of Aberdeen.

National Audit Office. 1986. *Review of the Forestry Commission's Objectives and Achievement, Report by the Comptroller and Auditor General*. London: HMSO.

NUFU (National Urban Forestry Unit). n.d. *Training the Long-Term Unemployed to Manage the Urban Forest*. Case Study 48. Wolverhampton: NUFU.

Office for National Statistics. 2006. *Social Trends N°36*. London: The Stationery Office.

PACEC (Public and Corporate Economic Consultants). 2000. *English Forestry Contribution to Rural Economies. Final Report*. Cambridge: PACEC.

Patmore, J. 1970. *Land and Leisure*. London: David & Charles.

Price, C. 2003. Quantifying the aesthetic benefits of urban forestry. *Urban Forestry and Urban Greening* 1 (3): 123–133.

Price, C. and M. Chambers. 2000. Hypotheses about recreational congestion: tests in the Forest of Dean (England) and wider management implications. In *Forest Tourism and Recreation. Case Studies in Environmental Management*, eds. X. Font and J. Tribe, 55–74. New York: CABI.

Rackham, O. 1990. *Trees and Woodlands in the British Landscape. The Complete History of Britain's Trees, Woods and Hedgerows*. London: Phoenix.

Rollinson, T. 2005. Introduction. In *Great Britain and England Annual Report and Accounts 2003–2004*. Edinburgh: Forestry Commission.

Roper, C. and A. Park. 1999. *Non market Benefits of Forestry. International Symposium Held in Edinburgh*. London: Stationery Office.

Rural Innovation. 2004. *An Economic Study of the Proposals for the Development of Mountain Bike Trails and Associated Trailhead Facilities at Whinlatter Forest*. http://mtb-lakes.com/whin_pdfs/app_2.pdf.

Slee, B. 2006. The socio-economic evaluation of the impact of forestry on rural development: a regional level analysis. *Forest Policy and Economics* 8: 542–554.

Smith S., J. Gilbet and R. Coppock. 2001. Forestry Commission. Great Britain: new forecast of softwood availability. *Forestry & British Timber* 30 (4): 20–22, 24–25.

United Kingdom Forest Products Association. 2007. *UK Softwood and Hardwood Sector Report*. http://www.ukfpa.co.uk/site/sawnews.html. Accessed 26 July 2007.

United Nations Forum on Forests. 2003. *National Report to the Third Session of the United Nations Forum on Forests. United Kingdom of Great Britain and Northern Ireland*. http://www.un.org/esa/forests/pdf/national_reports/unff3/united_kingdom.pdf. Accessed 17 November 2006.

Willis, K., G. Garrod, R. Scarpa et al. 2003. *Non Market Benefits of Forestry. Phase 1*. Report to the Forestry Commission. Newcastle: Centre for Research on Environmental Appraisal and Management, University of Newcastle.

Wolf, K. 1998a *Urban Forest Values: Economic Benefits of Trees in Cities*. University of Washington College of Forest Resources, Factsheet n°29.

Wolf, K. 1998b. *Trees in Business Districts - Positive Effects on Consumer Behaviour*. University of Washington College of Forest Resources, Factsheet n°30.

Wolf, K. 1998c. *Trees in Business Districts - Comparing Values of Consumers and Business*. University of Washington College of Forest Resources, Factsheet n°31.

Wolf, K. 1999. *Grow for the Gold. TreeLink* 14. Washington State Department of Natural Resources.

Wolf, K. 2000. *Community Image - Roadside Settings and Public Perceptions*. University of Washington College of Forest Resources, Factsheet #32

Wolf, K. 2003. *Public Response to the Urban Forest in Inner-City Business Districts. Journal of Arboriculture* 29 (3): 117–126.

Chapter 7
Phoenix Reborn: The Role of Forestry in Regeneration

The community and urban forests – that are literally growing around our towns and cities and breathing life into them – are good examples of how green spaces and woodlands can help to revitalise areas as part of an integrated approach to regeneration. They are as much a part of the infrastructure of success as transport links, housing, or business parks, and so are integral to all our plans for creating sustainable communities.

[ODPM 2004: 51]

Abstract The part played by forestry in rural and urban regeneration is one of the most advertised roles of multi-purpose forestry. This relates to the topic dealt with in the previous chapter, in so far as there are obvious financial returns expected from regeneration. But it also relates to changes in the economy that call for changes in land use. The discourse on regeneration is scrutinised in the seventh chapter, so as to elicit what missions forestry is supposed to fulfil, and compare with incentives and programmes, in particular in the 12 Community Forests and the National Forest. There is no doubt that tree planting for regeneration purposes makes sense as a cost-effective, sustainable tool, as well as a pleasant backcloth against which to develop business ventures or tourist sites. The indirect market benefits of forestry are substantial; yet, in spite of the progress made, the future may not be as rosy and the phoenix may find it increasingly difficult to be reborn out of the ashes of the Industrial Revolution.

In the 21st century, it looks as if the first of the two principles justifying forestry, i.e. that forestry must be profitable, still applies, albeit in a different form. How does the other proposal, i.e. that forestry only exists when there is no better use for the land, fare in the postproductivist era? The present circumstances are very different from those of past centuries, and official discourse is bent on demonstrating the multiple benefits of forestry as a given fact. Yet, we argue, it is the reality that generates the discourse, rather than land use being adapted to a new dogma according to which woodland would be more important than other land uses. Indeed, it is undeniable that most of the plantings and maintenance of woodland that are supported by the Forestry Commission are made in areas which are no longer necessary or possible options for other uses. One way of achieving indirect economic benefits in woodland since the 1990s has been through implementing forestry in areas largely unaffected by it before, in order to regenerate local and regional economies and attract inward investment: lowland forestry on agricultural lands, derelict industrial lands and urban areas. Regeneration probably provides the clearest example of this trend, in Britain as in other European countries, whether in rural, urban or industrial areas. In England, it is carried out mostly on land not belonging to the Forestry Commission, essentially in the 12 Community Forests, as well as in the National Forest, the Black Country Urban Forest and in woodlands belonging to various charitable organisations or private landowners.

7.1 Rural Regeneration

European agriculture has moved to what has been called 'the post-productivist transition' [Lowe et al. 1993: 205–222] due to the reforms of the Common Agricultural Policy (CAP) since 1992, the General Agreement on Tariffs and Trade (GATT) signed in 1993 and the increasing preoccupation for environmental policy. Accordingly, state intervention encourages the reduction of food production by decoupling subsidies from production, the creation of a more sustainable agriculture and the provision of environmental goods in a context of over-production relative to domestic demand. Extensification, dispersion and diversification are the dimensions of this transition, with obvious impact on land use and landscapes: increased emphasis on the maintenance of fragile ecosystems (heathlands, moorlands, wetlands and woodlands), encouragement of organic farming, introduction of set-aside schemes for 'surplus' farmland and increased management of farmland for public access [Bowler and Ibery 1999: 121–125].

7.1.1 Objectives and Incentives

Just as the *Wildlife and Countryside (Amendment) Act* 1985 had urged a balance between the productive and the protective missions of the Forestry Commission, the 1986 *Agriculture Act* made it a duty to balance the conservation of the countryside with the promotion of an efficient agriculture, thus putting an end to the productivist era of British agriculture. This was timely since, as the *Countryside Survey 1990* revealed, agriculture was responsible for the rapid disappearance of landscape features and animal and plant species [DoE 1993]. Following Regulation 1094/88, British farmers were encouraged to take land out of arable production voluntarily. Setting aside at least 20% of their lands for 5 years entitled them to subsidy, and the land could either lie fallow or be used for non-agricultural purposes or for woodland. A set of ten different agro-environment schemes were introduced by the Countryside Agency in the early 1990s to protect the rural environment and assist the rural economy, before being transferred to MAFF in 1996 (now the Rural Development Service of DEFRA).

Diversification has developed rapidly, often in connection with recreation and tourism (farm-based accommodation and/or recreation), and forestry is recognised as representing the alternative land use with the greatest potential [House of Commons Agriculture Committee 1990: 16-1]. A typical example is to be found in the National Forest Tender Scheme, whereby landowners choose to contribute to the creation of the Forest on their own lands. Beehive Farm (Derbyshire) has thus converted 18.2 hectares of its land into woodlands with two lakes, as well as a horse-riding path and a surfaced path for wheelchair access. There is full public access, as well as a car park, and some farm buildings complement the service to the users with a tea-room, craft units, a conference room and a falconry centre. Tree planting on farmland has taken various forms.

Although the Biomass Task Force's report of 2005 stressed that its aim was not to find use for redundant farmland, but to develop a biomass sector as such, its proposals are based on the assumption that one million hectares may eventually be available for non-food uses, out of the 9,166,900 hectares recorded in the *Agricultural Census* of 2004. Thus, biomass demand and farmland surplus combine to contribute to industrial uses or to the production of energy. Short rotation coppice (SRC) plantations of willow, miscanthus and poplar in particular, stimulate biomass fuelled heat and electricity generation. Although poplar is not very popular and mostly associated with the defunct match-making industry, thinnings after 7 years can provide kindling, while after 35 years the logs can be used for veneer or flooring: an example of such a successful venture is the Poplar Plantation, outside Swindon, part of the Great Western Community Forest.

There is a lot of scope to encourage biomass crops and make biomass fuel and electricity widely available, especially on account of the high cost of transportation of wood. The Energy Crops Scheme, run by DEFRA and the Forestry Commission between 2000 and 2006, which provided establishment grants for SRC, attracted farmers for two main reasons. First, energy crops offer the advantage of being short rotation crops, i.e. they are harvested on a 2- to 5-year cycle, the stools can be viable for up to 30 years before replanting is needed. This means that the farmers do not have to forfeit 'hope value' for their land, while the crops offer perspectives of income within the farmers' lifetime. And secondly, they are 'crops', i.e. they require a treatment similar to other agricultural crops (weeding between the rows, etc.), and are therefore more amenable to the farmers' professional culture. As one interviewee put it, looking at a young poplar plantation outside Warrington, 'well, *this* is farming. I can do this'.

Some ventures are already successful. In 2002, the Rural Energy Trust, a not-for-profit company, and its sister company, Rural Energy Ltd in the East Midlands, started to undertake turn-key boiler installation and supply wood fuels in order to 'develop the environmental, social and economic opportunities that a viable wood-heating industry offers' [Biomass Task Force 2005: 23]. By 2005, it had already installed 25 heating systems in the East Midlands with a combined capacity of 3 Megawatts. Annual fuel use by these systems is over 1,000 tonnes. Another success story is that of Trelowarren Estate in Cornwall which meets the heating and hot water needs of 38 timeshare houses as well as the tourist venues (restaurant, swimming-pool, leisure and craft centres), while burning annually 350 tonnes of coppice produced on the estate. As part of Kielder (Northumberland) Regeneration, there are plans to install a woodfired boiler in order to heat a school as well as the Youth Hostel and workshops. However, the Energy Crops Scheme has not met its original targets and was discontinued at the end of 2006. It was replaced by other support through the Environmental Land Management Fund, the single fund that incorporates Environmental Stewardship and other sources of environmental funding.

Fig. 7.1 Willow plantation for short rotation coppice on a Yorkshire farm (Photo courtesy of Forest Research)

Regeneration on farmland also aims at reversing the fragmentation of existing native woodlands for conservation purposes. Although over a quarter of England's woodland is part of a farm holding, in woods, hedgerows, waterside and individual field trees, they have often become patchy because of agricultural practice. In order to encourage their expansion and linkage by new native woodland, the Forestry Commission provided JIGSAW (Joining and Increasing Grant Scheme for Ancient Woodlands) Challenge Grants in some eligible areas for native woodland planting between 2000 and 2005.

For the same reason, regenerating and planting hedgerows and hedge trees was a priority. As part of the Forest Futures Programme (part of the Cumbria Rural Action Zone), 700 hectares should be brought into management every year so as to enhance and restore ancient semi-natural woodlands. According to the assessment made in 2001 by DEFRA of the Countryside Stewardship Scheme, over 6,000 miles of hedgerows have been restored [DEFRA/ERDP 2001: 11]. The *Countryside Survey 2000* is a bit more cautious, pointing out that if the loss of hedgerows registered during the 1980s had stopped, 'there was no significant difference in the 1990 and 1998 estimates of hedgerow length in England and Wales' [Haines-Young et al. 2000: 3]. On the whole, implementing forestry on agricultural lands has been far from easy.

Fig. 7.2 Historical continuity in Frodsham (Cheshire): a recently planted woodland, adjacent to Hobshaywood, an ancient woodland (Photo Sylvie Nail)

7.1.2 Resistances

Various studies testify to farmers resisting the conversion of agricultural land into woodland. The reasons put forward by the farmers include: the lack of a suitable area to plant; the inadequacy of the grants offered in relation to the income foregone; the uncertain financial returns of forestry as well as the decline in capital value of planted land in relation to agricultural land; the fear of possible complications arising from public access; the lack of attraction or expertise for forestry and; the long-term dedication of the land to forestry which might entail missing out on more favourable rates at some point in the future. The last argument is probably the most powerful one, even though accepting grant aid for tree-planting or managing implies a dedication of 30 years, thus not a permanent bind.

Some officials from the Forestry Commission argue, from past experience in Scotland for instance, that the farmers' professional culture is not really an obstacle to diversification, but what matters more is that the right amount of subsidies should be offered. As Community Forests rely a lot on farmland conversion to achieve their objectives, two of them have looked into farmers' attitudes and their findings concur with the ones mentioned above: the perceived irreversibility of woodland, which makes the land lose its 'development hope value' when it is close to urban areas, and also makes it impossible to respond rapidly to changes in agricultural policies; the insufficient financial incentives and finally the lack of expertise [Watkins et al. 1996: 169]. Similarly, a study into farmers' attitudes to woodland planting in Nottinghamshire shows that most farmers were keen on trees as amenity and planted them in hedgerows, field corners and around the farm house. They also willingly took up schemes to improve biodiversity and reduce agricultural intensity, but when it entailed planting or managing woodland, they reacted quite negatively [Watkins et al. 1996: 171].

This is in-keeping with the author's own findings in Cheshire and Merseyside. There, the few farmers who have planted trees on their lands as part of the Mersey Forest either are the laughing stock of their neighbours or are seen as farmers wealthy enough not to need revenue from their lands. Even for these, forestry has seldom been chosen as a land use; rather, it has come to them as the response to a problem. In one case, near St Helens, as a consequence of the closure of the local colliery, the farmers found themselves with redundant land on which they used to grow the potatoes which formed part and parcel of the miners' daily diet of fish and chips. As a consequence, they have devoted part of their land to Stewardship Schemes and woodland. In another case, the farmer had bought the land to build a house, and only when planning permission was refused did he envisage tree planting as an alternative. These findings coincide with those of Battershill and Gilg in the South West of England when they conclude that

> much of the environment-friendly activity observed was associated with poor quality farm land and difficult farming conditions [Battershill and Gilg 1996: 339].

Rejecting forestry on agricultural land may partake of the same unconscious cultural substract which rejects urban forestry: the 'progress' of mankind has been achieved through forest clearance, and it has been the agriculturists' task to achieve that result. Planting trees may be more alien to their culture than letting the land lie fallow, although some farmers interviewed feel ashamed at receiving subsidies for not doing their job and would rather plant trees than look at barren ground. For one interviewee outside Frodsham (Cheshire), it was a matter of pride to cultivate something on his lands, even though it was trees, but he did emphasise that he was isolated in this, all the more so as he had opened his land to public access. The question of access is indeed a tricky one, and at times a bone of contention between farmers and the Community Forests: farmers feel that their lands being part of the geographical remit of a Community Forest may send the wrong signal, i.e. that their private land is accessible.

One interviewed farmer summed up the evolution in mentalities engendered by economic changes by saying that his father had spent his life uprooting the hedges around the farm, and that he was spending his own planting them back!

7.1.3 Assessment

The rationale behind the Farm Woodland Scheme and its successors corresponded to various needs: remove arable land from agricultural production; improve landscapes and biodiversity; provide sites for recreation and tourism and; help rural economy and employment while providing timber and wood for diversified uses.

The assessment of the 12 English Community Forests reflects this agenda. At the beginning of the programme, their woodland cover was 6.5% (with wide variations from one Forest to the other), and the 30-year planting target was established at 65,820 hectares, bringing the total surface of the 12 Forests to just over 400,000 hectares [Forestry Commission 2001: 38], with 80% of that area on agricultural land. This objective was probably too ambitious, and the reality is that by March 2001, only 8,067 hectares had been planted, that is just over 12% of the objective, in a third of the time allocated. Moreover, the lands that were easiest to plant (public land, etc.) were converted to woodland first, so that it looks unlikely that the Forest teams will be able to meet the objectives. Indeed, the number of applications to the Farm Woodland Scheme declined sharply after the first few months: a third of all applications received between 1988 and the end of 1991 were made in the first 4 months of the scheme [Bell 1992: n.p.]. Keeping up the enthusiasm requires a lot of imagination.

An assessment of the first Farm Woodland Scheme was made in 1991. Improvements were noted as far as environmental objectives were concerned and in terms of reduction of agricultural production, but it seems that as far as rural income was concerned, the assessment was not conclusive. Furthermore, the grant system was too complex. These conclusions were taken on board when the second scheme, the Farm Woodland Premium Scheme (FWPS), was introduced in 1992. Under FWPS, over 3,000 hectares were approved per annum between 1992 and 2000, and over 4,000 between 2001–2004, most of them for woodland creation on arable land, where the payment rates were highest [Forestry Commission 2006b: 2–3].

It looks as though the swing of the pendulum which was identified as a constant feature of the relationships between agriculture and forestry persists to this day. In a context of agricultural surpluses, forestry represents an attractive alternative, even though the low prices of timber on international markets mitigate this attraction. The *Countryside Survey 2000* assessed the evolution since the previous report in 1990. It established that broadleaved woodland had progressed by 4% and that more broadleaved woodland was created on formerly developed land than was lost to new development.

Furthermore, the purpose of forestry has changed, being more attuned today to recreation and conservation purposes than to timber production as such. The extent to which agriculture has integrated forestry and conservation over the past 15 years, even if it is below expectations, is still quite remarkable. Regeneration in the countryside responds to various demands and to a mixture of visions of the countryside promoted by various stakeholders. Using Elands and Wiersum's analysis of socio-political discourses on the countryside [Elands and Wiersum 2001: 5–16], it can be said that the objectives of rural regeneration cut across four discourses on rural development. The *utilitarian discourse* or *agrarian tradition* is obviously present in rural regeneration objectives, to respond to farmers' needs to modernise and adapt to new markets so as to satisfy productive and consumptive needs. But, through schemes like the Energy Crops Scheme and the promotion of crafts, for instance, farmers are also expected to maintain the traditional values and face of the countryside, free from nature destruction, which is the tenet of the *agri-ruralist discourse*. The emphasis on recreation and tourism harks back to the *hedonist discourse*, whereby the aesthetics of the countryside is kept up to the standards expected by the urban elites' naturalist tradition. The ubiquitous 'quality of life' objectives, so present in the literature on woodland planting and regeneration, are to a certain extent a modern expression of this tradition, as are many of the forms taken by agricultural diversification (e.g. on-farm campsites, Bed and Breakfast accommodation, crafts shops and walking trails).

To these three discourses corresponding to the contemporary functions of the rural economy can be added another, emerging from the international context regarding conservation, namely the *nature conservation discourse*. Such schemes as the Countryside Stewardship Scheme, the Environmentally Sensitive Areas Scheme or the Organic Farming Scheme clearly partake of a concern to maintain ecological integrity in the countryside.

So it looks as if rural regeneration in England today is conceived from an anthropocentric perspective, in economic, social and cultural terms, but also from an intrinsic perspective, with the conservation and protection of nature *per se* being an aim. The need to optimise and integrate these various roles is obviously a challenge that is at times impossible to achieve. For example, the *Countryside Survey 2000* notes a decline in habitat quality in woodland planted on former agricultural land, which it puts down to nutrient enrichment [Haines-Young et al. 2000: 3–4], so that the conservation value loses out in these areas. Furthermore, forest owners and managers have their own agenda, while other stakeholders demand that forestry objectives should respond to other needs than economic ones. We shall come back to this in the next chapters.

7.2 Industrial and Urban Regeneration

Woodland planting associated with rural regeneration has drawn less attention than that linked to urban and industrial regeneration. The main reason for this is that, as mentioned previously, forestry is more traditionally associated with the countryside. Urban forestry, one use of which is regeneration, is more novel, therefore attracts more attention. Although there is more to urban regeneration than the fight against industrial decline, industrial and urban regeneration will be dealt with jointly here. The reason for this is historical. The Industrial Revolution gave birth to many of today's main English cities (e.g. Manchester, Liverpool, Sheffield, Birmingham, etc.), so the destiny of the urban fabric is often intertwined with that of the industrial sector. The very cities which grew astoundingly fast in the 19th century, making Britain the 'workshop of the world', have borne the brunt of industrial decline and lack of investment since the 1970s, and inner cities have been struck by unemployment and dereliction.

7.2.1 Rationale

One obvious reason why urban rather than rural regeneration figures prominently in official literature and academic research is statistical: most of the English population is urban. Urban decay therefore affects more people than the rural transition does. Beyond that, urban/ industrial dereliction is very visible (abandoned buildings, burnt cars, graffiti, litter), and this creates a much more negative impact on the viewer than set-aside land. Yet another potent reason underlying urban regeneration is that, as the renaissance of urban studies testifies [Amin and Graham 1997], town-life has become trendy. Perceived as the 'engines of the economy' (as Richard Rodgers put it in his hailed report *Towards an Urban Renaissance* in 1999), cities are critically important to national economies. Public policies since the 1990s have thus focused on stopping counterurbanisation and attracting inward investment by making urban living an attractive proposition with the provision of everything urban residents need to live, work and play, with green areas providing visual unity, continuity and connections.

Although urban regeneration is very much the product of the last years of the 20th and beginning of the 21st centuries, it would be wrong to associate greening policies with the New Labour governments only. Nine years before the publication of *Towards an Urban Renaissance*, the Conservatives published a White Paper on the Environment specifically aimed at bringing English people back to towns so as to protect the countryside [DoE 1990]. It announced the government's will to, first, regenerate brownfield land so as to build new housing; secondly, launch the *Greening the City Campaign* and; finally, create 12 Community Forests close to the main urban areas. Even though the White Paper was widely criticised for lacking scope, it included elements which have since become staples in urban policy: protecting the countryside by bringing open spaces close to where most people live, using brownfield land rather than undeveloped land/green belts to build the necessary houses, and finally improving the aesthetics of urban areas to attract the middle classes.

In 1993, the Department of Environment issued a *Survey of Derelict Land in England* (interestingly, the first issue of the *Journal of Land Contamination and Reclamation* was published the same year). It established that in spite of the high expenses incurred over the 5 previous years, the amount of derelict land had remained relatively static. The national regeneration agency, English Partnerships, was then launched, to advise the government on brownfield land conversion, especially to provide affordable housing and support the urban renaissance in some specific areas. From 1994 to 1999, it administered the Derelict Land Grant previously administered by the Department of the Environment. A lot of research investigating the possibilities for woodland planting on derelict sites was undertaken at the time by Forest Research's Land Reclamation Research group [among many others, Dobson and Moffat 1993; Bending and Moffat 1997; Moffat et al. 1997; Hutchings et al. 2001].

Since the publication of the Urban White Paper *Our Towns and Cities: the Future*, in 1999, various official documents have presented New Labour's vision about public space policies: among others, the government's answer to the report of the Urban Task Force, *Green Spaces, Better Places* (1999), and two publications by the Office of the Deputy Prime Minister (*Living Places: Cleaner, Safer, Greener*, 2002 and *Sustainable Communities: Building for the Future*, 2003) as well as the modified *Policy Planning Guidance Note 17: Open Space, Sport and Recreation* (revised in 2002). In *Living Places*, the authors assert that 'public spaces are a barometer of a community. As human beings we respond positively and instinctively to places that are welcoming'. They immediately add 'we want to spend time – and money – in such a community' [ODPM 2002]. Inward investment is indeed a *leitmotiv* of urban/industrial regeneration in England, and part of the success of attracting investors to a derelict area is to provide them with a green setting: the age of industrial and business 'parks' has come, with their bucolic imagery to distance themselves from the erstwhile forms taken by industry and commerce.

Urban renaissance is not to be confused with urban regeneration. Renaissance often results in dilapidated sites (old warehouses, waterfront factories, etc.) being 'gentrified' for a middle-class clientele, but does not necessarily address the core political and social issues of urban inequalities. Thus, urban renaissance does not make cities more sustainable, in the sense of being not only more compact, but also more inclusive and equitable [Mace et al. 2007: 51–65]. Put more bluntly, according to some critics, urban renaissance is for the affluent, while regeneration is aimed at the working-class [Amin et al. 2000]. However, urban forestry programmes partake of both renaissance and regeneration in their more 'cosmetic' dimensions, in so far as they improve the setting of urban life. Indeed, planting trees has been part of 'greening' urban policies for almost 2 decades, regardless of political trends, and the website of the Community Forests thus boasts 'Welcome to the website of England's Community Forests, the country's biggest environmental regeneration initiative' [Community Forest Partnership 2007]. The description of the first objective of the Community Forests clearly shows that the 'renaissance'-'regeneration' categories are not that far apart from a forestry point of view: the following paragraph, dedicated to 'Supporting regeneration and growth', could equally well apply to 'Renaissance':

> England's Community Forests are making towns and cities more attractive places in which to live, work and do business. They attract new investment and create new jobs and enterprises in environmental management, wood products, recreation, leisure and tourism [England's Community Forests n.d.].

Urban forestry thus aims at changing brown to green on the palette of towns and cities, as do the 245 new Millenium Greens, the 200 historic parks restored under the Urban Parks Programme, the new green spaces created/revamped under the Green Spaces and Sustainable Communities programme or the Barclays SiteSavers Scheme. The 12 Community Forests alone aim at converting the landscape from an initial 4% to 12% woodland cover in a period of 30 years. To achieve this requires substantial reclamation and tree planting to take place on brownfield land, whether vacant, derelict or contaminated.

But tree planting also remarkably fulfils the purpose of burying the traces of defunct industries and their associated land uses and social deprivation. Indeed, one cannot help being struck by the coincidence between the location of the 10% most deprived English wards [DCLG 2000: §6.15] and that of brownfield sites [ODPM 2004: 49].

With time, the roots of the newly planted trees will hopefully obliterate the scars of the Industrial Revolution from the face of England's 'green and pleasant land'. Behind this lies the hard economics of the postindustrial era, as the previous quotation makes clear. The key is to provide a pleasant environment as a means to an end, more precisely to break the vicious circle leading to counterurbanisation and inner-city deprivation, so as to attract inward investment and regenerate the local economy. The 'Woodlands by the Motorway' programme, sponsored by NUFU, with the Countryside Commission, Highways Agency and a number of private and voluntary sector resources as partners, was an example of such an initiative. It aimed at establishing new woodland on some public and private sites bordering the M5 and M6 in the West Midlands, to improve the view for potential investors. An additional bonus of land reclamation through woodland or other 'soft end uses' is that it is often the most economical option, estimated at a tenth of 'hard end uses' like built development. Finally, tree planting is at times simply the only option for land use on brownfield sites.

The Community Forests programme puts into practice three central tenets of New Labour's vision: performance management, democratic renewal and joined up partnership [Stoker 2004: 49]. This is possibly why the *England Forestry Strategy* asserts that 'the Community Forests in England have successfully pioneered a new approach to managing the countryside around our towns' [Forestry Commission 1998: 22], and also why they were thus vindicated by the Prime Minister Tony Blair in 2004:

> The success of England's community forests demonstrates that social and economic gain can be achieved hand in hand with environmental enhancement… I commend the community forests as a model and inspiration for others [National Community Forest Partnership 2004:1].

Leaving the question of democratic renewal for a later chapter, let us concentrate here on the strategies used to implement the principles mentioned above and their translation on the ground.

7.2.2 *Strategies*

One of the credos of New Labour's public policies, that of multi-level governance, is widely illustrated in the realm of urban policies. Its main constituents are public-private partnerships, a multi-agency approach, with different departments establishing joint collaborative presence in the area, and the involvement of the local community. These correspond to the recommendations made in 1993 by David Donnison, Emeritus Professor at the Centre for Social Housing in Edinburgh, probably as a result of the unsatisfactory urban policies of the 1980s under Margaret Thatcher. Many projects and initiatives illustrate this functioning, starting with the Community Forests.

When they were launched, the Community Forests were jointly supported by the Forestry Commission, the Countryside Agency, 58 local authorities and many other voluntary, public and private organisations. This meant in practice that the financing of the Community Forests was quite complex, with the core team of each forest being paid for half by the Countryside Agency and for half by the local authorities, with other staff financed by the British Trust for Conservation Volunteers, grants from the Forestry Commission or from a charitable organisation. Support from the Countryside Agency was meant to last for the first 10 years only, to be removed in 2005. However, the transition proved more difficult than anticipated and prompted the Rural Affairs Minister to announce in March 2004 that the Countryside Agency would go on financing the core teams for another 2 years [DEFRA 2004: 1], while each of the nine English regions prepared its *Regional Forestry Framework*, to guide future support from April 2007.

Added to this complex administrative setup, the Community Forests address issues relating to many public services (health, environment, education, housing, etc.) between which they form a bridge, in order to achieve their multiple benefits. Conversely, each individual public service includes an environmental part which links up with community forestry. It thus demonstrates what the *Strategy* calls 'integrated environmental planning'. These complex joined-up partnerships act either directly, coordinating initiatives of their own, or as part of specially designed schemes in which they themselves act as one of the partners. This is the case of Groundwork, the environmental regeneration charity founded in 1981. It is now a federation of over 50 trusts, each of which made up of a partnership of local authorities, voluntary organisations and the business world, but it is also often involved in 'greening' projects as one of the partners.

The projects, logically, have to be match-funded, and financed by *ad hoc* setups involving many partners, from European funds to the National Lottery. One example of this is the Capital Modernisation Fund, a £9 million project funded by the Treasury to create 1,000 hectares of new woodland within Thames Chase, the Mersey Forest and the Red Rose Forest. Another emblematic current initiative is Newlands (New Economic Environments – through Woodlands). Launched by the Deputy Prime Minister in July 2003, it is a regeneration project funded by the Northwest Development Agency, the first phase of which will transform over 5 years 435 hectares of derelict, underused or neglected land (DUNland) into community woodlands in the North West of England, with a budget of £23 million. It includes among its partners various Government agencies (the Forestry Commission, Forest Enterprise, the Northwest Development Agency) as well as two Community Forests, Groundwork and the British Trust for Conservation Volunteers. More recent still, the Land Restoration Trust was created in 2004 to deal specifically with brownfield sites unsuitable for development. The vision is to acquire within 10 years 10,000 hectares of derelict land, turn them into public open space and ensure their long-term sustainable maintenance. The Trust comprises a partnership between English Partnerships, Groundwork, the Forestry Commission and the Environment Agency.

The complexity resulting from organisation and projects means that a lot of time and energy are devoted within each regeneration project team to looking for funding, and to an officer sometimes being appointed especially to harness funds. To add to that difficulty, some funds are not accessible to public organisations. For instance, European Funds for rural development are not accessible to Community

Forests, but they can be accessed by individual landowners taking part in the Community Forest programme, so the teams help private farmers and landowners to apply for these funds; in return, the achievements allowed by the funds will go down in the Community Forests' assessment. Similarly, in order to avoid the limitations linked to the public status of Community Forests, most of them have lately created trusts. For instance, the two Community Forests of North East England have jointly established themselves as *North East Community Forests Limited*. This has the double benefit of conveying a different image, that is more attractive to some participants than that of a public body, and of allowing them to attract extra funding available for charitable organisations. The downside of these partnerships is that they can end up being in competition with each other for accessing the same funds.

Performance assessment is ensured through the implementation of scales. Among recent examples of these are the Public Record Benefit System (PRBS) and Additional Value Assessment, used for the Newlands Project. The former consists in assessing competitively the benefits of the sites against a number of indicators (social benefits, access benefits, economic benefits, environmental benefits). The attributes of each site are recorded in a Geodatabase and scored against benchmarks, then the sites which obtain scores above the benchmarks are submitted to an Additional Value Assessment to delve deeper into their potential before a decision is made.

7.2.3 Landfill Sites

If public space is going to be used as a cement between the various networks of the city, and to make the regeneration agenda progress in its human dimension, as Sharon Zukin and left-wing think-tanks advocate [Zukin 1995; Greenhalgh and Worpole 1995], then the choice of sites to establish these public spaces, including urban forests, is crucial. This brings us to the analysis of the forms which tree planting has taken to assist urban/industrial regeneration/renaissance. Some sites chosen for tree planting programmes do not strictly come under regeneration, so they will not be dealt with here, but they partake of landscape improvement strategies aimed at attracting inward investment. This is the case of the many initiatives to 'greenify' transport corridors, to make business environments more pleasant.

Since the Department of the Environment advised in 1986 against trees being planted on landfill sites because of the contaminants contained in the soil [DoE 1986], abundant research has been devoted to the topic. More recent results, including the Forestry Commission's own research in 1995 and the work of a team from the University of Liverpool, suggest the reverse, i.e. that a lot of tree species can actually thrive on such sites. An experiment on 11 old-type landfill sites took place within the Mersey and the Red Rose Forest in 1998, with 21 different tree species. Assessment after 3 years showed that 14 species had had a good survival rate, and that 'all sites appear to be suitable for some form of community forestry, although two sites would require additional soil amendment' [Rawlinson et al. 2004: 279], thus confirming another study on soil toxicity which revealed that trees frequently survive on soils containing much higher levels of contamination than was hitherto believed possible [Dickinson et al. 2000].

These results are very important for several reasons. First, closed landfill sites account for a large proportion of brownfield sites. It is estimated that they represent 3,000 hectares within Community Forest areas. Secondly, their regeneration is an essential tool to improve quality of life and attract inward investment to derelict areas. Thirdly, their conversion to woodland provides an opportunity to bring these brownfield sites into productive use, while it would be prohibitively expensive to restore them. And finally, tree planting may even contribute positively to the breakdown of organic contaminants, and so to the development of healthy sustainable soils at brownfield sites.

The North West of England is a case in point, since 1,200 out of the said 3,000 hectares are situated in the region between Liverpool and Manchester alone. An example of landfill sites among many others is to be found at Bidston Moss, a low-lying wetland marsh along the eastern coast of the Wirral Peninsula. This site, stuck between the M53 motorway, the railway line and the Wallasey Docks, was converted into a rubbish tip in 1936 and had become a 90-metre high eyesore by the time it was closed in 1995. It attracted media attention in the 1980s for the amount of unemployed people of Birkenhead and Wallasey who scavenged daily through the rubbish to complement their social security benefits. When Groundwork Wirral was founded in 1993, consultation with the local population revealed that this was the 'public enemy n°1' site and that an open green space with walkways and trees was the favoured option for restoration.

Fig. 7.3 From rubbish tip to community woodland: Bisdston Moss in 1996 before regeneration (Photo courtesy of Groundwork Wirral)

Restoration work, coordinated by Groundwork Wirral, started with landfill tax money, on behalf of Merseyside Waste Disposal Authority. A 3-metre layer of paper pulp and soil was spread over the site and trees were planted. The methane produced underground is used to generate electricity. The site, now restored, has been completed by a wooden bridge which gives access to the site to the local population.

Fig. 7.4 Bidston Moss in 2001 (Photo courtesy of Groundwork Wirral)

Moston Vale, in North Manchester, is the first project to have been completed as part of the Newlands programme in the Red Rose Forest. It was opened to the public in April 2006. A comparison of the site before and after the works shows dramatic improvements, with hardcore footpaths winding through the grass amidst trees and lit by solar-powered lamps. However, the social context remains difficult locally: in an ironical inversion of the times when the upper classes living in cities had a key which gave them exclusive access to leafy squares, a fence has been put up to separate the houses from the park, with a key giving the inhabitants access to their homes, so as to avoid burglaries from loiterers in the park.

Fig. 7.5 Moston Vale before regeneration. (Photo courtesy of Chris Robinson)

Fig. 7.6 Moston Vale after regeneration (Photo courtesy of Chris Robinson)

Also in the Red Rose Forest, the 'Green Tips' project turns landfill sites in Greater Manchester into community woodland.

7.2.4 *Mines and Industrial Sites*

Another development around tree planting has occurred in derelict mines and primary industry sites. The two Community Forests in North East England deal jointly with the derelict landscapes resulting from now defunct industries (energy production, heavy industry, manufacturing). One of these concerns the Mineral Valleys Project, the regeneration between 2003 and 2008 of an area of coalfields in West County Durham. A partnership of 50 organisations led by Natural England and supported by Heritage Lottery Fund aims both to regenerate the natural heritage of the area by restoring natural habitats and encouraging wildlife, and to increase access to the countryside. The town of Consett, sadly famous in the 1980s for being one of the worst unemployment blackspots in Britain after the closure of the steelworks, is part of the project. Reclamation of the derelict land and buildings associated with the steelworks was undertaken thanks to a Derelict Land Grant in order to prepare the economic conversion of the area. To reclaim the land and make the environment more attractive to new investors, the buildings were demolished, slag heaps were regraded, subsoiled and grassed. Woodland blocks were planted in areas set aside for recreation and new light industrial units. An urban forestry plan has been formulated as part of a strategy to attract inward investment and give a new sense of place and identity to the local population [Hughes 1991: 1]. The area now attracts visitors, it is part of the C2C (Sea to Sea) bicycle trail and the Consett and Sunderland Railway Path provides recreational access by walkers, cyclists and horse riders. Other coalfields in the area are being transformed into native woodlands as part of the Mineral Valleys Project.

In the Mersey Forest, regeneration of industrial sites and transport corridors will hopefully contribute to the economic buoyancy of an area still too often associated with primary industries. Among the industrial sites (and the place where the Mersey Forest was born), part of the Northwich community woodlands was established on 300 hectares of land used to deposit soda ash. In Saint Helens, Sutton Manor colliery became derelict after the mine closed down in 1991. Having undergone dramatic change, including the creation of the Sutton Manor Woodlands, it was selected as one of the sites for the Channel 4 public art initiative series, 'Big Art', for 2008. This was the very year chosen for Liverpool to be European Capital of Culture and so both events were bound to ensure a wide media coverage for the area. Reclaimed transport corridors, for their part, include canals, railway lines and motorways. For instance, on the Shropshire Union Canal, opened in 1795, boats were used to carry kaolin from Cornwall on their way to the Potteries, while other boats transported coal and iron ore to feed the factories adjacent to the canal. Nowadays, along the canal in Ellesmere Port, close to the Boat Museum, the factories have closed down, some dilapidated warehouses remain standing, waiting for conversion into commercial warehouses or into a museum, and the M53 overlooks the scene. The canal and its surroundings have now become a leisure landscape, with pleasure boats where the barges used to be, the railway line connecting the canal to Stoke-on-Trent converted into a footpath and the towing path transformed since 2000 into a pathway and cycle path running from Chester to Ellesmere Port, as part of the Millenium National Cycle Route. Various partners contributed to making the surroundings of the canal more amenable to visitors: the Shropshire Union Volunteers installed benches along the way; a local artist provided five sculptured signposts; young people from the Prince's Trust created a ramp for wheelchairs, while schoolchildren painted a fresco under a bridge where there used to be graffiti. The Mersey Forest, for its part, planted up various sites along the canal, where chemical waste or sludge used to be deposited, as well as around reservoirs belonging to a nearby oil factory.

Community Forests are one of the two main regeneration projects proposed by the Countryside Commission in 1987 [Countryside Commission 1987]. The other one is the National Forest, created in the Midlands from 1990 onwards, and which has been promoted by the Government as the largest UK environmental project of the 20th century.

7.2.5 The Case of the National Forest

The area that emerged as the Countryside Commission's choice after a complex selection process, is situated between Birmingham, Nottingham and Leicester and had a very low tree cover to begin with (6% on average). Although the area as such only harboured 200,000 people, a further ten million lived within an hour

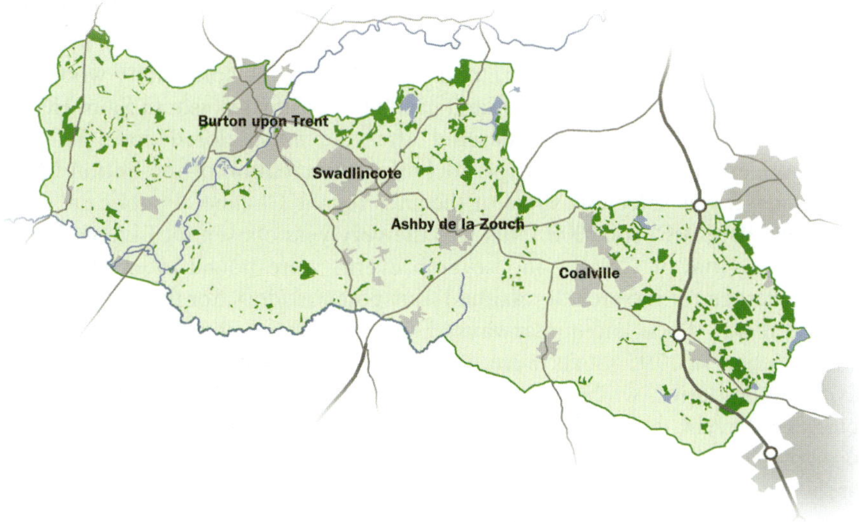

Fig. 7.7 Map of the site of the future National Forest, 1991 (Map courtesy of the National Forest Company)

and a half's drive. It included the remnants of two ancient Forests (Charnwood in the east and Needwood in the west). Between these, the landscape was dominated by the coalfield of North West Leicestershire, the coal and clay-workings of South Derbyshire and the industrial transport corridor of the Trent Valley, on top of farmland.

The project of transforming about a third of the area into woodland by 2020, with the rest a mosaic of farmland and open country, was inspired, official discourse has it, by the multifunctionality of the New Forest. It was founded jointly by the Department of the Environment and the Ministry for Agriculture, Fisheries and Food. A first Forest Strategy describing the vision was elaborated, and an independent National Forest Company, limited by guarantee, established, as a non-Departmental Public Body sponsored by DEFRA. It is in charge of implementing the vision thanks to its own budget and other partnerships for specific projects. The regeneration of the economy and the creation of jobs underlie the improvement of the landscape and the provision of sites for recreation and nature conservation. As the *National Forest Strategy* makes it clear,

> an improved environment will be a spur to new investment and a sound economy will be a foundation to ensure long-lasting environmental benefits [Countryside Commission 1994: 5].

The fact that indirect market benefits and non-market benefits dominate the agenda of the National Forest is logical, since the background studies used Cost Benefit Analysis as their basis [Beaverstock et al. 1997: 55]. The planned creation of 3,796 jobs will be mostly derived from inward investment as a result of regeneration, while tourism and forestry will only account for less than a third of jobs. The Forest Company is seen as a catalyst at the heart of multiple partnerships (Rural Development Commission, English Partnerships, EU sources such as RECHAR II, etc.).

The project straddles rural and industrial regeneration, since part of the 500-square kilometre triangle where it is being established was devoted to agriculture and part to the Midland Coalfield. However, we shall not insist on the rural regeneration aspect, in so far as it is in-keeping with what has been described above, especially with the objective of farm diversification. Rather, following the stress put by the Secretary of State John Gummer on the necessary regeneration after 'much of the land had been despoiled by mineral workings' [Sheail 1998: 20], we shall focus here on industrial regeneration, made necessary after the closure of the mines in the 1980s. In total, the regeneration of the former coalfield area had secured, by August 1996, £12.16 million, to be match-funded by around £43 million from the private sector [Beaverstock et al. 1997: 58]. On top of planting on land belonging to Forest Enterprise or to organisations like the Woodland Trust, planting on private land is encouraged by National Forest Company's grants (*Tender Scheme* and *Locational Supplement*), whether on arable land, derelict mineral workings, in hedges or as links between ancient semi-natural woodlands. By March 2007, some seven million trees had been planted, bringing wood cover to 17.5% of the area.

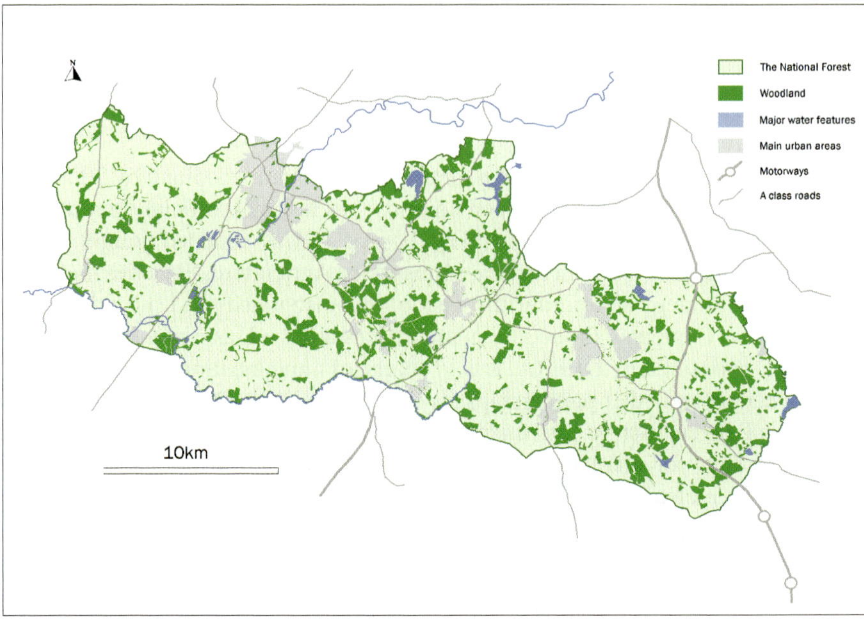

Fig. 7.8 The National Forest in March 2007 (Map courtesy of the National Forest Company)

The tree-planting programme is clearly subsidiary to the re-orientation of the sites formerly associated with the collieries, i.e. business and tourism. The colliery site in Rawdon should be transformed into a 1,000 acre wooded park by 2025; in Coalville, Whitwick Colliery has now been converted into Whitwick Retail and Business Centre while Snibston Colliery has been transformed into The Snibston Discovery Park, a museum of science and industry; a 60-hectare Urban Forest Park has been created on a former spoil heap in Swadlincote; a new National Forest Discovery Centre known as CONKERS has been set up in Moira, and the local blast furnace has been turned into a museum complex.

The experience of the National Forest is replicated on a minor scale at many industrial sites. Initiatives are under way to remediate 100 former colliery sites as part of the National Coalfield Programme, a partnership between the Forestry Commission, Forest Enterprise, English Partnerships, the Regional Development Agency *Yorkshire Forward* and the local authorities. At more than half of the sites targeted, derelict colliery sites have already been converted into business parks and community parks/woodlands. Bentley Colliery, a 102 hectare site in Yorkshire, has been hailed as a ground-breaking restoration for its use of new techniques to overcome the negative impact of soil compaction and poor soil quality on root-growing. The site is now covered by a semi-commercial pine and larch plantation and open to the public, and its ownership, like that of the other sites mentioned, has been transferred to the Land Restoration Trust.

Fig. 7.9 Beginning regeneration work at Bentley colliery (Yorkshire) (Photo courtesy of Chris Robinson)

Fig. 7.10 Community woodland and semi-commercial plantations now cover the site at Bentley colliery (Photo courtesy of Chris Robinson)

> considered to be preventable, and suggest a level of ignorance or disinterest by the managers of the land on which woodland is established [Moffat and Laing 2003].

The mortality rate of trees is high, as will be shown in the next chapter, mostly due to negligence and the lack of the minimum care required by the trees to become established. The assertion according to which

> the role of the forest as a restorative, if not redemptive, agent in old industrial landscapes (as with the National Forest and the England Forest Policy, for example), epitomises a transformed social construction of the forest [Mather 2001: 257],

summarises the wishful thinking that lies behind numerous regeneration projects.

As far as the National Forest is concerned, this controversy may be coming to an end, as the amount of coal extracted from opencast mining in the area has declined drastically, reducing the prospect of continuing competition between land uses in the area. According to *Opencast Coal Mining Statistics 2005*, opencast coal production in Derbyshire in 2004 was 332,240 tonnes on two sites, while in 2005 it had declined to 48,560; in Leicestershire, the corresponding figures were 412,422 on two sites in 2004 to 94,844 for 2005 [BGS 2005: table 1]. But the same process applies to landfill sites. The landfill sites which are being closed and regenerated are in many cases being replaced by others, in other places, which in due course will need to be closed and regenerated, too.

Much of the literature promoting regeneration projects toy with the idea of turning the clock back, possibly to a time before the Industrial Revolution, by restoring the environment. This is a myth, of course. Changes in the soil composition caused by extra nutrients, whether on arable or on derelict land, make it uncertain that the same life-forms can be harboured in the new woods as in the old ones. Supposing the same trees were planted, the complex ecosystems that go with them (native plants, birds, insects, for instance) would not necessarily come back. Frequent land-use change is not necessarily conducive to well established woodland. In other words, new woodlands does not mean the same woodlands.

The case of the Emscher Park, in the Ruhr Valley in Germany, suggests an innovative alternative to the burying of the past which, as we shall see, can create resentment from the local communities. There, the regeneration of the derelict industrial site has re-established the natural vegetation and riparian woodland where possible. But instead of getting rid of the traces of the industrial past (buildings and landscape), it has woven them, so to speak, into the urban forest and its green corridors, so as to vindicate the local population's creativity and skills in creating that particular layer of the palimpsest of local landscape history. By the same token, the local community has been invited to use these very qualities in contributing to the creation of the new layer, that of the postindustrial landscape of their area [Simson 2005: 490].

Trees are such powerful emblems that they lend themselves easily to fashions and multifarious discourses, elaborating thereby the 'new social construction of the forest' mentioned by A. Mather. A lot of the initiatives described above, and many others, start from the unquestioned assumption of the positive contribution of woodland to the landscape, an assumption which is greatly helped by the realisation of the cost-effectiveness of forestry in relation to other end-uses for dilapidated sites. They therefore use forestry as a background – one is tempted to say as an alibi in some cases – for other delivery objectives, more often than not linked to economic prospects. Until that bubble bursts, too?

Forestry has clearly become a means to an end, with economic benefits looming large. However, indirect market benefits have superseded the direct market benefits of forestry. This is partly due to the fact that the only worthwhile plantations in direct economic terms are conifers, which have been all but given up by the Forestry Commission and are poorly grant-aided, but also because the indirect benefits are as numerous and varied as the objectives of multi-purpose forestry, thus offering much more scope. However, the income to be expected by private landowners from planting for indirect market benefits is not sufficient to enlist their participation, leading to delays in fulfilling the objectives. A wider range of mechanisms is needed and is under study by the Forestry Commission, including promotion, advocacy, tech transfer, support for the development of markets.

The increase in the share of indirect market benefits is such that, due to a boomerang effect, it is having direct bearing on the management techniques of the forestry sector. The traditional clear-felling and replanting techniques associated with high plantations are now often being replaced by 'softer' management forms such as 'continuous cover' and 'low impact' forestry, which impacts less on the landscape [Forestry Commission 2006c: 15]. These regimes have prompted new research to develop adequate strategies [Forestry Commission 2004b]. Amusingly enough, devising methods to implement non-market benefits results in serving the market benefits, since regeneration instead of seeding turns out to reduce the cost of restocking.

Regeneration programmes invoke, mantra-like, sustainability to attract investors. But there is more to sustainability than economics, and the next two chapters will focus on the inclusion of the other two sides of the sustainability triangle.

References

Amin, A. and S. Graham. 1997. The ordinary city. *Transactions of the Institute of British Geographers* 22 (4): 411–429.

Amin, A., D. Massey and N. Thrift. 2000. *Cities for the Many not the Few*. Bristol: Policy.

Battershill, M. and A. Gilg. 1996. New approaches to creative conservation on farms in South-west England. *Journal of Environmental Management* 48: 321–340.

Beaverstock, J. et al. 1997. The National Forest and global environmental sustainability. *European Environment* 7: 54–62.

Bell, M. 1992. Woodlands on farms. In *New Forests for the 21st Century: The Scale, Nature and Location of Forestry Expansion in Britain*. Speakers' papers at the Forestry Commission Conference, 26 and 27 March 1992, Forestry Commission, n.p.

Bending, N. and A. Moffat. 1997. *Tree Establishment on Landfill Sites: Research and Updated Guidance*. Edinburgh: Forestry Commission.

BGS (British Geological Survey). 2005. *Opencast Coal Mining Statistics 2005*. http://www.bgs.ac.uk/mineralsuk/minequar/coal/occ/home.html#TABLE1. Accessed 20 March 2007.

Biomass Task Force. 2005. *Report to Government*. London: DEFRA.

Bowler, I. and B. Ibery. 1999. Agricultural land-use and landscape change under the post-productivist transition – examples from the United Kingdom. In *Land Use Changes and their Environmental Impact in Rural Ares in Europe*, eds. Kronert, R., J. Baudry, I. Bowler and A. Reenberg, 121–139. Carnforth Lancashire: Parthenon.

Cloke, P., P. Milbourne and C. Thomas. 1996a. The English National Forest: local reactions to plans for renegotiated nature-society relations in the countryside. *Transactions of the Institute of British Geographers* NS21: 552–571.

Cloke, P., P. Milbourne and C. Thomas. 1996b. From wasteland to wonderland: opencast mining, regeneration and the English National Forest. *Geoforum* 27 (2): 159–174.

Community Forest Partnership. 2005. *Community Forests Monitoring Report 2004–2005. Final Report*. Bury Saint Edmunds: Projects Partners Research. http://www.communityforest.org.uk/resources/monitoringreport_2004_05.pdf. Accessed 7 February 2007.

Community Forest Partnership. 2007. http://www.communityforest.org.uk/. Accessed 6 February 2007.

Countryside Commission. 1987. *Forestry in the Countryside*. CCP 245. Cheltenham: Countryside Commission.

Countryside Commission. 1994. *The National Forest. The Strategy*. CCP 468. Cheltenham: Countryside Commission.

DCLG (Department for Communities and Local Government). 2000. *Our Towns and Cities - the Future*. London: DCLG.

DEFRA (Department of Environment, Food and Rural Affairs). 2004. Alun Michael announces two year extension in funding for community forests. *News Releases 2004*, 113/04, 29 March 2004. http:/ww.defra.gov.uk/news/2004/040329a.htm. Accessed 15 April 2004.

DEFRA (Department of Environment, Food and Rural Affairs) and ERDP. 2001. *The Countryside Stewardship Scheme. Traditional Farming in the Modern Environment*. London: DEFRA.

Dickinson, M., J. Mackay, A. Goodman and P. Putwain. 2000. Planting trees on contaminated soils: issues and guidelines. *Land Contamination and Reclamation* 8 (2): 87–101.

Dobson, M. and A. Moffat. 1993. *The Potential for Woodland Establishment on Landfill Sites*. London: HMSO.

DoE (Department of the Environment). 1986. *Waste Management Paper 26: Landfilling Wastes*. London: HMSO.

DoE (Department of The Environment). 1990. *This Common Inheritance: A Summary of the White Paper on the Environment*. Cm 1200. London: HMSO.

DoE (Department of the Environment). 1993. *Countryside Survey 1990*. London: Department of the Environment.

Elands, B. and K. Wiersum. 2001. Forestry and rural development in Europe: an exploration of socio-political discourses. *Forest Policy and Economics* 3: 5–16.

England's Community Forests. 2007. *Supporting Regeneration and Growth*. http://www.communityforest.org.uk/whatwedo_cases1.htm. Accessed 13 March n.d.

Forestry Commission. 1998. *England Forestry Strategy. A New Focus for England's Woodlands*. Cambridge: Forestry Commission.

Forestry Commission. 2001. *Forestry Statistics 2001*. Edinburgh: Economics and Statistics Unit, Forestry Commission.

Forestry Commission. 2004a. *Great Britain and England Annual Report and Accounts 2003–2004*. Edinburgh: Forestry Commission.

Forestry Commission. 2004b. *Managing Light to Enable Natural Regeneration in British Conifer Forests. Information Note*. Edinburgh: Forestry Commission.

Forestry Commission. 2005. *Forest Life*. Spring-summer 2005.

Forestry Commission. 2006a. *England Forestry Strategy Progress Report 2006*. Edinburgh: Forestry Commission.

Forestry Commission. 2006b. *England Forestry Strategy Review. Topic Paper: Forestry and Agriculture - Interactions*. http://www.defra.gov.uk/corporate/consult/forestry-strategy/interactions.pdf. Accessed 12 April 2007.

Forestry Commission. 2006c. *Great Britain and England Annual Report and Accounts 2004–2005*. Norwich: The Stationery Office.
Greenhalgh, L. and K. Worpole. 1995. *Park Life. Urban Parks and Social Renewal*. London: Comedia/Demos.
Haines-Young et al. 2000. *Countryside Survey 2000. Accounting for Nature. Assessing Habitats in the UK Countryside. Executive Summary*. London: Department of the Environment, Transport and The Regions.
House of Commons Agriculture Committee. 1990. *Land Use and Forestry. Second Report, Vol. 1*. House of Commons Paper 16-1. London: HMSO.
Hughes, R. 1991. The role of urban forestry in environmental strategies and the economic regeneration of the post industrial town. *European Environment* 1 (2): 1–5.
Hutchings, T., A. Moffat and I. Stubbs. 2001. Woodland restoration of contaminated land. In *Forest Research Annual Report and Accounts 1999–2000*, Forestry Commission, 30–37. Edinburgh: The Stationery Office.
Lowe, P., J. Murdoch, T. Marsden, R. Munton and A. Flynn. 1993. Regulating the new rural spaces: the uneven development of land. *Journal of Rural Studies* 9 (3): 205–222.
Mace, A., P. Hall and N. Gallent. 2007. New East Manchester: Urban renaissance or urban opportunism? *European Planning Studies* 15 (1):51–65.
Mather, A. 2001. Forests of consumption: postproductivism, postmaterialism and the postindustrial forest. *Environment and Planning C: Government and Policy* 19: 249–268.
Moffat, A. and J. Laing. 2003. An audit of Woodland performance on reclaimed land in England. *Arboricultural Journal* 27 (1): 11–25.
Moffat, A., T. Hutchings and D. Elgy. 1997. *Tree Establishment on Landfill Sites. Interim Report*. Forest Research Report to DETR.
National Community Forest Partnership. 2004. *England's Community Forests – Shaping the Future, 1990–2004. Briefing note*. http:/ww.defra.gov.uk/news/2004/040329a-briefing.pdf. Accessed 15 April 2004.
ODPM (Office of the Deputy Prime Minister). 2002. *Living Places: Cleaner, Safer, Greener*. London: ODPM.
ODPM (Office of the Deputy Prime Minister). 2004. *Making it Happen – The Northern Way*. London: ODPM.
Rawlinson, H., N. Dickinson, P. Nolan and P. Putwain. 2004. Woodland establishment on closed old-style landfill sites in N.W. England. *Forest Ecology and Management* 202: 265–280.
Sheail, J. 1998. The National Forest: an historical context. *East Midland Geographer* 21 (1): 12–22.
Simson, A. 2005. Urban forestry in Europe: innovative solutions and future potential. In *Urban Forests and Trees. A Reference Book*, eds. C. Konijnendijk, K. Nilsson, T. Randrup and J. Schipperijn, 479–504. Berlin: Springer.
Stoker, G. 2004. *Transforming Local Governance: From Thatcherism to New Labour*. Beyond the Centre Series. Basingstoke: Palgrave Macmillan.
Watkins, C., D. Williams and T. Lloyd. 1996. Constraints on farm woodland planting in England: a study of Nottinghamshire farmers. *Forestry* 69 (2): 167–176.
Zukin, S. 1995. *The Culture of Cities*. Cambridge: Blackwell.

Chapter 8
The Contribution of Woodlands to the Environment

Some recent tree planting promotions would lead one to think that the moment
that a tree is planted, global warming will suddenly decrease and wildlife will appear

[Bradley 1995: 4]

Abstract One of the main functions of 21st century forestry is that linked to the environment, in relation to biodiversity and climate change. This is an area that has given birth to a wealth of publications and organisations, because an environmentally-friendly image entails, non only non-market benefits, but also clear indirect market benefits. It is therefore essential to bear in mind the context of the production of scientific knowledge in order to review the recognised scientific facts, and expose the fantasies and the exploitation of the tree image meant to elicit support and generate profits. Tree planting, especially deciduous native trees, is a plus for a company's image, as stressed in the testimonies of firms that claim to have become 'carbon neutral'. But is it really that efficient in environmental terms? The eighth chapter analyses these issues, focusing on the recognised benefits of trees and the exploitation of their image.

As far as sustainable forestry is concerned, environmental and social aspects lie half-way between the market and the non-market benefits. Indeed, even if these are not the essential bonus of contemporary forestry, there are obvious indirect market benefits, or 'invisible earnings', as regards improved health or improved community relations, if only in terms of deferred medical costs or policing budgets, for instance. However, in this chapter and the next, and without forgetting these categories, these aspects of forestry will be tackled, not through the market/non-market benefits grid, but from a wider perspective. In particular, one needs to cut through the categories used in forestry literature, so as to analyse the discourse and relate it to practice.

Until the 1980s, when mention was made in scientific journals of the relation between trees and air pollution, it was almost inevitably to stress the damage done to trees by airborne pollutants, and in relation to the productivity of forests. This obviously continues to be a matter of concern for foresters, as can be seen in the 2003 *National Report to the Third Session of the United Nations Forum on Forests. United Kingdom of Great Britain and Northern Ireland.* In the part devoted to forest health and productivity, the report stresses the threats that air pollution and climate change represent for the health and productivity of trees in Europe. National surveys on the condition of forests were integrated into a pan-European survey in 1987 and the European Commission passed a regulation requiring member states to undertake regular inventories of forest condition. This programme of monitoring, amended in 1994, permits us to assess the importance of pollutant depositions compared with the other factors (biotic and abiotic) which have a bearing on forest stability and productivity [United Nations Forum on Forests 2003: 9–10; Tagaki and Gyokusen 2004: 170–171].

At the same time as tree health has remained a source of preoccupation, another discourse emerged in the last 2 decades of the 20th century, which takes the opposite perspective, that is, the contribution that trees, urban trees in particular, can make to a healthy environment [Beckett et al. 1998: 357–359]. In the latter conception, the more trees are planted, the better it is for the atmosphere and for biodiversity. That is at least the view that the government and the National Urban Forestry Unit have chosen in order to promote tree-planting campaigns, as the following quotations exemplify:

> Tree planting is an important issue as trees contribute to the overall reduction in global warming by absorbing carbon dioxide [ONS 1999: 190];
> Trees are the air conditioners of the city, trapping particulate pollution, absorbing noxious gases and making the air in towns much healthier to breathe [NUFU a (n.d.): 4]
> Our urban forests also have a modest impact on global warming, since they convert carbon dioxide into timber [NUFU b (n.d.): 3]

Similarly, in the calculation of non-market benefits, the Forestry Commission has attributed a price to biodiversity. The first years of the 21st century seem to point to yet another reading of the functions of trees in the urban environment.

8.1 The Benefits of Trees

Margaret Thatcher's famous 'Green Speech' to the Royal Society in September 1988 is often taken as a U-turn in the politicisation of environmental issues. Indeed, the environment had been relegated to 'low politics', in spite of the creation of the Department of the Environment in the 1970s, and was even denigrated in the early 1980s [Burgess et al. 1991: 500]. On the contrary, studies on the environmental benefits of trees and woodlands have been numerous since the 1990s, and the benefits they identify come under several headings. They can directly contribute to the reduction of carbon emissions into the atmosphere, while fuel substitution (using wood instead of fossil fuels in energy production) and material substitution (using wood products instead of products made with fossil fuel) can contribute to the improvement of the environment.

8.1.1 Absorption

CO_2 is the most important greenhouse gas, responsible for 80% of carbon dioxide emissions in the UK [DEFRA 2005]. Growing trees, as well as mature ones, store

CO_2 as biomass above and below ground, and emit oxygen in the process of photosynthesis. Increasing carbon stocks in forests through expanding woodland area and managing existing woodland can therefore make a contribution towards meeting the UK's commitment under the Kyoto Protocol.

Dust and particulate pollution are quite high in urban areas. Studies have revealed that they are due essentially to road transport and industries. As well as being responsible for smog under certain atmospheric conditions and for tree disease, they can have adverse effects on health. The Department of Health manifested its concern as to the potential danger of this type of pollution, in the form of cardiovascular and respiratory diseases in particular [Department of Health 1995a, b]. The very small particulate matter, PM_{10}, in particular, are believed to be small enough to reach the lungs, and an increase in their concentration can increase mortality. Trees can trap and absorb some of these particles in their canopy [Beckett et al. 1998: 353], especially in the urban environment and especially if they are planted close to the source of emission (industries, roads, etc.). However, according to the Forestry Commission, the effects are 'relatively modest' [Forestry Commission 2006: 5].

8.1.2 Regulation

Cities are subject to what is known as the heat island effect, that is to say to temperatures usually warmer than in the open countryside, by 0.5–1.5°C. In some places, asphalt parking areas for instance, temperatures can be as much as 20°C warmer than in areas shaded by trees. Trees can help regulate excessive heat in two ways. They absorb solar radiation on their leaves and reflect it, thus reducing radiation on buildings and on the ground. Added to that, provided they find enough water in the soil, trees can emit through transpiration about 400 litres of water per day, which creates a cool microclimate deemed to be equivalent to five average air-conditioners. Thus, through the combined effect of reduced solar radiation and evapotranspiration, urban trees can limit the heat island effect as far as 150 metres from where they are planted. Moreover, at night, the tree canopies slow down the loss of heat stored during the day, leading to a warmer night temperature under trees (5–8°C) [Grey and Deneke 1978: 67–76]. So, if judiciously planted, trees can contribute to reducing the amount of fossil fuels used to warm or to cool, buildings.

8.1.3 Protection

Energy savings can also be induced by planting appropriate trees to act as wind shields to insulate houses from the prevailing winds. Fuel savings have been estimated at 23% [Grey and Deneke 1978: 48–53]. The shade cast by trees is particularly appreciated in summer as protection against strong sun radiance which is responsible for a growing number of skin cancers: according to Cancer UK, sunburn can double the risk of skin cancer. Noise has become a form of pollution, which even if it is invisible, can have important psychological effects. Trees absorb sound waves through their leaves and branches and, used in wide belts (30 metres), can act as noise barriers. Additionally, although this is probably a more valid argument in the open countryside and in mountain regions than it is in cities, tree roots help fix the soil, and the canopy intercepts the rain and slows its descent into the soil, thus decreasing runoff and preventing the washing away of humus and the erosion of soils in case of strong rainfall. Finally, trees and shrubs harbour numerous species of fauna and flora which contribute to biodiversity as well as to the quality of life: the first benefit people mention in relation to woodland walking is the presence of birdsong. The disappearance of hedges in the countryside makes it even more important to provide shelter for wildlife in cities.

Once put in parallel with the costs induced by trees (planting, irrigation, pruning and maintenance), these benefits generate a per tree net benefits figure. But beyond economic calculations, the environmental benefits of trees have attracted the attention of organisations wishing to improve their image by putting forward tree-friendly policies.

8.2 The 'Field of Dreams'

Numerous supermarkets plant trees in their car parks and thus underline their contribution to their customers' quality of life. Shopping centres often play on the tree/woodland image in their name to attract customers or borrow the name of the land that has been cleared to make space for their establishment: Cheshire Oaks Outlet Park shopping centre outside Chester or the Glades in London, among many others. Similarly, business parks, complete with lawns, trees and fountains, have become imperative for a company wishing to create a positive impact on the public.

8.2.1 Trendy Trees and Corporate Images

Although forestry is not currently part of the UK Emissions Trading Scheme, the private sector has begun to develop markets for carbon trading. Possibly the most daring use of the positive environmental association of tree planting is that made by the company Future Forests, renamed in 2005 the CarbonNeutral Company. The company, the first limited company to offer carbon offsets in the UK and now the largest one, was set up in 1997 by Dan Morrell, a popular DJ, and Sue Welland, a specialist in integrated marketing communication. It describes itself as 'one of Europe's leading environmental businesses helping companies and the public take action on climate change' [Future Forests 2004a]:

> Future Forests calculates the level of carbon dioxide (CO2) produced by various activities and provides 'offsets' to balance this out, and make that activity Carbon Neutral [Future Forests 2004b].

The firm sells trees through the Internet, to be planted so as to offset the customers' polluting activities (going on holiday, using their cars, travelling by plane, etc.): a variation on the theme of the 'polluter pays' principle. The trees are planted in the UK or in developing countries, in particular through the Plan Vivo projects in Uganda, Mexico and Mozambique. The calculations are made by putting in parallel the amount of carbon released in the air by one of the customer's activities, with the estimated amount of carbon stored in a tree for the whole of its lifetime, estimated (optimistically) at 99 years. The calculation is deliberately kept simple: five trees per car per year, 15 trees per person per year [Wylie 2000]. The company thus boasts planting in 2001 more than 2,000 trees, at £7 apiece. Each customer receives a certificate indicating where 'his'/'her' tree was planted. In 2005, CarbonNeutral planted a little under 300 hectares [Randerson 2005].

Among the first customers were a number of individuals or groups from show-business who could thus boast that the CO_2 produced in making their latest CD had been 'soaked up' with forests in Mozambique or England, or that their tour was 'carbon neutral', since they had paid for a number of trees to be planted to offset the environmental costs of their activities. Bands even have their own Celebrity forests, where fans are welcome to buy trees. The change of name to CarbonNeutral in 2005 was the result of a market study among the customers, who thought that the image of the company was too closely associated with tree planting and wanted a wider array of strategies to fight against climate change and to enhance brand value.

Other carbon-offsetting commercial ventures have subsequently been created, and it is safe to assume that the trend will continue for a while. Indeed, public support is strong for such actions, as can be seen from the results of the 2005 *Public Opinion on Forestry Survey*. One of the three main reasons why people support forestry with public money is that trees contribute to 'prevent the greenhouse effect and global warming' (56%), and among the people who had heard of trees and woodlands in the media over the previous 12 months, tree-planting was the dominant topic they had heard of (23%) [Forestry Commission 2005: 3, 7]. More than the environmental benefits, the testimonies of companies underline the impact of their move in terms of brand image. Presenting one's firm as carbon neutral (especially when one is connected with transport) has a positive ring with customers [Parkinson 1995: viii]. An illustration of this can be seen at the Liverpool John Lennon airport. Collecting boxes in the waiting lounge boast 'Liverpool John Lennon airport is the first UK airport to instigate a CarbonNeutral flight initiative'

From the point of view of the companies, it is well worth investing in offsetting actions in order to make a 'a brand differential' in a saturated market. Put in the words of CarbonNeutral,

> companies who have recently operated in the ethical brand space have seen their profits soar. Better yet, they're proving that what's good for the shareholder has synergy with environmental objectives [Carbon Neutral Company n.d.].

Fig. 8.1 Flying 'carbon-neutral' from the Liverpool John Lennon airport (Photo Sylvie Nail)

Commercial ventures like CarbonNeutral and other environmental service providers sponsored by oil companies have been criticised by environmentalists: through tokenistic gestures ('greenateering' [Katz 1998: 52]) and 'guilt-free option(s)' [Randerson 2005], they offer corporate capitalism the equivalent of 'medieval indulgences' [Honigsbaum 2007], i.e. a convenient way of buying the patronage of much of the population while shirking their responsibilities and changing next to nothing to a lifestyle based on consumption.

One characteristic that many tree-planting projects have in common is that their scientific backing is often scarce, which has led to environmentalists protesting against the use and abuse of the tree image.

8.2.2 Tree Planting Versus Climate Change: Myths and Realities

In an American article dated 1991, the authors did point out that planting trees could help compensate for human activities, and they calculated how many trees a company would need to plant in specific places in order to save on heating and air conditioning bills [Rowntree and Nowak 1991: 269–275]. They went on to suggest how good-willed US citizens could do their share in order to offset their carbon emissions, and it can be supposed that this reasoning lies behind such ventures as that of Future Forests/CarbonNeutral. However, two essential points were made in the demonstration which never appear in the arguments of such companies. First, the number of trees to be planted per person varies enormously according to the person's age. According to the authors' reckonings, a newborn baby would only need 45 seedlings to be planted in order to offset all the carbon of his/lifetime, while a 20-year old would need 80 and a 50-year old 550. In all cases, the figures are given provided the trees lived as long as the person did and provided carbon consumption did not increase.

The other point is that the effectiveness of the process depends entirely on the care one gives to 'one's' trees. Stewardship is an essential part of the process and there is no question in their argument of planting the trees and walking away to let them fend for themselves. A brochure by the authors entitled 'How many trees does it take to store the carbon you produce?' underlined the need for responsible behaviour for one's whole lifetime, 'to counter the misconception that one can do their part by just planting a seedling and not following up with tree care' [Rowntree and Nowak 1991: 273–274]. However, this is not part of the process induced by commercial ventures: once one has paid for one's trees, no follow-up is expected. In the case of CarbonNeutral, the monitoring of UK sites is only done every 5 years, which is far from sufficient to ensure the welfare of the trees.

Gordon Bradley was also warning in 1995 against excessive enthusiasm, in particular concerning the role of trees in environmental terms:

> some recent tree planting promotions would lead one to think that the moment that a tree is planted, global warming will suddenly decrease and wildlife will appear. This is a kind of 'field of dreams' effect: we plant the tree and benefits will come [Bradley 1995: 4].

The fact that trees do not necessarily receive the care they need is only one aspect of the criticisms which can be addressed to commercial projects, the environmental benefits of tree-planting are themselves subject to discussion, for several reasons.

The first one concerns the use of polluting chemicals in forestry itself. Whether to keep the plantations 'clean' by killing weeds or to reduce the consequences of diseases and pests on trees and thus ensure better productivity, pesticides are used and this practice is likely to contribute to pollution and the degradation of the environment. The Forestry Commission became aware of the issue and, on top of guidelines on the subject in 1995 [Willoughby and Dewar 1995; Forestry Commission 1996], it published in 2004 a booklet entitled *Reducing Pesticide Use in Forestry*. This provides alternative methods and ways to reduce chemical pest and weed control and encourages landowners to prioritise risks (to operators, neighbours, forest users, wildlife and the environment in general), before using pesticides [Willoughby et al. 2004: 5]. Similarly, in January 2006, a new *Code of Practice for Using Plant Protection Products* was published by DEFRA, urging the reduction of chemicals in amenity, agriculture, horticulture and forestry, both to comply with national and European regulations and to protect the environment [DEFRA Health and Safety Commission 2006]. Both stress the necessity of assessing the need for pesticides before using them.

Concerning the role of trees in atmospheric pollution abatement itself, Grey and Deneke, in their classic book on urban forestry, were very cautious. They asserted that some plants had the capacity to absorb specific pollutants such as hydrogen fluoride, sulphur dioxide, nitrogen dioxide, but stated that the least absorbed was the very noxious carbon monoxide which accounted for more or less half the total weight of air pollutants in the United States [Grey and Deneke 1978: 77–78]. Worse, trees can emit volatile organic compounds (VOC) which contribute to the formation of ozone and carbon monoxide [Nowak 2000: 145].

There are also huge gaps between the amount of carbon trees can store and what is emitted. The Academy of Sciences defined in 1994 the average storage of carbon in temperate forests at 2.5 tonnes per hectare per year [Hermeline and Rey1994]. Bradshaw and Walmsley calculate that a mature oak tree captures 18 kilos of carbon per year, and put this in parallel with the average driver travelling about 10,000 miles (about 16,000 kilometres) per year, producing about 2,500 kilos of carbon. In order to fix the CO_2 emitted by that driver alone in a year would require about 140 trees [Bradshaw et al. 1995: 13]. In order to sequester emissions from the 23 million cars of the UK in 1999, an area of '3.7 million ha of poplars or 8.5 million ha of conifers or 12.6 million ha of hardwoods' would need to be planted, bearing in mind that the entire rural land area of the UK is 23 million hectares [Cannell 1999: 244].

Furthermore, various studies point out the difference between capturing carbon and absorbing it. In a lot of publications, indeed, trees are made to appear as carbon-eaters (the concept of 'carbon-sink' is ambiguous in that sense), while they in fact 'store' the carbon for as long as they live, but release most of it when they die. Their function thus consists in slowing down the emission of CO_2, not eliminating it.

8.2.3 The Limits of 'Green Power'

The Intergovernmental Panel on Climate Change calculated in 1995 the rate of planting that could realistically be achieved worldwide and their environmental consequences on CO_2. They concluded that slowing deforestation and promoting a global forestation programme worldwide may amount to a 12–15% reduction in CO_2 emissions by 2050, but that this would not stabilise levels or keep them below a level likely to be dangerous for climate change in the 21st century. So, even though the Kyoto Protocol, in its articles 3.3 and 3.4, acknowledges the role of forests as carbon sinks, trees cannot be expected to do more than their share. David Nowak, a researcher at the US Forest Service, has demonstrated in over one hundred publications that the simple presence of trees is far from enough to capture carbon in urban areas if other accompanying measures are not implemented. He modeled the planting of ten million urban trees each year for 10 years in the United States [Nowak and McPherson 1992]. Fifty years later, these judiciously planted trees would have stored 85 million tonnes of carbon and prevented the production of 315 tonnes, provided none of them had died in the meantime and not including the carbon emissions when they eventually died. Even with this generous calculation, this amount is hardly equivalent to 1% of the carbon emissions forecast over that 50-year period in the United States. Not only is planting trees not enough: they must be judiciously planted. The reduction in carbon emissions due to appropriate planting is estimated at four times higher than that of carbon storage by trees [Nowak 2000: 146].

The UK share of fossil fuel carbon emissions is about 2.5% of the global total and a sustained programme of tree-planting is estimated to result in the sequestering of less than 2% of the UK fossil fuel carbon emissions. A 1999 study by the University of Lancaster with the Centre for Ecology and Hydrology used Urban Tree Air Quality Scores (UTAQs) to weigh the benefits of trees to air quality against the costs in the Midlands. Their computer model calculations from a sample tree survey allowed them to calculate that

> the total amount of carbon stored in the West Midlands tree population is equivalent to only 6% of the carbon dioxide emitted to the atmosphere from the West Midlands in a single year. In other words, all the trees in the West Midlands hold the equivalent of three weeks worth of emissions of CO2 from the conurbation [Stewart et al. 2002: 9].

Cuts in emissions are therefore indispensable to improve the situation.

In an article published in 2002, Nowak and his colleagues stress another fact, seldom mentioned, namely that trees have to live to a certain age in order to justify the carbon emissions produced by using fossil fuel driven machines and vehicles in the planting and maintenance process (chainsaws, trucks, etc.). If the trees die young, the process of tree-planting can turn out to worsen CO_2 emissions [Nowak et al. 2002]. Hence the importance of providing the newly-planted trees with a minimum care, if possible without using fossil-fuel machines. The sad reality is that in England, 30% of the newly planted trees die within 12 months of being planted, and 50% (up to 90% in some areas of England) within the first 10 years, for lack of the minimum care they require [Thurman 1997: 31–33; Rival 1998: 18]. This means that far from offsetting atmospheric CO_2, tree-planting has actually contributed to carbon emissions. This is also true if the trees are left to decompose and release the carbon they have accumulated, so making sure that the trees are not left to decompose, in other words managing the forest, is another important recommendation. Attention must be paid to the use made of timber and wood, so as to 'seal' the carbon (by using it for building or furniture purposes, for instance) or use it in replacement of other fossil fuel (burning wood instead of oil for heating purposes). This is difficult to reconcile with conservation and biodiversity purposes which advocate leaving nature to its own devices because dead wood provides niches and habitat for wildlife.

An example of this ambivalence can be found in the publications of the Forestry Commission. The *UK Forestry Standard*, the official British government's programme of sustainable forestry, stresses the role of trees in CO_2 reduction [Forestry Authority 1998: 1], while the booklet *Life in the Dead Wood* insists on the need not to remove deadwood so as to sustain woodland species. It goes as far as to say that 'dead and dying trees play a key role in the functioning and productivity of forest ecosystems through effects on biodiversity, carbon storage, (...)' [Forest Enterprise 2002: 1], which is surprising, to say the least: on occasions, policies can try to kill too many birds with one stone, it seems.

Fig. 8.2 Good for biodiversity, bad for CO_2: the dilemma of dead wood (Photo courtesy of the Mersey Forest)

8.2.4 Caring and Choosing

The need for tree maintenance has increasingly been taken into account in recent years. 'Trees Love Care', for instance, a programme set up by the Tree Council, was made much of during the 30th anniversary of the Council in 2004. It aims at making people aware of the need to follow up the planting with an increased sense of responsibility, doing such simple things as loosening the ties that hold the saplings or cleaning the ground of the weeds around them every spring.

All in all, the sources agree as to the greater role played by trees in the cleaning of the air than in their function in CO_2 storage, which is mostly a short-term measure to buy time while inventing new solutions to combat climate change.

Another controversial issue is the choice of species to be planted. Trees in towns are not without their drawbacks, such as root invasion of pipes, foundations and pavements, the danger of falling branches or the need for the cleanup of falling leaves.

Hazards can include allergy to pollen particles, or even the contribution of trees to the formation of ozone by the release of allergens and emission of biogenic volatile organic compounds (VOC) by some species. So the choice of species to maximise positive environmental effects is crucial, but this is hardly ever mentioned, apart from articles in specialised magazines aimed essentially at planners. Scientific evidence points to the superiority of conifers to achieve environmental advantages (shade excepted, since they do not provide much). Indeed, broadleaves shed their leaves every autumn, while conifers can operate as fixators all year round. Moreover, their needles are better at fixing particulate dust than broadleaved trees, since they offer a rough and sticky surface [Beckett et al. 1998: 354]. Conifers are 'more efficient scavengers of particulate lead than broadleaves' [Fergusson et al. 1980: 293–310] and in general they are better at cleaning the air. In terms of noise reduction, when planted in belts at least six metres wide, conifers are at a definite advantage over broadleaves – although sycamores are surprisingly good at this – since their foliage and branches go down to ground level. The Lancaster study estimates that the species more likely to remove the most pollutants from the air without contributing to the formation of VOCs were Scots pine, common alder, larch, Norway maple, field maple, ash and silver birch. On the other hand, those classified as high biogenic hydrocarbon emitters, thus conducive to ozone formation include oaks, poplars and willows [Stewart et al. 2002: 10], a result confirmed by David Nowak [Nowak 2000: 145]. All in all, as an article published in 1998 put it, 'the information reviewed here suggests that future urban planting should focus on the increased use of conifers' [Beckett et al. 1998: 357].

Yet, the immense majority of trees being planted in Britain since the 1985 *Broadleaves Policy* are natives, more often than not including oak, as well as poplar for short rotation coppice. Concerning carbon storage, the oldest trees in the UK are broadleaves, and since larger, mature trees store more carbon than small ones, broadleaves presently store the largest amount of carbon, hence the need to conserve them. In the short term, faster-growing trees, such as poplar (*Populus* spp.) or conifers store carbon faster, therefore are more interesting than slow-growers, but in the longer term (100–200 years), the choice of species is less critical. In short, 'life-span and size' are identified as the two key-factors in carbon storage, factors that should also be taken into account in the selection of trees species.

Some even go as far as to say that the existence of forests is not essential for fixing the carbon. Without denying the interest of trees in reducing the particulate air pollutants by diluting them or by trapping them so that they will be washed off by the rain, the fact is that as far as oxygen and CO_2 are concerned, the overwhelming majority of oxygen produced on earth through photosynthesis comes not from the trees, but from the seas, and that the oxygen produced by 1 acre of forest per year represents only about 0.03% of the total oxygen produced on that 1 acre of land [Grey and Deneke 1978: 77–78]. Noël Decourt adds that the disappearance of the forests would not even lead to the doubling of CO_2 in the atmosphere [Decourt 1999: 15]. However, such is the love of trees, particularly deciduous trees, and the power of their image, that should one wish to do so, it would be quite difficult to go against the grain. Bearing all these scientific factors in mind, one may wonder about the success of a commercial venture such as CarbonNeutral, whose environmental contribution is far from unequivocal, to say the least.

The key may lie somewhere else, both in the development of environmental law which has put increasing pressure on companies to become greener, and in the English cultural context.

8.3 Tree Planting and Environmental Law

Tree planting, especially deciduous native trees, is a plus for a company's image, as stressed in the testimonies of companies that claim to have become 'carbon neutral'. Whether this is just a change in packaging meant to pursue 'business as usual' while putting up an environmentally-friendly face in line with social expectations, or whether it is a commitment to more, probably varies from one company to the other.

8.3.1 Complying with Regulations and Expectations

Some customers' testimonies show that joining the Future Forests programme has raised their awareness of environmental problems and made them re-think their relationship with consumption: one car-rental company, for instance, says that it has started promoting car-sharing clubs. Heather Voisey and Tim O'Riordan note that businesses try to reduce the 'spread of their ecological footprint' partly to comply with tough regulations, but also to satisfy customer expectations. According to a study by a consulting organization (EFTEC) in connection with Green Alliance, an environmental organisation, more than half of the companies studied take initiatives beyond what is strictly required [Voisey and O'Riordan 1998: 170]. Becoming the customer of a carbon-offsetting company can therefore respond to external market pressure, or go beyond this and be part of an internalised environmental governance.

Undoubtedly, companies have become increasingly screened against environmental criteria in Europe and the USA, and international environmental law has become very influential. Companies' responsibilities are now expected to extend, not only to shareholders, but to stakeholders too, who include the customers and the protection of the environment [Ong 2001: 688]. In Britain, further to Lord Marshall's Task Force report, *Economic Instruments and the Business Use of Energy,* published in October 1998, the Climate Change Levy was introduced in the *Finance Act* of 1999 and came into operation in April 2002. The purpose of this tax on energy consumption in the industrial, commercial and public sectors is to act as an incentive to promote renewable energies.

In order to encourage businesses to integrate environmental concerns into their agenda, and not just to comply with rules imposed from the outside, the European

Community introduced Environmental Agreements in 1996, that is to say agreements between ‘industries and public authorities on the achievement of certain environmental objectives’ [Ong 2001: 690]. After the publication by the British government of its Climate Change Programme in November 2000, DEFRA negotiated *Climate Change Agreements* with 44 industrial sectors in 2001 [Huddleston 2003: ii]. Participants in the *Agreements* can be entitled to up to 80% reductions in the Climate Levy if they comply with energy savings targets. Tax incentives, funded by the Levy, were further added in 2001 (Enhanced Capital Allowances), to aid companies investing in renewable energies. To top it all, since 2000, investors have been encouraged to take into account in their reports the social, environmental and ethical achievements of the companies they invest in, hence the attractive reference to shareholders in the quotation above.

These changes in the legislation as part of the climate change debate provide a more rounded picture of the backdrop of tree-planting projects at the beginning of the 21st century. They partake of a shift from considering environmental duties as a liability to viewing environmental policies as an asset.

8.3.2 Has the Bubble Burst?

The scientific debate, so far limited to scientific journals, came to the forefront of the media at the end of 2005. The debate was on the possible inability of the British government to reach its targets regarding greenhouse gas emissions, and of climate change more generally. The articles were fuelled by the publication of three reports. The first of these, *Leading by Example? Not exactly…* by the Sustainable Development Commission, was the first report published since the government’s will to be an example in sustainable development was expressed in *The UK Government’s Sustainable Development Strategy* published in March 2005. The second report, published a month later in January 2006, was commissioned by the Department for Transport and entitled *Looking Over the Horizon: Visioning and Backcasting for UK Transport Policy* [Banister and Hickman 2006]. The third one, the title of which is *Avoiding Dangerous Climate Change*, was the report of an international conference held in February 2005 at the Met Office as part of the emphasis that the government wished to give to climate change at the beginning of its presidence of the G8 [Schnellnhuber 2006].

The press articles, for their part, focused on three complementary aspects: the need to think of different sources of energy to cut fossil fuel emissions to curb climate change [Clark 2006; Black 2006; King 2005]; the incapacity of the governments to reduce the CO_2 emissions of its own departments [Mullholand 2005; Wintour 2006] and the very limited efficiency of tree-planting projects [Randerson 2005; Honigsbaum 2007]. Furthermore, unless tree-planting projects are well thought-out, they may have negative side-effects, such as loss in stream flow, soil acidification and salinisation [Jackson et al. 2005: 1944–1947].

The publication in January 2006 of the results of a meeting of experts gathered at Exeter in February 2005 added fuel to the debate and attracted media attention.

The book stressed different interesting facts. Trees were viewed from the perspective of the negative impacts they suffered from climate change and pollution rather than as carbon-sinks. Concerning the latter point, the contributions pointed out that trees were likely to become saturated in carbon at some point [Schnellnhuber 2006: 149]. This acceleration of alarming information shows how perceptive Robin Grove-White was when he warned in the early 1990s of what might happen with regards to environmental issues [Grove-White 1991]. Using the case of the 1990 White Paper *This Common Inheritance*, he stressed the danger of the institutions' snug confidence in scientific information without concern for the human and cultural context in which the said scientific information was produced. In other words, he deplored the lack of investment in the sociology of science and saw in this a potential cause for inaccurate information on environmental risks.

The events at the end of 2005 appear just like one of these cultural turns in the production of scientific research on climate change and the result is a series of alarmist newspaper articles and government declarations, after a decade of apparent control of the situation.

The present position may be less naïve than the previous ones, in the sense that it does not condone semi-measures to protect the environment, such as tree planting, which can never hope to offset the increase in the emissions in the atmosphere. It may also reveal a new political agenda in Britain, with the promotion of nuclear energy by the British government at the end of 2005. Different arguments were put forward, but once again, transparency was less obvious than orthodoxy, and information seemed to be used so as to make nuclear energy the unavoidable solution. Indeed, if convincing the population was not the intention, one would expect to see the arguments put forward by David Nowak on tree-planting used in relation with the building of the nuclear plants.

Nowak argues that in order to form a real assessment of the benefits of trees in the absorption of carbon dioxide, the amount of fossil fuel used in tree planting and maintenance should be taken into account. In the government's arguments, as relayed by the press, only the environmental benefits of nuclear energy are mentioned. Yet it would make good environmental sense to take into account the fossil fuel used in the huge and protracted work involved in the building of the nuclear plants in question. And yet, the time needed by nuclear plants simply to offset the input of fossil fuel that went into their construction is never mentioned. The high number of nuclear plants needed to actually limit the amount of emissions to 500 parts per million (the government's objectives) is also passed over, as is the question of nuclear waste and its disposal. Yet, to take only the first point, the figure of 500 parts per million is already an excessive figure to protect the planet against climate change according to climatologists, who estimate that 400 parts per million are the maximum beyond which climate change might become irreversible [Elliott 2006]. Information has shifted, but it is still incomplete information.

To draw this discussion to a close, now that climate change is not some distant possibility for future generations, but is already showing its effects on the ski slopes of Switzerland, one can suppose that the discourse on trees and forests as the miracle cure for carbon emissions is going to recede somewhat. The British government published a draft *Climate Change Bill* on 13 March 2007, the first one of its kind at an international level, which aims at introducing in 2008 a legally binding framework to reduce carbon dioxide emissions.

Moreover, it seems that public opinion is becoming more aware of the issues, as the success of the campaign 'Your world. Your say', initiated by *The Independent* to gather public opinion at the end of March 2006 testifies. Yet, the myth is still pervasive: the preliminary report of *Public Opinion of Forestry 2007: England* reveals that an overwhelming majority of the population agrees with the assertion: 'the UK could offset all its greenhouse gas emissions by planting more trees' [Forestry Commission 2007: 12].

In all logic, the discourse on trees as a miracle cure for urban environments and for the environment in general should be replaced by more sober advice, namely how to plant the right species to provide the maximum range of effects, how to plant and maintain trees with as much care and using as little fossil fuel as possible, and how to reduce one's consumption of fossil fuel. Now that even the advantages of 'cleaning the air' by fixing fine particles are being contested, because particles might protect the earth from stronger solar radiation, the tide seems to have turned [Sample 2005]. The response to the consultation set up by DEFRA in 2006 in preparation for the new Forestry Strategy yet again stresses the complexity of the issues and the impossibility of giving a straightforward answer as to the role of trees. To give just one example related to climate change, the stakeholders present at a seminar organised by DEFRA in May 2006 underlined the difficulty of reconciling environmental objectives with climate change ones, the latter requiring for instance shorter rotations and different species choice than what environmental priorities would demand [Land Use Consultants 2006: 19].

In sociological terms, however, the last 3 decades of the 20th century will remain central to the positive perception of tree-planting, both from an anthropocentric and an ecocentric perception. What is being lost on the one hand is being regained on the other: environmental benefits may appear less miraculous than what was previously thought, but the social construction of reality has now pushed forward the social agenda of woodland (health, amenity, community). The next chapters will analyse to what extent this agenda is tuned to people's preoccupations and more likely to produce tangible benefits than the environmental one.

References

Banister, D. and R. Hickman. 2006. *Looking Over the Horizon: Visioning and Backcasting for UK Transport Policy*. London: Department for Transport.

Beckett, K., P. Freer-Smith and G. Taylor. 1998. Urban woodlands: their role in reducing the effects of particulate pollution. *Environmental Pollution* 99: 347–360.

Black, R. 2006. Stark warning over climate change. *BBC News* 30 January. http://news.bbc.co.uk/1/hi/sci/tech/4660938.stm.

Bradley, G. (ed.). 1995. *Urban Forest Landscapes: Integrating Multidisciplinary Perspectives*. Seattle: University of Washington Press.

Bradshaw, A., B. Hunt and T. Walmsley. 1995. *Trees in the Urban Landscape: Principles and Practice*. London: E&FN Spon.
Burgess, J., C. Harrison and M. Limb. 1991. Contested meanings: the consumption of news about nature conservation. *Media, Culture and Society* 13 (4): 499–519.
Cannell, M. 1999. Growing trees to sequester carbon in the UK: answers to some common questions. *Forestry* 72 (3): 237–247.
Carbon Neutral Company. n.d. *Why it's good for business*. http://www.carbonneutral.com/pages/goodforbusiness.asp. Accessed 6 January 2006.
Clark, A. 2006. Green fuel 'not enough' to cut transport pollution. *The Guardian*, 27 January.
Decourt, N. 1999. *La forêt*. Paris: Presses Universitaires de France.
DEFRA. 2005. Guidelines for Company Reporting on Greenhouse gas Emissions. http://www.defra.gov.uk/environment/business/envrp/gas/envrpgas.pdf. Accessed 27 January 2006.
DEFRA. 2006. *England Forestry Strategy Review. Topic Paper: Natural Resource Protection*. http://www.defra.gov.uk/corporate/consult/forestry-strategy/interactions.pdf. Accessed 12 April 2007.
DEFRA, Health and Safety Commission. 2006. *Pesticides. Code of Practice for Using Plant Protection Products*. London: DEFRA.
Department of Health. 1995a. *Committee on the Medical Effects of Air Pollutants: Health Effects of Non-Biological Particles*. London: HMSO.
Department of Health. 1995b. *Committee on the Medical Effects of Air Pollutants: Asthma and Outdoor Air Pollutants*. London: HMSO.
Elliott, L. 2006. Boardrooms feeling the heat. *The Guardian* 16 February.
Fergusson, J., R. Hayes, T. Yong and S. Thiew. 1980. Heavy metal pollution by traffic in Christchurch, New Zealand. *New Zealand Journal of Science* 23: 293–310.
Forest Enterprise. 2002. *Life in the Dead Wood. A Guide to Managing Deadwood in Forestry Commission Forests*. Edinburgh: Forestry Commission.
Forestry Authority. 1998. *The UK Forestry Standard. The Government's Approach to Sustainable Forestry*. Edinburgh: Forestry Commission.
Forestry Commission. 1996. *Herbicides for Farm Woodlands and Short Rotation Coppice*. Field Book 14. Edinburgh: Forestry Commission.
Forestry Commission. 2005. *Public Opinion of Forestry 2005: England*. Edinburgh: Forestry Commission, Economics and Statistics.
Forestry Commission. 2007. *Public Opinion of Forestry 2007: England (preliminary report)*. Edinburgh: Economics and Statistics.
Future Forests. 2004a. http://www.littlebigman.com/pr/LBM/LBM-020917-FutureForests.pdf. Accessed 11 January 2006.
Future Forests. 2004b. http://www.littlebigman.com/pr/LBM/LBM-020917-FutureForests.pdf. Accessed 11 January 2006.
Grey, G. and F. Deneke. 1978. *Urban Forestry*. New York: Wiley.
Grove-White, R. 1991. The emerging shape of environmental conflict in the 1990s. *RSA Journal* 139 (5419): 437–447.
Hermeline, M. and G. Rey. 1994. *L'Europe et la forêt*. Paris: Eurofor; Luxembourg: Parlement européen/Office des publications des Communautés européennes.
Honigsbaum, M. 2007. Is carbon offsetting the solution? (or part of the problem?). *The Observer*, 10 June.
Huddleston, J. 2003. *Climate Change Agreements – Results of the First Target Period Assessment*. Harwell: Future Energy Solutions.
Jackson, R. et al. 2005. Trading water for carbon with biological carbon sequestration. *Science* 310 (5756): 1944–1947.
Katz, C. 1998. Whose nature, whose culture? Private productions of space and the 'preservation' of nature. In *Remaking Reality. Nature at the Millenium*, eds. B. Braun and N. Castree, 374–402. London: Routledge.
King, D. 2005. The nuclear option isn't political expediency but scientific necessity. *The Guardian* 16 December.

Land Use Consultants. 2006. *England's Trees, Woods and Forests. A Consultation Document. Analysis of Responses to the Consultation.* Bristol: DEFRA. http://www.defra.gov.uk/corporate/consult/forestry-strategy/responseanalysis-final.pdf. Accessed 7 February 2007.

Mullholland, H. 2005. Government fails to hit carbon emission targets. *The Guardian* 16 December.

Nowak, D. 2000. Impact of urban forest management on air pollution and greenhouse gases. In *Proceedings of the Society of American Foresters, 1999 National Convention, Portland, Oregon*, 143–148. Bethesda: Society of American Foresters.

Nowak, D. and E. McPherson. 1992. Quantifying the impact of trees: the Chicago urban forest climate project. *Unasylva* 173. http://www.fao.org/docrep/u9300e/u9300e08.htm. Accessed 12 November 2007.

Nowak, D., J. Stevens, S. Sisinni and C. Luley. 2002. Effects of urban tree management and species selection on atmospheric carbon dioxide. *Journal of Arboriculture* 28 (3): 113–122.

NUFU (National Urban Forestry Unit) a. (n.d.) *Trees and Woods in Towns and Cities. How to Develop Local Strategies for Urban Forestry: A Guide*. Wolverhampton: NUFU.

NUFU (National Urban Forestry Unit) b. (n.d.) *Trees Matter. The Benefits of Trees and Woods in Towns: A Review*. Wolverhampton: NUFU.

ONG, D. 2001. The impact of environmental law on corporate governance: international and comparative perspectives. *European Journal of Environmental Law* 12 (4): 685–726.

ONS (Office For National Statistics). 1999. *Social Trends 29*, London: The Stationery Office.

Parkinson, J. 1995. *Corporate Power and Responsibility. Issues on the Theory of Company Laws*. Oxford: Oxford University Press.

Randerson, J. 2005. Tree-planting projects may not be so green. *The Guardian* 23 December.

Rival, L. 1998. *The Social Life of Trees*. Oxford: Berg.

Rowntree, R. and D. Nowak. 1991. Quantifying the role of urban forests in removing atmospheric carbon dioxide. *Journal of Arboriculture* 17 (10): 269–275.

Sample, I. 2005. Pollutants ward off global warming, study finds. *The Guardian* 22 December.

Schnellnhuber, H.J. (ed.). 2006. *Avoiding Dangerous Climate Change*. Cambridge: Cambridge University Press.

Stewart, H., S. Owen, R. Donovan et al. 2002. *Trees and Sustainable Urban Air Quality*. Lancaster: University of Lancaster with Centre for Ecology and Hydrology.

Sustainable Development Commission. 2005. *Leading by Example? Not exactly...* London: SDiG.

Tagaki, M. and K. Gyokusen. 2004. Light and atmospheric pollution affect photosynthesis of street trees in urban environments. *Urban Forestry and Urban Greening* 2 (3): 167–172.

Thurman, P. 1997. The end of green cities? *Landscape Design* June: 31–3.

United Nations Forum on Forests. 2003. *National Report to the Third Session of the United Nations Forum on Forests. United Kingdom of Great Britain and Northern Ireland.* http://www.un.org/esa/forests/pdf/national_reports/unff3/united_kingdom.pdf. Accessed 17 November 2006.

Voisey, H. and T. O'Riordan. 1998. Sustainable development: the UK national approach. In *The Transition to Sustainability*, eds. T. O'riordan and H. Voisey, 156–173. London: Earthscan.

Willoughby, I. and J. Dewar. 1995. *The Use of Herbicides in the Forest*. Field Book 8. Edinburgh: Forestry Commission.

Willoughby, I. et al. 2004. *Reducing Pesticide Use in Forestry. Practice Guide*. Edinburgh: Forestry Commission.

Wintour, P. 2006. Carbon emission targets delayed by government row. *The Guardian* 31 January.

Wylie, I. 2000. We recycle your air. *Fast Company.Com* 40: 334. http://www.fastcompany.com/magazine/40/wf_morrell.html. Accessed 10 November 2007.

Chapter 9
Social Forestry and the Health and Education Agenda

> *The chronology of [Forest Research] programmes reflects the development of forestry in the 20th century. Consequently there are many developed research branches with economic considerations as the primary focus (...), fewer more recent branches with environmental considerations (...) and, until recently, none with social considerations*
>
> [Hislop 2001: 11]

Abstract The cornerstone of sustainability is the balance between its three dimensions: economic development, environmental protection and social equity. Moving on from the economic returns and environmental objectives of multi-purpose forestry, the ninth chapter introduces its social dimension, by focusing on the health and education agenda. They are presented together because they have a lot in common in terms of objectives and of strategies. As in the previous chapters, it is shown that economic benefits are not absent from the agenda (in terms of savings in health or better trained youths). Moreover, official programmes as well as those implemented by charitable organisations reveal new definitions of health and education and a new conception of the missions of the State in these areas, to which social forestry can contribute, to a certain extent.

The cornerstone of sustainability is the *balance* between its three dimensions: economic development, environmental protection and social equity. The latter, however, is very often neglected to the benefit of the other two, at times to the sole benefit of the former.

9.1 The Birth of Social Forestry

In forestry, the Helsinki and the Lisbon conferences provided a clearer focus for action in favour of multiple benefit forestry, and in return called for more adequate knowledge in order to deliver best practice. At the European level, a 4-year programme of research, COST Action E39, started in 2004 to 'increase the knowledge

S. Nail, *Forest Policies and Social Change in England.*

about the contribution that forests, trees and natural places make, and might make, to the health and wellbeing of people in Europe' [COST E39 2004] and a number of European projects are under way. The need to understand the values and meanings attached to woodlands and people's needs in relation to them has given birth to a new realm, social forestry. An oxymoron for some, since forestry means producing timber for economic purposes while social implies spending money for the general public's welfare, this new discipline has nonetheless entered the vocabulary of research and practice.

9.1.1 Forest Research

After the restructuring of the Forestry Commission in 1992, the Policy Studies Division became responsible for forest surveys, within the Policy and Resources Group. After being broken up in 1995, its branch in charge of surveys finally became Forest Research, an Executive Agency of the Forestry Commission, in 1997. The aim of Forest Research is to provide the forestry sector with appropriate research, and the government with advice to support its forestry policies.

Research on the relationships between people and trees began to appear within the Silviculture Branches of the Forestry Commission Research Programme, normally more prone to research on biodiversity and the management of woodlands. Two of them focused on urban forestry and community woodlands. By 1998, the multiple benefits of trees and wooded spaces seemed to be taken for granted, so much so that the Forestry Commission could state in the introduction to the *England Forestry Strategy*, without further explanation, that

> Our concern is with trees and woods both in rural areas, and also in towns and cities, where amenity, environmental improvement and quality of life benefits of trees *are so important* [Forestry Commission 1998: 4, italics added].

9.1.2 The Social Research Unit

Yet, a review of social forestry research undertaken by Forest Research that same year revealed a gap 'between the traditional biophysical research undertaken within Forest Research and the social sciences' [Hislop 2001: 11], and as a consequence research budgets in social forestry were increased and two project leaders were appointed: the Social Research Unit was born. In 2000, it organised a 1-day seminar to draw together all the research done until then within the Commission and by academics. It was followed by a 2-day conference in 2001 on social science and research into woodlands. These and the publications that followed [O'Brien 2001; O'Brien and Claridge 2002] constitute an important milestone in the opening of the Forestry Commission to sustainability in its widest sense. The Social Forestry Unit's programme of research has followed the evolution of the agenda of forestry in general, becoming increasingly wider, as can be seen from the following list of themes tackled by the unit:

Governance and public involvement
Social and cultural values
Health and well-being
Education and learning
Recreation, access and tourism [Forest Research 2004: 41].

In other words, there seems to be very little that forestry cannot do for society, whether in individual or in collective terms. And yet, as was shown in the introductory chapter, forests have often been associated with danger, fear and injustice, and seen as places which could harbour undesirable members of society and illegal activities. How did the idea that woodlands are good for health and education come to be accepted to the point of being at the heart of many policies?

9.2 Health and the Natural World

It has been one constant source of amazement to the author how much research has been invested to demonstrate what was accepted, commonsensical knowledge until a century ago, i.e. that nature had positive effects on people, physically and psychologically.

9.2.1 Bidding Farewell to the Enlightenment

This probably results from what Raymond Williams describes as the separation of human beings and their actions from nature over the last 200 years, consequence of the Enlightenment and of the development of physical sciences [Williams 1972]. Technological advances in the 20th century, in the field of medicine among others, have contributed to severing people from their natural needs, with positive but also negative consequences to body and mind: in Orstein and Ehrlich's words, 'there is now a mismatch between the human mind and the world people inhabit' [Ohrstein and Ehrlich 1989: 9]. The 'biophilia' hypothesis [Wilson 1984], according to which humans are attracted to other living organisms, and contacts with the natural world may benefit health, has risen to prominence in recent decades.

An article published in 1984 by a behavioural scientist is regularly quoted as ground-breaking in that it established a direct relationship between recovery from surgery and natural views [Ulrich 1984]. Its author, Roger Ulrich, endeavoured to show that patients with a window overlooking trees recovered more rapidly and needed less analgesics than other patients whose window overlooked a brick wall. Since then, his research has widened to show the incidence of good architecture and planning on medical outcomes (he himself is the director of the Texas A&M University Center for Health Systems and Design, and professor in the Department of Landscape Architecture and Urban Planning). His work has been hailed as having influenced the architecture of health establishments in the United States.

An interesting phenomenon has been at play here. While the title of the so-oft cited article of 1984 explicitly said 'patients *may* recover…', and the content did not pretend to be a full-fledged demonstration, the hypothesis of the author has progressively been taken for proof and the 'may' of the title has often become something like 'Ulrich has demonstrated' when his work is cited. The numerous authors who quote the article hardly ever discuss it either, as if the simple mention of Ulrich's research was enough to give it – and the authors who quote it – credibility. The only exception is a witty comment whose provoking dimension shows how much its author goes against the tide when he declares that

> Ulrich's much-quoted paper shows that people recover faster from surgery if there are trees in view from the hospital bed (though the hypothesis was not tested, that people hate trees, and if forced to confront them from a hospital bed, will strive to recover quickly in order to escape from them) [Price 2003: 124].

9.2.2 The Thrust of Environmental Psychology

The widespread reference to Ulrich's hypothesis may have to do with the fact that his were among the first studies of its kind. A lot of research had been devoted to leisure and recreation in the 1960s and 1970s, but little to the relationships between human well-being and nature, and it was noted in the mid-1980s that further education was needed to understand the importance of the natural environment [Kaplan 1984: 138]. 'During the last years of the last millennium a complementary interest in what came to be known as 'therapeutic landscapes' developed within the usually unrelated disciplines of medicine and geography' [Muir 2005: 118].

A quantitative analysis of publications on the topic shows that in fact it was mostly anecdotal before the 1980s. It started developing in the United States in that decade, with the work of Roger Ulrich and that of Stephen and Rachel Kaplan, associated with cognitive responses to nature and landscape preferences, and it developed in the 1990s essentially, as the following, not exhaustive, table of published references shows:

Table 9.1 Table of published sources on environmental psychology in the United States

Year	Author and title
1974	Tuan, Yi-Fu. *Topophilia: A Study of Environmental Perception, Attitudes, and Values*. New York: Columbia University Press
1979	Gibson, P. Therapeutic aspects of wilderness programs. *Therapeutic Recreation Journal* 13 (2): 21–33
1982	Getz, D. A. Karow and J. Kielbaso. Inner city preference for trees and urban forestry. *Journal of Arboriculture* 8 (10), October: 258–263
1983	Kaplan, S. A model of person - environment compatibility. *Environment and Behavior* 15: 311–332

(continued)

Table 9.1 (continued)

Year	Author and title
	Altman, I. and J. Wohlwill (eds.). *Human Behaviour and Environment 6: Behaviour and the Natural Environment*. New York: Plenum
1984	Wilson, E. *Biophilia*. Harvard: Harvard University Press
	Kaplan, R. Human needs for renewable resources and supportive environments. In *Land Use and Forest Resources in a Changing Environment*, ed. B. Gordon, 133–140. Seattle and London: University of Washington Press
	Ulrich, R. View through a window may influence recovery from surgery. *Science* 224: 420–421
1986	Dick, R. and J. Hendee. Human response to encounters with wildlife in urban parks. *Leisure Sciences* 8 (1): 63–77
1987	Altman, I. and E. Zube (eds.). *Public Places and Spaces*. New York: Plenum
1989	Shroeder, H. Environment, behavior, and design research on urban forests. In E. Zube and G. Moore (eds.). *Advances in Environment, Behavior and Design*, 87–117.
	Hull, B. and A. Harvey. Explaining the emotion people experience in suburban parks. *Environment and Behavior* 21 (3): 323–345
	Kaplan, R. and S. Kaplan. *The Experience of Nature: A Psychological Perspective*. Cambridge: Cambridge University Press
1991	Shroeder, H. The Psychological value of trees. *The Public Garden* 6: 16–19
	Shroeder, H. Preference and meaning of arboretum landscape: combining qualitative and quantitative data. *Journal of Environmental Psychology* 11: 231–248
	Dwyer, J., H. Shroeder and P. Gobster. The significance of urban trees and forests: toward a deeper understanding of values. *Journal of Arboriculture* 17 (10): 276–284
	Ulrich, R., R. Simons B. Losito E. Fiorito, M. Miles and M. Zelson. Stress recovery during exposure to natural and urban environments. *Journal of Environmental Psychology* 11: 201–230
	Parsons, R. The potential influences of environmental perception on human health. *Journal of Environmental Psychology* 11: 1–23
	Hartig, T., M. Mang and G. Evans. Restorative effects of natural environment experience. *Environment and Behavior* 23: 3–26
1992	Gesler, W. Therapeutic landscapes: medical issues in light of the new

(continued)

Table 9.1 (continued)

Year	Author and title
	cultural geography. *Social Science and Medicine* 34 (7): 735–746
	Stokols, D. Establishing and maintaining health environments: towards a social ecology health promotion. *American Psychologist* 47: 6–22
	Relf, D. (ed.). *The Role of Horticulture in Human Well-Being and Social Development: A National Symposium*. Portland: Timber
1993	Hartig, T. and G. Evans. Psychological foundations of nature experience. In *Behavior and Environment. Psychological and Geographical Approaches*, eds. T. Garling and R. Golledge. Amsterdam: Elsevier
	Kellert, S. and E. Wilson (eds.). *The Biophilia Hypothesis*. Washington, DC: Island
1994	Perlman, M. *The Power of Trees: The Reforesting of the Soul*. Dallas: Spring
1995	Friese, G., J. Taylor Pittman and J. Hendee. *Studies of the Use of Wilderness for Personal Growth, Therapy, Education, and Leadership Development: An Annotation and Evaluation.* Moscow (Id): University of Idaho Wilderness Research Center
	Kaplan, S. The urban forest as a source of well-being. In *Urban Forest Landscapes: Integrating Multidisciplinary Perspectives*, ed. G. Bradley. Seattle: University of Washington Press
	Kaplan, S. The restorative benefits of nature: toward an integrative framework. *Journal of Environmental Psychology* 15: 169–182
	Roszak T., M. Gomes and A. Kanner (eds.). *Ecopsychology: Restoring the Earth, Healing the Mind*. San Francisco: Sierra Club Books
1996	Hartig, T., A. Böök J. Garvill T. Olsson and T. Gärling. Environmental influences on psychological restoration. *Scandinavian Journal of Psychology* 37: 378–393
	Laitakari, J., I. Vuori and P. Oja. Is long-term maintenance of health-related physical activity possible? An analysis of concepts and evidence. *Health Education Research* 4: 463–477
	Grifo F. and J. Rosenthal. *Biodiversity and Human Health*. Washington, DC: Island
1998	Tyson, M. *The Healing Landscape. Therapeutic Outdoor Environments.* New York: McGraw-Hill
	Freeman, H. and S. Stansfeld. Psychosocial effects of urban environments, noise, and crowding. In *Environment and Mental Health*, ed. A. Lundberg, 147–173. Mahwah: Erlbaum–

(continued)

Table 9.1 (continued)

Year	Author and title
2000	Dwyer, J., D. Nowak M. Noble and S. Sisinni. *Connecting People with Ecosystems in the 21st Century. An Assessment of Our Nation's Urban Forests.* General Technical Report PNW-GTR-490. U.S. Department of Agriculture, Forest Service, Pacific Northwest Research Station, Portland, OR
2001	Kuo, F. and W. Sullivan. Aggression and violence in the inner city: impacts of environment via mental fatigue. *Environment and Behavior* 33: 543–571
	Frumkin, H. Beyond toxicity.Human health and the natural environment. *American Journal of Preventive Medicine* 20 (3): 234–240

In 1994, a British review of literature on human well-being and nature in urban areas assessed that

> It would appear that exposure to nature can have benefits for human beings which may enhance their well-being at all stages through the life cycle – childhood, adulthood and old age. However, although this conclusion is based on some empirical evidence, the scarcity of the latter makes it rather tentative [Rhode and Kendle 1994: 136].

Indeed, environmental psychology became prominent in European publications only in the late 1990s, as the following table testifies.

Table 9.2 Table of published sources on environmental psychology in Britain

1994	Rhode, C. and A. Kendle. *Human Well-Being, Natural Landscapes and Wildlife in Urban Areas. A Review*. English Nature Science report N°22. Peterborough: English Nature
	Lewis, G. and M. Booth. Are cities bad for your mental health?. *Psychological Medicine* 24 (4): 913–915
1995	Godbey, G. *Parks and Health: A Coming Alliance.* Paper presented at the *Wildlife, People and Parks Conference*. London Zoological Rooms: Wick Consultants
1998	Burns, G. *Nature Guided Therapy. Brief Integrative Strategies for Health and Well Being*. London: Taylor & Francis
1999	Honari, M. and T. Boleyn (eds.). *Health Ecology. Health, Culture and Human-Environment Interaction*. London: Routledge
2000	Countryside Agency. Evaluation of a health walks scheme: led walks in the Thames Valley. The Countryside Agency
	Department of Health. Minister Sets Out Plans to Extend Exercise on Prescription. Department of Health Press Release 8th June, 2000, Ref. 2000/0336
	Edensor, T. Walking in the British countryside: reflexivity, embodied practices and ways to escape. *Body and Society* 6 (3–4): 81–106.

(continued)

Table 9.2 (continued)

	Macnaghten and P. Urry. Bodies in the woods. *Body and Society* 6 (3–4): 166–182
	Health Walks Research and Development Unit. *Proceedings. Health Walks and Research and Development Unit Symposium*. Oxford: Oxford Brookes University/HWRDU
	Countryside Agency. *Better Mental Health from the Countryside*. Countryside Agency
2001	Countryside Agency. *Walking for health: the First Randomised Control Trial*. CRN 18. The Countryside Agency
	Henwood, K. *Exploring the Linkages Between the Environment and Health: Is There a Role for Environmental and Countryside Agencies in Promoting Benefits to Health?* Report to the Forestry Commission
	Coles, R. and M. Caseiro. Social Criteria for the Evaluation and Development of Urban Green Spaces, in *Comparisons Report to the European Commission*, Project URGE – Development of Urban Green Spaces to Improve the Quality of Life in Cities and Urban Regions, EVK4-CT-2000–00022, Part B Annex B1
2002	British Trust for Conservation Volunteers. *Well-Being Comes Naturally: Evaluation of the Portslade Green Gym. Research Summary*
	Nilsson, K. Urban forestry: where people meet trees. *Forests for the Community* http://www.communityforest.org.uk/tpsn.html
	Lindon, S. and J. Grut. *The Healing Fields: Working with Psychotherapy and Nature to Rebuild Shattered Lives*. London: Frances Lincoln
2003	Morris, N. *Health, Well-Being and Open Space. Literature Review*. Edinburgh: Openspace Research Centre, Edinburgh College of Art
	Special issue of *Built Environment*, Perspectives on Urban Greeenspace in Europe 29 (2)
	Seymour, L. *Nature and Psychological Well-Being*. English Nature Research Report N°533. Peterborough: English Nature
	Grahn, P. and U. Stigsdotter. Landscape planning and stress. *Urban Forestry and Urban Greening* 2: 1–18
2004	Land Use Consultants. *Making the Links. Greenspace and Quality of Life* Scottish Natural Heritage. Commissioned Report No. 060 (ROAME No. F03AB01)
	Miligan, C., A. Gatrell and A. Bingley. "Cultivating health": therapeutic landscapes and older people in northern England. *Social Science & Medicine* 58: 1781–1793

Environmental psychology, which highlights the restorative powers of the natural environment, is to a large extent the scientific study of the century-old awareness of the relaxing effects which are part of the visual amenity benefits of plants and trees (cf. the positive effect of the colour green mentioned in a previous chapter). To describe the restorative power of nature, the Kaplans devised the term 'involuntary attention', meaning the reduction of fatigue directed attention through the simple contemplation of natural scenes.

9.2.3 Lifestyle Diseases and Green Therapies

The development and success of this approach testify to the renewed appreciation, through academic research, of the importance of the environment to human life. One of these effects concerns the benefits of a nice, green surrounding on physical and mental health, more particularly in the case of recuperation from illnesses. There is not a shadow of a doubt to anyone who has ever walked into a hospital, especially as a patient, that the architecture of the building, the presence of comfortable single-occupancy patient rooms, with soft music and nice paintings on the walls, overlooking well-groomed gardens, as Ulrich advocates, is conducive to greater personal satisfaction and lifts more spirits than a drab mineral environment aimed purely at hygiene and efficiency. This environment is bound to have direct consequences on the patient's recovery. Environmental psychology, focusing mostly on the positive visual impact of plants on the psyche, has been put into practice by a number of companies which have set up their offices in nice grounds with trees, to lift their employees' spirits and increase their productivity [Hull 1992: 322–324].

The growing literature on environmental psychology based on empirical evidence also testifies to the changes in the conception and the provision of healthcare over the last 2 decades, and as such is not devoid of political and economic implications. On the one hand, demand for healthcare has kept escalating, and containing costs has become a priority of the British governments since the 1980s. On the other hand, modernising the National Health Service and improving its efficiency have been objectives of the Labour government since 2000. In parallel to cost-containment, the provision of healthcare has become more consumer-oriented, as patients have increasingly been viewed as customers. A Patients' Charter was introduced in 1991, a Patients' Forum was created in 1999, and these novelties have contributed to raising patients' expectations with regard to their health service, hence to increasing public dissatisfaction when these were not met. If 'therapeutic landscapes' can contribute to quicker recuperation and to customer satisfaction while alleviating the costs of treatment, they are worth looking into and investing in, as Colin Price points out:

> There are, at minimum, clear market benefits in earlier return to the productive work-force [Price 2003: 124].

The benefits of plants in patient recuperation, but also in cases of depression, have been explored beyond simple visual effects. Through active engagement with nature, people's cognitive functioning may improve and a higher sense of self may arise [Rhode 1983]. One of the earliest examples of such a hands-on approach to horticultural therapy – or therapeutic horticulture – in Britain has been the work of the charitable organisation Thrive since 1978. It operates directly in four gardens in England, and indirectly through support to numerous projects around the country. Patients recovering from accidents or heart surgery or requiring special needs are encouraged to practise gardening to help recuperation and inclusion. In 2003, together with the Centre for Child and Family Research, it published the results of an in-depth study entitled *Growing Together: Promoting Social Inclusion, Health and Well-Being*, covering more than 800 garden projects, and accompanied by a practical guide [Sempik et al. 2005a and b].

Fig. 9.1 Cultivating health in Thrive gardens (Photo courtesy Thrive)

Lastly, the concept of health itself has become much broader. No longer limited to being the contrary of disease (the 'pathogenic', or medical, model), health now follows a 'salutogenic', or health, model [Antonovsky 1979]. This view, adapted by the World Health Organisation in 1996, implies a holistic approach to health, perceived as a state of complete well-being, and not merely the absence of diseases.

The modern way of life is often seen as conducive to what is called 'lifestyle diseases'. These include well-known illnesses such as diabetes, cardiovascular diseases and certain types of cancer, but also stress or obesity, which can lead to the diseases aforementioned. These affect an increasing proportion of the population and have economic repercussions in terms of lower productivity, absence from work and healthcare costs, not to mention lesser quality of life and higher morbidity. Beyond strictly physical ailments, mental health and peace of mind have been identified as influencing quality of life. Stress in particular, described as resulting from the excess of demands on the individual and the loss of control over one's life, has been the object of abundant research over the past 20 years, because its cost to society has risen significantly. 'Stress-related conditions and musculoskeletal disorders are now the commonest reported causes of work-related sickness absence', states the White Paper *Choosing Health: Making Healthy Choices Easier* [Department of Health 2004: 161], while the cost of workplace stress is estimated at £3.7 billion each year [Carvel 2005].

Rightly or wrongly, stress has been associated with urban living, although there seems to be no evidence to support the idea that living in a city is in itself more stressful than living in the countryside. This is again probably derived from the positive cultural associations of the countryside and the long-rooted negative connotations of cities. In the early 20th century, L. Wirth coined the term 'anomie' to express the fact that the city alters people's psychology for the worse as a result of excessive sensory stimuli [Wirth 1938]. Accordingly, publications underline the positive effects of the contact with the natural world in cities, which can help to compensate the psychic overload, restore mental peace and fight against aggressiveness and violence [Kuo and Sullivan 2001]. Ian Douglas links the presence of extensive grounds around mental asylums in the 19th century with therapeutic effects and he summarises mental health benefits of the contact with nature in the city as comprising

- Improved self-awareness, self-esteem, self-concept, and positively effected mood state.
- Reduction of negative feelings such as anger, fear, anxiety and frustration.
- Improved ability to recover from stressful episodes.
- Effective alleviation of the symptoms of anxiety, depression and psychosomatic illness (including irritability, insomnia, tension, headaches and indigestion).
- Improved psychological health, especially emotional and cognitive aspects (including reduced symptoms of ADD).
- Restored capacity for concentration and attention [Douglas 2005].

However, one has to beware of rosy clichés linking leafy environments to improved mental health. While proportions of mental illness are similar in town and country [Leighton et al. 1963], studies in rural estates and New Towns even reveal that moving from the city to a greener environment could lead to higher rates of mental illness and 'suburban neurosis' [Martin 1957; Ineichen 1993]. By the same token, connecting stress with population density, hence with urban living, seems to be simplistic, as the contradictory results of experiments on reactions to crowding in public open spaces have shown [Fischer 1976; Alterman and Amir 1983; Price and Chambers 2000].

9.2.4 Implementing a New Conception of Health

The British government has made health one of its public policy priorities since 2001. In that year, the Chancellor of the Exchequer commissioned a research to find out about the resources needed by 2022 for the National Health Service to provide high quality service. The Interim Report, published in November 2001, showed that Britain had fallen behind other countries in terms of health outcomes, while the Final Report stated that health requirements had changed and underlined the need for more investments to be made (around 5% per year over the next 20 years). It also established a close link between healthcare and social care and advocated increasing spending in health promotion and disease prevention, so as to reduce future healthcare costs [Wanless 2002: 6–14].

Other official publications stressed the dire consequences of sedentary lifestyles and prompted action at the beginning of the 21st century. Such was the case of the annual report of the Chief Medical Officer for 2002, which devoted a whole chapter to obesity, 'the health time bomb', especially among children. It stressed the importance of encouraging people to participate in sports and recommended 30 minutes of moderate exercise at least five times a week for adults, and 60 minutes of physical activity per day for children [Chief Medical Officer 2002: 42–43]. A parallel between obese children and elderly people, two categories particularly exposed to health problems, is drawn in a cartoon by Mac Lachlan, published in *The Times* in April 2007: the attention of a man driving past a residential home and a junior school is drawn by markings on the road warning 'SLOW' in front of the residential home, and 'VERY SLOW' in front of the junior school, while road signs on the pavement explain: 'elderly people' and 'obese children'.

The minimum time requirement for physical activity was taken up in a 2004 report by the same Chief Medial Officer addressed to NHS health professionals, which stated that

> effective solutions need the engagement of a wide range of agencies. No single organisation will have sufficient impact alone. We will need concerted effort from a range of key partners – Government (at national, regional and local levels), leisure and sports services, schools and colleges, town and regional planners, transport planners and providers, architects, countryside agencies, the NHS and social care, voluntary and consumer groups, employers and the media. All will need to work in a coordinated and comprehensive way to influence the way we live [Chief Medical Officer 2002: iv].

In recognition of the enlarged definition of health, and of research showing that sustainable health implied not only adequate medical provision for health restoration, but also healthy environments and lifestyles for the prevention of diseases, the government has included this in its policies. PPG 17 underlines the need for adequate provision to ensure frequent recreation in natural greenspaces to promote health benefits [ODPM 2002: 66]. The 2004 White Paper, *Choosing Health: Making Healthy Choices Easier* was followed by two delivery plans which provide support and information on how to improve the nation's health: *Delivering Choosing Health: Making Healthier Choices Easier* and *Choosing Activity: A Physical*

Activity Action Plan. The latter aims to promote activity for all, whether by making public spaces more amenable to physical activity easier (cycle lanes, paths, woodlands) or by providing organised activities. Experimental studies in Sweden and in the Netherlands have shown that walking for only 20 minutes helps lower blood pressure and lift spirits, and that the more often a person visits open spaces, the less often he/she is likely to report stress-related diseases [Hartig 2002; Grahn and Stigsdotter 2003].

According to the Forestry Commission,

> a study in the West Midlands has revealed that a walk in the woods can cut the risk of heart attacks. It is estimated that the health benefits of walking and cycling in woodland beauty spots could be saving the health sector up to £45 million per annum in the region [Forestry Commission 2005a: 52].

Health issues combined with the rise in leisure and recreation have logically led to a new focus on outdoor recreation as a source of well-being. Because of research in environmental psychology and the diversification of forestry, woodlands have increasingly been perceived as the ideal venue for healthy leisure recreation pursuits.

Fig. 9.2 Walking one's way to health in the new plantations of the Mersey Forest (Photo courtesy of the Mersey Forest)

In 2002, the Forestry Commission hosted three expert consultations in England, Scotland and Wales. They brought together environment and health professionals to devise ways to improve public health and well-being, and gave birth to a publication, *Health and Well-Being*. Research and policy culminated in an Outdoor

Health Concordat, signed by the Forestry Commission, the Countryside Agency, English Nature, Sport England and the Association of National Parks Authorities. Through piloting projects, participation in seminars and events and coordination of programmes of activities, the Concordat aims at encouraging outdoor activity for everyone for health and enjoyment, regardless of age, gender, race or ability. *Trees and Woodlands: Nature's Health Service*, published by the Social Forestry Unit in 2005, took stock of the research into the health agenda of forests and of the recent programmes and publications, and aimed at providing environmental and health professionals with 'examples of how the woodland sector can contribute to people's health and well-being' [O'Brien 2005: 4]. Among the various recent campaigns, *Active Woods. Naturally Good for You* was devised by the Forestry Commission in 2005 and supported by the British Heart Foundation to establish a link between health, well-being and woodlands and promote physical activity in the woods.

Several types of projects have been implemented, catering for the needs of various categories of users. First, infrastructures have been built, thanks to lottery funding from Active England: Haldon Forest Park, near Exeter, was inaugurated in April 2006 and provides 42 kilometres of purpose-built trails over 1,400 hectares, while Roslington Forestry Centre in Swadlincote is now promoted as the physical activity centre of the National Forest.

Woodland recreation and health pursuits have also led to a collaboration with Primary Care Trusts. A Community pilot programme launched in May 2002 aims at evaluating the effectiveness of Primary Care Trust-led approaches to physical activity programmes for the community. An example of this type of approach is the Green Gym Fund, a joint initiative of the Primary Care Trust and the British Trust for Conservation Volunteers, in association with local authorities, which is being implemented in many English and Scottish towns. It revolves around the double objective of undertaking conservation work and providing physical health as well as 'exercise-therapy' for distressed or depressed people. As a consequence of the study in the West Midlands mentioned above, a woodland-based pilot providing grant-aid has been established in relation to the Walking for Health Initiatives.

Another initiative is entitled General Practitioner Referral Scheme, whereby GPs can refer voluntary patients suffering from overweight, mild depression, stress or high blood pressure, to a Health Promotion Worker for a programme of activities including walks and other physical activity in the woods, over a number of weeks Such pilot projects have been experimented at Chopwell Wood, near Newcastle, and in the Forest of Dean, and are deemed to have been popular and successful at reducing obesity and creating social networks. The REACT (Regeneration through Environmental ACTion) projects, implemented in seven English cities to tackle poor health and exclusion, include programmes to reduce coronary heart disease and obesity within the ethnic minorities, young people and disabled people of three communities of Central Liverpool.

Fig. 9.3 Tree-planting as part of the REACT project in Liverpool (Photo courtesy of the Mersey Forest)

Although these alternative treatments are taking time to percolate through to patients and doctors alike, the Mental Health Foundation, the charitable organisation which recommends government investment to develop the scheme, claims that there is evidence that it helps alleviate patients, reduce drug intake (in particular antidepressants the consumption of which has soared to unknown heights over the last decade) and hence save on medical costs [Mental Health Foundation 2006]. In the same line, the 'Route to Health' Project at Cannock Chase, a partnership between Cannock Chase Council and the Forestry Commission, with support from South Staffordshire Primary Care Trust, consists in a 1-mile trail displaying artworks inspired by mental and physical health topics. Such an initiative harks back to two essential aspects of human-forests relationships today: on the one hand, the close association of these natural surroundings and health and, on the other hand, the role of forests as a source of inspiration for artists. This project contributed to Cannock Chase Council's success in gaining the 'Beacon Award for Healthier Communities' in 2005. Art is also central to the Emscher Park project in the Ruhr region (Germany), described in a previous chapter. The *Trees and Woodlands Vision for the London 2012 Games*, beyond its symbolical nature as marker in the landscape, is completely in line with the health agenda, as the following presentation stresses:

> By creating a landscape that accentuates health, fitness and physical activity, the very embodiment of the Games ethos, London 2012 will be quite literally sowing the seeds of healthy, prosperous and vibrant communities in east London and across the UK [Forestry Commission 2005b].

Children have become a privileged target of forest programmes, for various reasons.

9.3 Education, a Newly-Born Target

The education agenda is the latest addition to the wide range of social forestry objectives. Indeed, with the new policies aimed at preserving health, it is felt that inadequate provision for children's mental and physical health is going to lead to adults who are not fulfilling their potential. Some, like the Vice president of the Health Visitors Association, advocate the creation of a new Social Institution, bringing together through woodland programmes the departments of Health and of the Environment, to tackle children's 'health' issues as diverse as bullying, attention deficit disorder, depression and suicide, dyslexia, obesity or lack of emotional resilience [Rigler 2001]. Among these serious problems, the 'obesity epidemic' has become a major cause of concern for the medical professions and the State, since it is believed that obesity-related diseases (heart attack, diabetes, high blood pressure, bowel cancer) could lead to today's children having a shorter life expectancy than their parents. It is estimated that 25% of 11–15 year olds are overweight or obese, a figure which is on the increase. Fighting against this 'lifestyle disease' has thus become a priority, for example by reducing inactivity and changing eating habits. So the issue of health is very closely related to that of education, in the widest sense of both words. We would argue that there are two basic ways in which the educational agenda relates to woodland in today's practice. Woodland can be used as a resource for content-learning and for character-forming.

9.3.1 Woodland as a Tool for Content-Learning

The National Curriculum, introduced in 1988 with the Baker Act, includes four key stages and, for each of these, achievement targets in various subjects: English, mathematics, religious education, design and technology, information and communication technology, geography, history, music, art, physical education and a modern foreign language. Environmental education is therefore not part of the curriculum as such. Yet, the number of initiatives, programmes and organisations revolving around woodland and education has soared, and they very often use the National Curriculum as their guideline, sometimes with a clear connection, sometimes less so.

Obvious connections can be made in science and geography for example. In a class the author attended one winter afternoon in a rural primary school of Cheshire, an officer of the Woodland Trust took a Key stage 2 class, in preparation for tree planting with the class the following week. He thus tackled such notions as the food chain, photosynthesis and recycling. Similarly, a walk through the woods can act as a stimulus for story-telling (English), or can be used for map-drawing (geography). Batboxes or bird nesting-boxes can be built within the design and technology course. In Prescot High School (Merseyside), an ancient woodland, fenced off from the school grounds but belonging to the school, has been used as a case study in information technology and business studies. A group of Key Stage 4 students was asked to assess how to raise finances in order to develop the land. What had started as an academic exercise soon turned out to be much more. The students and their teacher, with the help of the Mersey Forest Team, developed a management plan to turn the wood into an open classroom with an amphitheatre. Parents and children came one Saturday morning to clear the litter and cut back brambles, and a bid was put in by the Mersey Forest for European money to make a reality of the vision.

Fig. 9.4 Leaping over the fence: woodland at Prescot High School (Merseyside) (Photo Sylvie Nail)

In Chopwell Wood, already mentioned, school trips supported by the governmental initiative Healthy School Standard have also been organised in relation to health and nutrition education.

Through grants to assist projects, the Forestry Commission also supports the Forest Education Initiative, a programme which aims to increase children's understanding of trees and forests, both in relation to their uses in industry and to the environment. The FEI sets up 'Forest Schools' and provides teachers with material for the various stages in the key subjects. Based on the Danish model established in the 1950s, a Forest School England Network was created in 2002, organised in geographical 'clusters', most of which are to be found in Wales and Scotland so far. In a typical example of social forestry, these offer children and adults a hands-on approach to the woodlands to improve their confidence and self-esteem: for example, the Forest School in Wyre Forest (Worcestershire), in partnership with the probation service, has developed special action for pupils at risk of exclusion from school. Similarly, the Forest of Mercia offers twenty placements in urban parks and countryside sites to children experiencing difficulties at school.

Another programme, 'Forest Classrooms', based on a model created in the United States, also exists in various Forestry Commission sites, like Delamere Forest or Cannock Chase. They are normally targeted at young disadvantaged children and give them the 'ownership' of a particular wood where they work under the supervision of their teachers. The objectives include developing team work, motor and coordination skills, as well as self-confidence.

This type of projects, involving field trips or participation of forest specialists within the classroom, has increased markedly over the last few years, in a way which does not appear to be directly linked with changes in the National Curriculum. What has changed over the first few years of the 21st century is two-fold: the stress put on activities related to the importance of the woodland resource in economic and ecological terms, and what we might call the extra-curricular objectives of woodland, in connection with a belief in the wider benefits afforded to schoolchildren and young people through contact with woodland.

9.3.2 School Grounds, Grounds for Growth

Among the benefits sought through schoolchildren's participation in woodland activities, the main ones relate to physical and psychological development. Many experiments have been undertaken to change the school grounds and make them, not only more pleasant, but also more attuned to the different demands made on space, and therefore possibly more favourable to learning. This is the vision of 'Learning Through Landscapes', the national school grounds charity founded in 1990 which campaigns for better outdoor environments for schoolchildren.

Sometimes, the school is lucky enough to have a piece of woodland within or adjacent to the school grounds, which can be used as a teaching medium. This has become rarer, as the sale of school playing fields was encouraged in the 1980s to favour development. In the primary school of Barnton (Cheshire), all the pupils have been involved over several years in the conversion of part of the school grounds into a wooded play area which will eventually provide an extension to the adjacent ancient woodland, renowned for its bluebells in the spring. Assisted by the Mersey Forest officer in charge of school ground events and by a local gardener, the older pupils have planted the trees while the smaller ones have been busy planting bulbs, brought from home or provided by the Wildlife Trust. In the same school, a lunchtime workshop, gathering a dozen children under the supervision of two teachers, is developing plans for converting an unused corner of the school grounds into a school garden. One of the specialist areas which the Mersey Forest has developed is precisely related to this. Entitled 'School Grounds Development', it is a scheme based on the results of the Advisory and Inspection work in schools, as well as a lot of other research. The Mersey Forest has produced a *Schools Pack* for the creation of woodland spaces for play.

The rationale of improving school grounds is threefold. First, better grounds improve pupil behaviour. Some schools have experimented the plantation of small gardens with trees to delineate different spaces for different groups of children to enjoy during the breaks.

Fig. 9.5 Developing schoolgrounds to enrich the educational experience (Photo courtesy of the Mersey Forest)

The teachers claim that the result has been conclusive, in so far as different groups, of girls in particular, have been able to pursue different activities without suffering from the hegemony of the football-playing boys. As a result of their needs being better taken into account, the children are less frustrated. Secondly, and as a consequence of the previous point, better school grounds improve pupil's opportunities to learn. The pupils are more relaxed, and the teachers report that they need not dedicate so much time after the breaks to quarrel-solving due to bullying or boredom. Finally, better school grounds, on top of contributing to improved relationships within the school by learning how to share space, also improve the relationships between the school and the wider community, not least by presenting a more amenable appearance.

An assessment of 700 schools having undertaken improvements in their school grounds, was made by 'Learning Through Landscapes' in 2003 [Learning Through Landscapes 2003]. Half of them responded, and the results confirm the three basic tenets of the programme. Concerning the first point, three results can be connected: 90% off respondents state that the number of children who enjoy being in the school grounds has increased, 85% note an increase in the number of activities provided to pupils at break and lunchtime, and 73% have seen an improvement in pupil behaviour. In relation to better learning opportunities, 64% note a reduction in the number of incidents linked to bullying, 54% estimate that the amount of time taken to settle disputes after break and lunchtimes has decreased, and 52% note improved academic achievement. Concerning the third point, that of the improvement of the relationships between the school and the wider community, 87% of the schools report that the aesthetic improvement to the school grounds has contributed to a better perception of the school environment, by insiders (pupils and staff) and outsiders (parents) alike, 84% say that it has improved social interaction, which we can interpret as relations within the school, rather than with the outside community, and 66% note an improvement of the community/parental involvement in the school as a consequence of the programme (a significant proportion of parents were involved in the implementation of the projects).

The results which are most significant (and also the easiest to measure) concern children's enjoyment of the school grounds and the increased opportunities for varied activities. Additionally, it is important to note that in 85% of the cases, improved school grounds have stimulated increased active games and play, which in turn means that the children's physical shape and health is likely to improve as a consequence.

To finish with, let us mention another international initiative run by the Foundation for Environmental Education and entitled Eco-schools, which provides schools with a programme to enable them to become more sustainable, first by becoming aware of their ecological impact, then by trying to reduce it, by recycling, collecting rainwater, walking to school, collecting litter, switching off lights, etc. Through an award scheme, children are encouraged to become responsible citizens, which is also part of the curriculum.

9.3.3 Sowing the Seeds of Citizenship

Other initiatives based on the same vision of the child's development take place outside the school grounds. Enlisting the help of schoolchildren to measure local trees, count hedges and collect seeds in the woods is a frequent example of these, originating in individual initiatives (a Tree Warden, a ranger in an urban forest, etc.) or as part of national programmes such as 'Trees of Time and Place', educational charities like Green Light in Bury Saint Edmunds, or the National Tree Week. Trees and hedges are thus measured year on year to assess their growth, and seeds are collected to be sown in pots, ready for the next planting campaign 1 year or 2 later.

Educating the future adults is a key stake for all those involved in Community Forests and community woodland groups. Two simple ideas underlie these initiatives. First, one takes one's children to the places where one went to as a child. Thus, training children to know, respect and love woodlands helps to ensure that the next generation will also care for them. Secondly, the depredators of the woods are not those who maintain them, so the more the children are taught to care, the less the woodlands will suffer from vandalism when the children grow up. In Mill and Alder Wood, an ancient wood now straddling Liverpool, Knowsley and Halton, where vandalism is a daily challenge, the chairman of the Friends Group regularly organises seed collection and planting days with schoolchildren, to give them a sense of ownership of the wood. This may explain the increasing number of initiatives around woodland as part of the educational agenda.

Fig. 9.6 Children are taught that woodland has to be managed (Photo courtesy of the Mersey Forest)

Introducing the children to the woods, whether to undertake activities or on the occasion of a teddy bear picnic, is part of the education to 'citizenship' that has progressively entered the curriculum over the recent years. Together with other 'subjects', not traditionally associated with school education (balanced diet, sex education, etc.), it is part of the state's growing intervention into spheres traditionally reserved to parents. In that sense, schools' action reinforces the parents' education, or at times avoids leaving the initiative to parents who may or may not feel the urge to teach these values closely linked with a 'hidden agenda' concerning citizenship. This is one of the manifestations of the ever-growing intervention – some might say interference – of public policies into the private sphere, which at times smacks of a mutual distrust between parents and the State.

These projects also have to do with the reappropriation of the woodlands by the local inhabitants, who have often moved away from them on insecurity grounds. A good number of guided walks have been set up and are followed by lone women and/or people who otherwise would not have visited the woods on their own. Many testimonies point to a change: whereas interviewed people remember going to play in the woods as children, they also tell that they don't let their own children play there any more, for fear of assault or unsocial behaviour. Woodlands have become no-go areas in a number of urban areas, leading to a vicious circle: the least frequented they are, the more likely they are to attract unsociable behaviour. Thus inviting city-dwellers to rediscover their woodlands is also a way of promoting a virtuous circle, where civil frequentation is going to attract more civil frequentation.

Finally, planting is also a way of caring for other living organisms, with the discipline that this implies (the lack of watering of a newly planted tree has immediate consequences on its survival), but without the constraints or apparent threat of relating to other people. This is why planting is used in training programmes, not only for children, but also for some adults, for instance ex-prisoners and long-term unemployed people. Not only are they helped by the therapeutic virtues of gardening, but also caring for plants and trees is a way of reconciling them progressively with the discipline of work and the constraints of a regular schedule, without the human conflicts they are not yet prepared to handle. Groundwork often employs ex-prisoners or long-term unemployed people precisely to get them back on the track on an active professional life

9.3.4 Bodies in the Woods

Finally, on top of children's cognitive and social development, there is another dimension to woodland discovery by children and teenagers. Although it has been studied less [Hart 1979; Valentine and McKendrick 1997; Kong 2000], it is no less important and has to do with their relation to natural spaces.

Many elderly interviewees, when talking of the main differences between their childhood and that of their grandchildren, mention that they, as children, were allowed to go out and about to discover woods and fields, only coming back home at mealtimes, while 21st century children are not allowed to go anywhere on their own. There is certainly a very strong tendency to protect children much more now than 50 years ago. Few are the children who walk to school or who are allowed to go to the neighbouring wood on their own. In some places, 'walking buses' have been set up (e.g. in Knowsley, Liverpool) to get primary school children to school in groups, thus reconciling safety and health. One anecdote is symptomatic of parents' fears and children's consequent limitations as far as 'adventure' is concerned. An interviewee (a retired schoolteacher) told this author of how one of her neighbours had told her of her son's wish to camp in the (enclosed) family garden for one night, and how the mother had refused on security grounds. The interviewee convinced her neighbour of granting her son his wish. At their next encounter, the latter told her of the delight of her son, who had slept under the tent in the back garden, with a mobile phone by his pillow, while his father slept in the kitchen with the back door open, in case his son should call him.

Not only do many children not do sufficient exercise to ensure a harmonious physical development, but they also often grow without knowing how their bodies relate to the space around them, through climbing trees, etc. At school, for fear of accidents and/or lawsuits, 'dangerous' games like playing with conkers are sometimes prohibited (or allowed with goggles on), and for fear of hurting the self-esteem of the less gifted children, competition is sometimes banned in sports. As a result of what the Confederation of British Industry terms the 'risk-aversion culture' [Paton 2007] physical contact with the reality of competition and with the dangers of the natural world are often very limited. These measures, taken with the best of intentions to protect the children's physical and emotional development, may lead to the very opposite. This is why the Commission for Architecture and the Built Environment (CABE), the government's adviser on architecture, urban design, and more recently, on public space (CABE Space) has launched a 10-point manifesto which invites all British people, from ordinary citizens to politicians and businesses, to sign it through the Internet, calling them to 'work together to reclaim Britain's city parks, squares and open spaces for the people' [O'Brien and Tabbush 2005: 20].

A CBI report has also stressed the negative long-term economic consequences of the lack of recruits with a sense of entrepreneurship, initiative and competition. The education charity 'Heads, Teachers and Industry' has recently launched a campaign entitled 'Go4it' to reward the participating schools for creating the best initiatives to reintroduce competition and activity days. There are other consequences to children and teenagers growing up in a risk-free, but also challenge-free, environment. Because they do not have the opportunity to pit themselves against nature, in a boomerang effect this may end up producing exactly the type of behaviour which it tried to avoid, namely the taking of uncalculated risks. This paradox is encouraged by the advent of sophisticated commercial devices, often boasting the adjective 'extreme' to attract youngsters. These new 'extreme' activities sometimes reach the headlines, because they can result in deadly accidents. Such is the case of the 'human catapult' for instance, a typical example of the mediation of technology between the human body and the space around it, which in the end leaves the body extremely vulnerable, at the mercy of any technological incident [Dodd 2002]. The phenomenon has recently risen to prominence in the national press, as the consequences of such inability to cope with the real world are perceived as potentially disastrous. Another consequence is the declining sense of orientating oneself which, even though it has less dramatic consequences, impacts on the perception of woodlands as dangerous places, as we shall see.

So taking the children to the woods as part of their school activities is a way for them to rediscover spaces which have become no man's lands, to enjoy the physical and emotional freedom of moving about in a natural setting, and to experience 'safe danger' [Burgess 1995: 22] – and hopefully to take their parents to the woods at the weekend. Freedom of movement goes beyond health, exhilarating the senses and developing the imagination. It is difficult to talk of the importance of physical presence in natural space without referring to the late Roger Deakin, and to his passionate writing on swimming. Symptomatically, his last book, published in June 2007, is entitled *Wildwood. A Journey Through Trees*. The virtues he attributes to permission to swim could equally well apply to his vision of *being* in the woods:

> To deprive large sections of the population of swimming, either because there is no pool close enough to where they live to make regular swimming practical or because the rising costs involved put it out of reach, is to limit the imaginative and mental capacities of the people (especially children and young people), as well as their ability to become healthy and fit. As a participation sport and recreation, swimming has the potential to combat the tendency of television spectator sports to nurture a nation of couch potatoes. It is perhaps significant that a remarkable number of writers are regular, passionately committed swimmers. They find, as I do, that a daily swim is essential to imaginative as well as physical fitness [Deakin 2001].

9.3.5 Assessment

Programmes and initiatives revolving around the involvement of children in woodland activities are in general engineered by dedicated members of organisations. Results are difficult to measure, in so far as this initiation will only bear fruit in the long term. Children on the whole (up to a certain age, that is) seem enthusiastic, if only because wood planting or woodland walks provide a nice change from the classroom atmosphere. There is no doubt that for Community Forests, child participation is very important in their assessment of public involvement, even though in all fairness it would be difficult for the children involved to refuse to participate. Activities where the children are supposed to bring the parents along for a family woodland activity are not always successful, as some parents see the rangers/officers as child-minders and drop the children off to pick them up a couple of hours later.

Lastly, one cannot emphasise too much the necessity to plan very carefully the activities meant for children, otherwise they can be counter-productive. In the New Town of Runcorn, the author took part in tree-planting in the Town Park one November morning. This gathered a small number of children with their parents under the supervision of three rangers, to plant trees (grown from seeds gathered the year before) on a hill essentially made up of the gravel extracted to make the nearby Expressway. Several factors were identified which made of that morning a poor example. First, the seedlings had been collected by other children, thus there was no continuity with which the children might identify. Then, the ground was extremely difficult to work, and clearly beyond the possibilities of small children. Third, the opportunity was not seized upon to explain to the children what trees need in order to grow – at the very least, careful planting in relation to the root system, then water and weeding until they can manage without protection. Quite the reverse, they walked away probably thinking that it was difficult to dig, but easy to grow trees! Finally, a lack of coordination can have quite negative consequences on the appropriation of the trees by the children. In this particular instance, another 1,000 trees had been planted on the same site the year before, but less than 50 survived, due both to lack of care and to the fact that the landscaping services had not been told of the planting and therefore mowed the area in the summer. This event clearly smacked of tokenism in child and community participation.

From the actions and programmes both in theory and in practice, two types of objectives emerge. On the one hand, the woods and the initiatives around them have an educational content and are cognitive tools to get acquainted with the natural world in all its aspects (economy, conservation, biodiversity, etc.). On the other hand, the woods are tools for achieving other objectives, whether health, therapeutic, social. So we find in social forestry the same dichotomy as in the other realms tackled before, with nature being used as a medium while the discourse revolves around the benefits of nature *per se*.

The chances are that the latest functions of social forestry, related to health and education, will go from strength to strength, for several reasons. First, they can contribute to improving the nation's physical and mental health. They can help to reintroduce in a safe context a sense of risk and adventure in the 'cotton wool' society denounced by some. They may also contribute to the development of children's imagination, freedom and self-esteem. A UNICEF report published in February 2007 establishes that, out of 21 'economically advanced countries', the United Kingdom fares very poorly regarding children's welfare, coming bottom of the league on a number of indicators. Without taking the report at face value, for the same reasons as were stressed before concerning the social production of scientific knowledge, there are alarming trends. Let us leave aside, as there is little that woodland can contribute to this issue, the damning results concerning child poverty, a blow to Tony Blair's governments in spite of the policies implemented since 2001. The study asserts that British children are more exposed to the risks from alcohol or drug-taking or unsafe sex, and have poor relations with their parents. Considering the scope of social forestry to provide 'soft' answers to self-confidence, the Government and social forestry partners may well see more scope to social forestry than ever.

What remains to be seen is whether the agenda of multi-purpose forestry and social forestry and the policies that have been described and analysed in this part corresponds to the aspirations and perceptions of the users. This will be the object of the last part.

References

Alterman, R. and S. Amir. 1983. Neighbourhood physical form and use of public open space: Haifa, Israel. *Landscape Research* 2: 145–154.

Antonovsky, A. 1979. *Health, Stress and Coping*. San Francisco: Jossey-Bass.

Burgess, J. 1995. *Growing in Confidence. Understanding People's Perceptions of Urban Fringe Woodlands*. Technical Report CCP457. Cheltenham: Countryside Commission.

Carvel, J. 2005. Part-timers lose out on pay and potential. *The Guardian* 15 September.

Chief Medical Officer. 2002. *Health Check: On the State of Public Health*. London: Department of Health.

Cost Action E39. 2004. *Memorandum of Understanding for the Implementation of a European Concerted Research Action Designated as COST E39 "Forests, Trees, Human Health and Wellbeing"*. http://www.e39.com.ee/en/m-9/c-11/. Accessed 14 November 2007.

Deakin, R. 2001. Extract from the supplementary Memorandum to the House of Commons Select Committee on Culture, Media and Sport submitted by Mr Roger Deakin, 6 December 2001. http://www.hamishhamilton.co.uk/nf/shared/WebDisplay/0,,215073_11_1,00.html. Accessed 27 March 2007.

Department of Health. 2004. *Choosing Health: Making Healthy Choices Easier*. London: Stationery Office.

Dodd, V. 2002. Student killed as human catapult sport goes wrong. *The Guardian* 26 November.

Douglas, I. 2005. *Urban Greenspace and Mental Health*. Prepared for the UK MAB FORUM. http://ukmaburbanforum.org.uk/Publications/UF%20Publications%20and%20Reports/ukmabgrnspcepap2.htm. Accessed 16 May 2006.

Fischer, C. 1976. *The Urban Experience*. New York: Harcourt Brace Jovanovich.

Forestry Commission. 1998. *England Forestry Strategy. A New Focus for England's Woodlands*. Cambridge: Forestry Commission.

Forestry Commission. 2005a. *Great Britain and England Annual Report and Accounts 2003–2004*. Edinburgh: Forestry Commission.

Forestry Commission. 2005b. *The London Trees and Woodland Framework*. http://www.forestry.gov.uk/forestry/infd-6qhdva. Accessed 11 April 2007.

Forest Research. 2004. *Annual Report and Accounts 2002–2003*. Edinburgh: The Stationery Office.

Grahn, P. and U. Stigsdotter. 2003. Landscape planning and stress. *Urban Forestry and Urban Planning* 2: 1–18.

Hart, R. 1979. *Children's Experience of Place*. New York: Irvington Publishers.

Hartig, T. 2002. *Psychosocial Forestry and the Health of Urban Populations*. Paper presented at the *Urban Woods for People* Congress *La Perception de la Forêt*. Paris, November. http://www.onf.fr/europe/life-urbanwoods/pdf/Hartig_T.pdf. Accessed 10 November 2007.

Hislop, M. 2001. The Social Forestry Programme of Forest Research. Past, present and future. In *Social Forestry: Questions and Issues*, ed. E. O'Brien. Farnham: Forestry Commission.

Hull, R. 1992. Brief Encounters with urban forests produce moods that matter. *Journal of Arboriculture* 18 (6): 322–324.

Ineichen, B. 1993. *Homes and Health: How Housing and Health Interact*. London: Spon.

Kaplan, R. 1984. Human needs for renewable resources and supportive environments. In *Land Use and Forest Resources in a Changing Environment*, ed. G. Bradley, 133–140. Seattle and London: University of Washington Press.

Kong, L. 2000. Nature's dangers, nature's pleasures: Urban children and the natural world. *In Children's Geographies*, eds. S. Holloway and G. Valentine, 257–271. London: Routledge.

Kuo, F. and W. Sullivan. 2001. Aggression and violence in the inner-city. Effects of the environment via mental fatigue. *Environment and Behavior* 33 (4): 543–571.

Learning Through Landscapes. 2003. *National School Grounds Survey 2003*. Winchester: LTT.

Leighton, D. et al. 1963. Psychiatric findings of the Stirling Country study. *American Journal of Psychiatry* 119: 1021–1026.

Martin, E. et al. 1957. Incidence of neurosis in a new housing estate. *British Journal of Preventive and Social Medicine* 11: 196–202.

Mental Health Foundation. 2006. *Up and Running. How Exercise Can Help Beat Depression*. London: Mental Health Foundation.

Muir, R. 2005. *Ancient Trees, Living Landscapes*. Stroud: Tempus.

O'Brien, E. (ed.). 2001. *Social Forestry: Questions and Issues*. Farnham: Forest Research.

O'Brien, L. 2005. *Trees and Woodlands. Nature's Health Service*. Farnham: Forest Research.

O'Brien, L. and J. Claridge. 2002. *Trees are Company. Social Science Research into Woodlands and the Natural Environment*. Edinburgh: Forestry Commission.

O'Brien, L. and P. Tabbush. 2005. *Accessibility of Woodlands and Natural Places. Addressing Crime and Safety Issues*. Farnham: Forest Research.

ODPM (Office of the Deputy Prime Minister). 2002. *Assessing Needs and Opportunities: A Companion Guide to PPG 17*. London: Her majesty's Stationery Office.

Orstein, R. and P. Ehrlich. 1989. *New World, New Mind. Changing the Way We Think to Change Our Future*. London: Methuen.

Paton, G. 2007. The danger from our 'cotton wool kids'. *The Daily Telegraph* 7 February.

Price, C. 2003. Quantifying the aesthetic benefits of urban forestry. *Urban Forestry and Urban Greening* 1 (3): 123–133.

Price, C. and M. Chambers. 2000. Hypotheses about recreational congestion: tests in the Forest of Dean (England) and wider management implications. In *Forest Tourism and Recreation: Case Studies in Environmental Management*, eds. X. Font and J. Tribe, 55–74. Wallingford: CABI.

Rhode, C. 1983. *An Investigation of Reinforcement Mechanisms in Depressed Psychiatric Patients*. Unpublished PhD Thesis: University of London.

Rhode, C. and A. Kendle. 1994. *Human Well-Being, Natural Landscapes and Wildlife in Urban Areas. A Review*. English Nature Science report N°22. Peterborough: English Nature.

Rigler, M. 2001. Sharing knowledge and moving boundaries: towards a new social institution for Health and Environment. In *A Way into Woodland. Making Access Work Well for All Concerned*.

Conference report of the National Small Woods Association, Small Woods Association, 15–20. Telford: Small Woods Association. http://www.smallwoods.org.uk/wayintowoodland.doc. Accessed 8 November 2007.

Sempik, J., J. Aldridge and S. Becker. 2005a. *Growing Together: A Practice Guide to Promoting Social Inclusion Through Gardening and Horticulture*. Bristol: Policy.

Sempik, J., J. Aldridge and S. Becker 2005b. *Health, Well-Being and Social Inclusion: Therapeutic Horticulture in the UK*. Bristol: Policy.

Ulrich, R. 1984. View through a window may influence recovery from surgery. *Science* 224: 420–421.

Valentine, G. and J. McKendrick. 1997. Children's outdoor play: exploring parental concerns about children's safety and the changing nature of childhood. *Geoforum* 28 (2): 219–235.

Wanless, D. 2002. *Securing Our Future Health: Taking a Long-Term View*. London: HM Treasury.

Williams, R. 1972. Ideas of nature. In *Problems in Materialism and Culture. Selected Essays*, Raymond Williams 1980: 67–85.

Wilson, E. 1984. *Biophilia: The Human Bond with Other Species*. Cambridge: Harvard University Press.

Wirth, L. 1938. Urbanism as a way of life. *American Journal of Sociology* 44: 1–24.

Part III
The Tree-Led Solution to Empowerment

Chapter 10
Forestry Paradigms and Resource Provision

> *Modern day planners in the UK are locked into the concept of 'visual amenity', i.e. the concept of 'if you can see it, it has amenity, if you can't it doesn't'.[...] However, when 'amenity' is viewed in the broadest possible context [...,], it means 'pleasant circumstances, advantages and all that flows from that'. It becomes obvious that 'amenity' is much more than just the visual component*
>
> [O'Callaghan 2003: 100]

Abstract The last part of this study moves away from the 'objective' benefits of woodlands into the realm of perceptions and social issues. It explores how policies may differ from people's expectations, what woodlands mean to people, independently from official programmes to promote them, and what they reveal of society. The tenth chapter revolves around the values associated with the forest resource and the question of access/ibility. In spite of the fact that values are presented as immutable, they in fact vary according to the forestry paradigms in fashion. Behind apparently trivial questions like forestry aesthetics lie assumptions about who defines the criteria, who uses the resource and how. Recent research shows that people's preferences may be at variance with what the Forestry Commission has for a long time assumed them to be according to an expert paradigm. Linked to this question is that of access. This two-fold question is fundamental, in so far as inadequate understanding and provision can stand in the way of official objectives being reached. Furthermore, the issue of access/ibility is a culturally rooted one and its treatment today is indicative of social relationships.

The forestry agenda has adapted to the new economic context, as well as to sociological changes. This has that meant transforming the ways in which profitability, health or education are defined, adapting the resources and changing the way outcomes are measured. On top of these changes often related to the all-important concept of 'quality of life', forestry has taken on board political missions relating to the fight against democratic deficit and political apathy, and the search for 21st century versions of the local/regional/national identity.

The first aspect of the question of forestry as a tool for empowerment revolves around looking into values associated with the forest resource. In spite of the fact that these are always presented as immutable, they in fact vary according to the forestry paradigms in fashion at a particular point in time. Behind apparently trivial questions like forestry aesthetics lie assumptions about who defines the criteria, who uses the resource and how, and also about the limits to access.

10.1 Questioning the Expert Paradigm

The provision of resources derives from interpretations of/assumptions about needs, both in quantitative and qualitative terms. Throwing light on these reveals a lot about the various paradigms that are at play in the policies to provide woodland resources for recreation, and about their adaptation to social needs.

10.1.1 Quantitative and Qualitative Valuation

The National Playing Fields Association (NPFA), a charitable organisation dedicated since 1925 to the protection and promotion of open recreational spaces, defined these along quantitative lines. Its 'acre standard', widely used by local authorities, thus advocated a minimum of 2.4 hectares of open space per 1,000 inhabitants, i.e. 1.6 hectares for outdoor sports and recreation, including parks, and 0.8 hectares for children's play, of which 0.25 hectares of equipped playgrounds [NPFA 2001]. Qualitative analysis of what was deemed valuable came later, with landscape appraisal research starting in the 1960s. It aimed at eliciting environmental and landscape values in order to provide protection for landscapes of great quality.

Landscape appraisal methods were at first strongly marked by a positivist approach, i.e. based on impartial measurements by a 'neutral' observer. Techniques of evaluation through numerical scales were expected to lead to an unquestionable, qualitative assessment of landscape value [Penning-Rowsell 1981: 25]. The assessment of landscape value was left to the 'experts' who chose the variables, very often based on environmental value [Hill-Tout 1999: 166–167]. The search for objective landscape assessment criteria informed a lot of research in the 1970s and 1980s, which was carried out essentially by landscape architects, geographers and planners. The criteria, logically enough, reflected the researchers' approaches. Judith Tsouvalis has pointed out that it is no coincidence if these approaches used environmental economics as the standard for environmental valuation in the 1980s and early 1990s, at a time when the accent lay on market mechanisms [Tsouvalis 2000: 166]: the underestimation of other than economic criteria reflected the thrust towards economic growth.

However, as public benefits, including all social groups, were becoming the cornerstone of public forest policies, revisiting the definition of visual amenity criteria became crucial. Indeed, there is a lot to be said about the criteria that make a landscape amenable to different categories of the population, as has recently been investigated.

10.1.2 Challenging the 'Landscape Value' Approach

Under the influence of research in cultural geography and environmental psychology, the idea that numerical scales of evaluation would lead to an unquestionable, qualitative assessment of landscape value began to be superseded by a new model, which acknowledged the need to take into account people's values and preferences, rather than take their preference for 'quality landscapes' for granted. Even while the statistical approach to landscape evaluation was fashionable, criticism about its validity rose in the 1970s, first in the United States, then in Britain. The harmonious juxtaposition of various visual components was deemed insufficient to constitute landscape value. The latter became inseparable from the social perceptions of landscapes, including cultural plurality. Taking these values into account entailed eliminating judgments as to the 'rightness' or the 'wrongness' of such or such perception [O'Brien and Guerrier 1995: xiv]. Thus arose the 'landscape preference approach', which entailed reintroducing the subjectivity which had been so painstakingly eliminated [Penning-Rowsell et al. 1977]. Through questionnaires and direct questioning, the Contingent Valuation Method began to elicit people's preferences from their willingness to pay to keep an area of vegetation from development, or their willingness to accept compensation for the loss of goods or services. What came out is that for the interviewees, landscape value was not necessarily related to objective landscape quality as described by experts.

The people preference approach is not without its shortcomings, however. The profusion of attitudinal surveys, consisting mostly of questionnaires, one-time only interviews, or opinions based on photographs, is far from delivering reliable and consistent data. Another objection is that the respondents do not necessarily have adequate knowledge to estimate the value of unpriced goods. As Colin Price puts it,

> it is extraordinary that members of the public, armed with intuition, a few scraps of real information and a lifetime's accumulation of folklore, should be deemed capable of making a proper decision about the appropriate balance of interest for forestry in Britain [Price 1997: 185].

More fundamentally, this method, like the previous ones, tries to price the priceless along monetary lines, not in terms of equity, thus again reducing value to valuation. Still another problem with the Contingent Valuation Method has to do with the fact that it brings to the surface individual likes and dislikes which can be incoherent or contradictory, and in any case do not necessarily lead to a clear picture on which to base policies for the public good. One instance of this is the type of emotion that is sought in the contact with nature. Provision of natural spaces is often tuned to the need for peace and quiet, 'to get away from it all', as a lot of interviewees put it. This is the type of preferences that the National Trust for instance provides for, mostly in rural and peri-urban areas:

> Our speciality is tranquillity. Our brand image is quiet enjoyment of the countryside. We make places available, provide car parks where appropriate and maintain paths free from hazards, so far as possible.(...) Our aim is to intrude as little as possible in people's experience of the place [Russell 2000: 251].

Fig. 10.1 Burton Wood, a mature woodland now owned by the National Trust, overlooks the Dee estuary on the Wirral peninsula (Photo Sylvie Nail)

However, there are other demands, often from the very same people, for interest, mystery and exhilaration, which cannot necessarily be found in settings designed for tranquillity [Rhode and Kendle 1994: 79–80]. Another example of ambivalence can be found in relation to people's preferences for tree density. A higher density of trees and thick undergrowth, providing habitat for a lot of wildlife, seems to bring more satisfaction to users. This attitude is coherent with a search for adventure and mystery in outdoor recreation. But on the other hand, a feeling of personal safety is closely associated with a rather low-density canopy, open views and clearly marked paths [Hull and Harvey 1989]. So providing for both types of experience requires diversity within the same space, or in two different spaces not too far apart, so that visitors can choose according to their personal requirements or mood.

10.1.3 Translating Values into Policies

Åsa Ode and Gary Fry argue that visual amenity which takes people's preferences into account should be a criterion of management, as it enriches the quality of daily life and contributes to the good physical and mental health of the population. Yet, their comparative study of official guidelines on visual aspects of urban forest management in Sweden and the UK shows that visual preferences are not much taken into account in British policies [Ode and Fry 2002].

They selected 12 publications on urban woodland management and correlated them with the dominant visual concepts used in landscape literature, namely scale, diversity, naturalness-continuity, stewardship, visual accessibility and coherence. The authors found that guidelines varied greatly in their inclusion of these elements and concluded that the 'British guidelines discuss visual concepts at a more general level' than their Swedish counterparts, and that visual preferences, although recognised, are not given a high priority in the UK. Moreover, the justification of 'the importance of visual aspects and aesthetics [is] based on an expert paradigm' whereas in Sweden, people preferences prevail.

Sustaining and increasing 'the public benefits given by existing woodlands', as well as helping 'create new woodlands to deliver additional public benefit' [Forestry Commission England 2007] lie at the heart of the English Woodland Grant Scheme (EWGS), introduced as part of the *Government's Rural Strategy* in 2004. It is therefore particularly relevant to look into what is advocated in these new and improved spaces in terms of visual amenity, to establish a comparison with Ode and Fry's findings. Do official guidelines take users' preferences into account, or does expert opinion still prevail? After stating that society's preferences are for a number of 'values', including 'producing healthy and pleasing living and working environments for people', as well as 'providing safe areas for recreation and sport', the introduction to the *General Guide to EWGS* states that 'improving the environment of disadvantaged urban communities' [Forestry Commission England 2006: 3] figures among the key targets. This can be perceived as a measure against environmental injustice; yet, surprisingly, nothing more is said within the document on that topic or on how to achieve it.

The leaflets which present the six distinct grants which make up the EWGS (Woodland Planning Grant, Woodland Assessment Grant, Woodland Regeneration Grant, Woodland Improvement Grant, Woodland Management Grant, and Woodland Creation Grant) do not provide any details as regards the ways and means to achieve 'pleasing' environments either, whereas they are very precise on technical details of all kinds. The vocabulary used in the various leaflets refers to 'information' and 'promotion' of the amenity benefits to the population, not the inclusion of the population's points of view in the definition of the term or its content: a top-down approach prevails, based on supposedly consensual criteria. Requirements concerning public access can be quite stringent if one applies for a grant, while visual amenity, when it is mentioned, seems to be left to the appreciation of the applicant. In one particular activity eligible for one of the six new grants, i.e. access, visual amenity is mentioned. Access is defined as 'action to support good quality permissive public access. Action to maintain the *visual amenity* along public rights of way and permissive access paths' [Forestry Commission England 2003: 11. Italics added]. What is meant by this is made more explicit in the 'guidance' section of the *Woodland Management Grant Guide*:

> *Clear evidence* of recent and cyclical management of woodland edges and open spaces appropriate to the particular biodiversity values of the woodland.
>
> - A diverse and *unregimented* shape and structure.
> - A developing or developed range of heights and ages of vegetation.
> - Maintenance of, or the process of forming, predominantly graded edges for major rides, glades and riparian zones.
> - Rides are wide enough to allow light to the ride surface and scrub edge.
> - Light being allowed into the edges to support the shrub layer.
> - No *significant damage* to the diversity of vegetation and shrub edges from grazing and browsing.
> - significant damage to ride surfaces from vehicles [Forestry Commission 2005: 9. Italics added].

Aesthetic and environmental criteria are intertwined: the insistence on maintenance clearly implies that visual amenity depends on the environmental quality of the site. The woodland environment is to be manicured, protected from damage due to grazing animals and to visitors; width and light are important for the growth of the nearby vegetation. What is presented here is the 'scenic aesthetic' that Gobster traces back to three 'cultural legacies': 'a preference for idealised nature; the predominance of a static, visual mode of landscape appreciation; and an uneasiness associated with change' [Gobster 1995: 6–10]. Below the test quoted above, there is a drawing that reinforces the importance of neatness and orderliness. Along a riding path, the vegetation grows in three tiers, from the shortest one close to the path to the highest one inside the woodland, and the caption underlines the maintenance required. The central ride should be mown regularly, while the tall grasses on either side should be cut every 2–4 years, and the coppice adjoining the path cut every 8–20 years [Forestry Commission 2005: 9]. This evokes a naturalistic setting, with variety being a crucial element of visual pleasure. The general layout is very reminiscent of paths in a country house park or in a public park.

There is an obvious correlation between these remarks and Ode and Fry's findings. Little clear and precise information is provided on visual guidelines, in comparison with, for example, ecological guidelines, which are often clear and concrete. This leads one to conclude that they are assumed to be known, or supposed to derive from environmental quality, which the focus on maintenance and orderliness confirms. It may also imply a consensus on landscape appreciation, at least among experts, on what makes woodlands visually agreeable to the visitors. The failure to reassess visual amenity criteria along more open lines results in a reinforcement of expert opinion, which in turn can stand in the way of social inclusion.

The criteria used to define visual amenity have taken time to take on board people's aesthetic preferences and the fact that experience of the landscape can extend well beyond the visual. However, faced with the shortcomings of existing methods, innovative approaches have been tried.

10.2 Fresh Approaches to Environmental Valuation

One of the Social Forestry Unit research reports lists the reasons why it has taken so long for people's values to be taken into consideration in public policies:

- difficulties of incorporating numerous values into decision-making;
- problems with obtaining information from people;
- forest professionals sometimes viewing the public as unknowledgeable/irrational;
- members of the public's ideas seen as impractical;
- the public often viewed as overtly emotional in response to environmental issues [O'Brien 2003: 6].

To these could be added the sheer difficulty of defining the attitudes and values associated with woodlands along non-economic lines.

10.2.1 Bridging the Gap

An illustration of the gap between what experts would define as a pleasing environment and the negative way in which it can be perceived by local communities appears, ironically enough, in a programme for social inclusion implemented in two Community Forests in the North of England. Entitled 'Bridging the Gap', it revolves around the discovery/appropriation of local woodlands by underprivileged urban groups (women, ethnic minorities, children and handicapped people). As the programme developed, it was realised that a number of mental frameworks stood in the way of people visiting these amenity woodlands on a regular basis. They had to do with personal and social barriers, and one of the latter was 'judgements about the social class of those who use the countryside' [Davies and Vaughan 2002: 6]. In other words, these landscapes are still perceived by some as landscapes of exclusion. This can be explained by the fact that today's 'landscapes of beauty', with their high environmental value, are very often yesterday's landscapes of power, reflecting both the economic supremacy and the sense of beauty of the social and cultural elites of the past. These past landscapes are often promoted today as landscapes of beauty and enjoyed as such by many, but in turn, this leads to their association with today's elite.

The perception of beauty in landscapes as adding economic value to a site may concur to this feeling of exclusiveness, hence of exclusion. Colin Price provides an illustration of this phenomenon in his discussion of the Hedonic Pricing Model. Contrary to what is normally concluded from this method, i.e. that nice views lead to an increase in real estate prices, he argues that it is the willingness of some well-off people to pay high prices for them, and the resulting snowball effect leading to other well-off people choosing these 'nice neighbourhoods', 'for social concourse and a sense of security' [Price 2003: 130] at least as much as for the view, which contributes to keeping house prices high and the neighbourhood socially homogeneous.

From his analysis, Price concludes that planting more trees in poor areas would allow people of lesser means to share the beauty of these landscapes and would contribute to the reduction of what he calls the 'separation value' between the wealthy and the poor. Bringing urban forests and parks to the less well-off would decrease the gap between landscapes of power and landscapes of destitution. The social construction of woodlands by minorities, exemplified in the 'Bridging the gap' experience, had been surprisingly little taken into account until recently – or maybe not so surprisingly, as it is difficult to assess. Yet, these minorities are a key target of Community Forests for instance:

> a priority focus is on people who have traditionally been socially excluded from environmental regeneration programmes and countryside activity [Davies and Vaughan 2002: 1].

There is good reason to believe that the frequentation by minorities of urban and peri-urban forests displays the same characteristics as those found in the countryside by Bill Slee et al., namely low participation rates and fear of stigmatisation, due to their minority status [Slee et al. 2001]. This may be reinforced by the characteristics of the population who most habitually frequents the woodlands. As early as 1978, a survey of 3,000 visitors to 16 greenbelt sites around London had found out that there was a preponderance of car-owners and an over-representation of A, B and C1 social groups[1] among them [Harrison 1981]. Twenty-five years later, *Leisure Day Visits 2004* reports similar data on the profile of visitors to woods and forests in comparison with the whole population sampled: visitors to woods tend to belong to social classes A,B, C1 (66% vs 51% for the whole population sampled); they have access to a car in a higher proportion (91% vs 80% for the whole sample) and they are more likely to have a full time job (51% vs 45%) [Countryside Agency 2004: 4]. The activities provided in the woodlands, mostly walking and cycling, do not necessarily cater for the needs of all social groups, either. According to the 2006 edition of *Social Trends*, even though walking is the most popular sport activity (involving 46% of the British population in 2003) participation rates vary widely from one socio-economic group to the other: walking is much more practised by the higher socio-economic groups than by the lower ones [ONS 2006: 196].

The portrait that emerges from these data is confirmed by the author's field experience during woodland walks and plantation days: woodland visitors apparently dominantly belong to the white middle class.

10.2.2 Multiple Value Forestry

Researchers with a background in social geography and ecological economics have elaborated new ways of attributing value to unpriced goods, taking into account

[1] According to this occupation-related classification, most often used by market researchers, groups A, B, C1 are middle-class and groups C2, D, E are working-class.

new, broader, criteria: different types of knowledge, equity, opportunities for learning, transparency and public participation [Sustainable Development Research Network 2007]. Among the copious research into people's reactions and preferences in natural spaces of urban areas, the work of social geographers Jacquelin Burgess, Melanie Limb and Carolyn Harrison is crucial to understanding the popular values of the natural world. In 1988, they noted that

> neither the 'new conservationists' nor the conventional leisure managers, have made any sustained attempt to discover whether the beliefs, attitudes and behaviours of urban residents accord with the newly advocated role for public open spaces [Burgess et al. 1988a: 456].

According to them, empirical social research into environmental values was still limited and quite insufficient to make allowances for people's multiple and complex appreciations of landscapes. These evoked a 'kaleidoscope of sounds, smells, sights and tactile experiences of nature' [Harrison et al. 1986: 20], in which visual perceptions were only one component. As part of a 2-year research project on green open spaces in 1985–1986, Burgess et al. developed a deliberative approach to environmental valuation through small group discussions, using the principles of group analytical psychotherapy for geographical purposes. This entailed paying particular attention to the language, concepts and beliefs expressed by the members of the groups. In their fieldwork in Greenwich, they held discussions over 6 weeks with three different groups coming from various socio-economic backgrounds. Not only did the group-work serve to elicit personal sensitivity, coloured by experience and memories, but it also acted as a forum where ideas could be debated and common values expressed, which could then validly inform public policies based on the common good. They discovered that the values attributed to the open spaces of cities were far more diversified than had hitherto been gathered, and that aesthetics was only part of the equation. A point common to the three groups was the wealth of sensory experience and pleasures derived from contact with the natural world. A walk in the countryside or in the woods was seen as an antidote to the hurly-burly of urban life, a way of finding oneself, away from the others, but in a safe environment where civilisation was just round the corner. A sense of adventure was identified as being valuable, particularly for children, to develop their imagination and independence. Interaction with nature in open spaces, whether formal or informal, was particular important.

They concluded that open natural spaces on the urban fringe acted as 'gateways to a better world' for people from all walks of life [Burgess et al. 1988b: 322]. In order to provide for a wider range of recreational pursuits that are adapted, not only to a wider public, but also to different moments or moods, they recommended countryside policies covering both agriculture and recreation, wider access to different types of recreation sites, and the creation of natural areas and wildlife corridors in housing estates as well as suburban developments.

Focus groups have also been used in Wales [Henwood and Pidgeon 1998; Bishop et al. 2002] in order to find out about cultural values and the relationships that local people have with the woodlands. Similarly, a long-term study has examined the criteria which define social outputs of urban green spaces, which 'reflect the pleasure that people derive from the use of local green sites, the social interaction that occurs, and the factors that contribute to successful interaction' [Coles and Caseiro 2003: 1]. These studies stress the importance of feelings and emotions and advocate the provision of spaces of a certain size likely to ensure quality of experience. More than multiple *use* forestry, multiple *value* forestry might well be the order of the day [Bengston 1994: 519].

One of the background papers, focusing on community forestry, accompanying the *England Forestry Strategy Review*, shows that, progressively, multiple value forestry is being integrated. Particular value is placed on 'the everyday, often unspectacular natural environment that is, or ideally would be, nearby' [Forestry Commission 2005]. Accordingly, a range of areas, from the most highly managed to areas of wildness, should be available locally to all users, to provide for the whole range of experiences sought.

10.2.3 Reconciling Aesthetic and Biodiversity Values: Towards a New Paradigm

Over the last 2 decades, woodlands have come into their own as user-oriented spaces associated positively with amenity, leisure and recreation. According to *Leisure Day Visits* published in 2004, 40% of English and British adults had visited a wood/forest between March 2002 and March 2003, and visits to woodlands accounted for one out of six visits to the countryside [Countryside Agency 2004: 4]. Yet, new achievements raise new challenges. Finding new economic outlets to multi-purpose forests while catering for people's various visual preferences and the multiplicity of their requirements for a fulfilling experience of the woodland is not enough: these have to be compatible with the other dimension of sustainability, ecology.

The Peter Wall Institute of Advanced Studies (PWIAS) held an Exploratory Workshop at the University of British Columbia (Vancouver) in 1999. The first one of its kind, in that it gathered an interdisciplinary panel (ecologists, forest resource scientists, landscape architects, philosophers, psychologists, sociologists and computer scientists), it used forest landscapes to explore the relationship between ecologically sustainable landscapes and aesthetic quality. Their definition of aesthetics builds upon previous research and embraces 'the full range of aesthetic and perceptual qualities received by the senses and appreciated by the mind, including the meanings to be found in the landscape, such as the symbology of timber harvesting practices and forest stewardship' [Sheppard and Harshaw 2000: 6]. The conclusions of the workshop provide a few answers, but also raise a number of questions.

Even though there is still considerable doubt as to the exact relationship that exists between visual and ecological parameters, the authors argue that there seems to be a positive aesthetic perception of sustainable landscapes from the population. This is even more the case if the viewers are aware of looking at a sustainable landscape. Yet, there are limits to this, as some sustainable landscapes are not scenic in the traditional sense of being neat and tidy, and conversely, some scenic landscapes are far from being sustainable.

Questions raised include the sustainability of management methods used to preserve the aesthetics of scenic landscapes (e.g. helicopter logging). They also address values, since in cases where aesthetics and ecological sustainability are not compatible, choices have to be made. In the PWIAS Workshop, participants agreed that the need for sustainability was paramount, but that may not be the case in all circumstances. Values are contingent, as has amply been shown through forest history. Moreover, if the viewers' perception of the aesthetics of landscapes can be positively influenced by their knowledge that they are sustainable, this points to the role of mental mediation between perception and a value system. The participants to the Workshop agreed that there was a need to educate the public, so as to improve its acceptance of ecological practices. This is not without its difficulties, however, since it entails going against the grain of the cultural legacies which have shaped the present 'scenic aesthetics'. If ecology is paramount, then public preferences for manicured landscapes have to be transformed into an 'ecological aesthetic'.

Not only does this raise practical questions into the methods to be implemented to reach the required result; it also raises ethical questions concerning the manipulation of public preferences involved in the process. Can one disapprove of it in some cases and approve of it in others, depending on the current value system? It is difficult not to see a parallel between the 1960s–1970s and today, albeit in opposite directions. Then, forest managers wanted to educate the public 'to like timber production forestry because it was deemed expedient for the public to learn to like the landscape that pursuit of production objectives promoted' [Price 2007: 20]. Now, although the previous attempt is condemned, it is seen as justifiable to do the same to adapt to the new credo of sustainability. The participants to the Workshop with a social sciences or arts background questioned the ethics of so doing, and Colin Price puts forward an alternative:

> There is, at the least, an option to be explored, that we should reserve the right to maintain our preferences unsullied by expediency; and, where indulging our preferences demonstrably leads to unsustainability, accept that preferences will be flouted, into the indefinite future. [Price 2007: 20]

Taking into consideration a whole range of values and preferences is not enough to ensure full public benefits: access is an essential element of multi-purpose forestry. The authors of the comparative study on Sweden and the UK relate their results to the structure of land ownership in the two countries. In Sweden, public preferences are an obvious criterion since public property dominates, leading to wide access. In the UK on the contrary, land ownership is overwhelmingly private, which can explain why people preferences are secondary to expert opinion. Specialists of rural sociology like Howard Newby, Philip Lowe and Marion Shoard have also related restrictive planning policies in relation to the recreational use of the countryside to the rights of private property in British law.

10.3 The Key Issue of Access

The *England Forestry Strategy* asserts that the government's priority lies in creating accessible woodlands, preferably next to where people live, on account of their social and environmental benefits. This priority has geographical as well as social implications. From a geographical point of view, as the *Strategy* points out, there is a gap between the places where woodlands are to be found and the places where people live. Woodlands in the South of England are often over-used: such is the case of the Forest of Dean, receiving over one million visits per annum. In the North, woodlands are often lacking, hence a lot of the regeneration projects focus on tree-planting, as was shown in a previous chapter. From a social point of view, there is a crying need to provide access for a more numerous and more varied public than has hitherto been the case, in order to promote social inclusion, one of the government's objectives. What do people want from their woodlands and where? What sort of access is provided to these spaces? In order to answer these questions, it is important to adopt a different perspective on the three objectives generally grouped together under one heading by the Forestry Commission: 'recreation, access and tourism', in order to focus specifically on questions of access, at times neglected to the benefit of the other two.

10.3.1 Research into Needs

If regular contact with natural spaces is crucial to people's physical and mental well-being, then the question of mobility and accessibility becomes a priority. Indeed, the benefits of regular contact with nature can be more than offset by the stress of driving to open spaces [Frumkin 2002]. Moreover, more than 60% of visits to urban open areas are made on foot [Burgess et al. 1988a]. This makes the provision of open spaces close to where people live indispensable, so as to make frequent, stress-free contact possible.

With recent research taking into account not only minimum area, but also type of space, diversity and proximity, new norms have been suggested. English Nature, the government-funded body in charge of promoting the conservation of England's wildlife and natural features, commissioned the UK Man and the Biosphere Committee's Urban Forum in 1992 'to produce and publish a discussion paper discussing targets and guidelines to be used in providing adequate accessible natural greenspace in urban areas' [Harrison et al. 1995: i]. One of the authors of the resulting paper was Carolyn Harrison, whose background research into qualitative assessment of people's needs in relation to natural spaces has already been referred to. Unsurprisingly, the paper's definition of natural greenspace reflected the authors' research, and considerably widened the NPFA's definition of greenspace to include

> land, water and geological features which have been naturally colonised by plants and animals and which are accessible on foot to a large number of residents [Box and Harrison 1993: 232].

It took stock of various nature conservation strategies which tried to pinpoint the areas lacking in such spaces – 'urban deserts' in the West Midlands, 'areas of deficiency' in London, 'areas of nature deprivation' in the Black Country or 'areas lacking in natural habitats' in Bristol. But it stressed the fact that there was still little agreement on the adequate distance between people's dwellings and accessible greenspace. The authors suggested that

- an urban resident should be able to enter a natural greenspace of at least 2 hectares within 0.5 kilometres of their home.
- provision should be made for Local Nature Reserves in every urban area at the minimum level of 1 hectare per thousand population [Box and Harrison 1993: 234].

They also suggested that the provision of natural open space should be no less than 20 hectares within 2 kilometres of all residents; at least 100 hectares within 5 kilometres and 500 hectares within 10 kilometres of all residents. These figures were taken up in the *Accessible Natural Greenspace Standard* by English Nature [Handley et al. 2003], where provision is linked to accessibility, not to population density as in the NPFA standard.

For woodlands, the correlation between frequentation and distance was confirmed by Richard Coles and Shelagh Bussey, who studied how users in Redditch, in the Midlands, interact with the various woods which make up the Community Forest. For them, distance is a key factor for the frequentation of a wood. Five minutes, or a 100–400 metres range, is ideal [Coles and Bussey 2000: 187]. These figures coincide with other research: Harrison et al. found out that 280 metres was the optimum distance from the home to make the most of local woodlands on foot [Harrison et al. 1995], while Richard Coles and Maria Caserio show that use declines dramatically when sites are located at more than 5 minutes on foot from where people live [Coles and Caseiro 2003: 1].

Even though they have to be taken with some caution on account of the method used (questionnaire and photographs) and of the low response rate (only 8% of the population targeted), the results of a social survey made at Telford (Shropshire) as part of the NeighbourWood project confirms these trends: 77% of the respondents said that 'the woodland closest to them was the one they visited most frequently and that 90% of them normally reached the woodland on foot' [Simson and Ryan 2003: 28]. The Woodland Trust, the leading UK woodland conservation charitable organisation, determined that

> 59 per cent of woodland visits are undertaken within a total round-trip distance of under 5 miles (8 kilometres). Beyond 8 kilometres, the frequency of visits drops markedly. Furthermore, the main mode of transport to woodland is on foot (more than half), with only 32 per cent of visits made by car [Woodland Trust 2004: 15].

The Forestry Commission has taken these data into account and has embarked on a policy of selling its distant properties and concentrating on those closer to the population, especially if there is already a wooded area nearby, so as to increase its area. But this is not always easy: in spite of the numerous woodlands created over the past decades, space is often at a premium and woodlands are not necessarily to be found where the population lives. Some 'Woods on your Doorstep' can be rather far away from anyone's doorstep, in fact. To facilitate walking to the woods, special attention is also being paid to corridors. On top of being good for wildlife, which they help to go from one wooded zone to another, corridors help pedestrians travel from one woodland to the next. The 2,800-hectare 'Green Network' established in Telford in 1991 thus includes corridors within the New Town and between it and the surrounding countryside.

Fig. 10.2 Green corridors between Telford New Town and the surrounding countryside (Photo courtesy of Alan Simson)

Apart from distance, other crucial components are size and composition. Requirements to obtain a grant for woodland establishment from the Forestry Commission have become more flexible, possibly to attract applicants: although 0.25 hectare is considered as a minimum surface and 30 metres a minimum width, there is no minimum size to create a new woodland. Public benefits are not connected with size according to the Forestry Commission, but rather with proximity to urban areas, in particular to disadvantaged urban areas.

Although the proximity factor is essential, this does not mean that size is unimportant in the quality of the experience, far from it. According to qualitative research, proximity is not sufficient, a minimum size is required to provide an escape from the views and sounds of the city. Woodlands, in order to be attractive, to act as refuges from the pressures of urban life and to attract visitors in search of peace and quiet, must to have a minimum of 2 hectares, according to the various research mentioned. This means that blocks are more appreciated than linear plantations such as roadside plantations, and that below a certain size, they may not fulfil their functions optimally. Concerning the age and composition of the woodlands, expert assessment can be at variance with recent research. If Coles and Bussey state that in amenity value terms, mature woodlands (especially of mixed deciduous species) score higher than recent plantations, yet age and composition do not seem to matter that much to the visitors. In their findings as in the author's, visitors have no objection to walking the dog or taking a walk in young poplar plantations for example.

Not only is there no systematic preference 'for truly natural woodland' [Coles and Bussey 2000: 190]: plantations are at times preferred because they are less likely to harbour attackers. The Telford social survey confirms these results. Although none of the respondents stated that they preferred young woodland, 32% said they had no preference (about a third preferring mature woodland and the other third preferring a mixture). Similarly, a majority of respondents does not have any preference in terms of tree species [Simson and Ryan 2003: 22].

If age or composition is not all-important to users, what does matter is the percentage of canopy cover. Open woodlands are preferred, with dappled light and glades. According to a 1985 study, a density of 100–160 trees to the hectare was the most attractive [Shroeder 1989: 94]. However, as was stated before, this varies according to whether exhilaration or personal safety is the preferred criterion, hence the need to take into account various needs for various places.

10.3.2 Measures to Facilitate Access and Diversify Supply

As Wendy Darby puts it in relation to accessibility in the English countryside, access is made up of at least two components: 'being able to reach it' and 'being able to walk in it'. Having reviewed what measures have been implemented to fulfil the first part of the definition, let us turn to the second one.

The Forestry Commission's estate has been, in its vast majority, open to the public for a long time, and its woodlands receive over 50 million visitors per year. On private lands, access has been encouraged through grants. The Forestry Commission's Woodland Grant Scheme Community Woodland Supplement supported between 1992 and 2005 the creation of around 7,300 hectares of new woodlands with public access [Forestry Commission 2006: 18]. Woodland access is now being promoted through three grants as part of the Woodland Grant Scheme: a Community Woodland Contribution, an Annual Maintenance Grant and a Woodland

Improvement Grant. Additionally, a *Walkers Welcome* package, made up of signs and discs, has been made for landowners who agree to open their woodland to inform the public that their land is accessible.

Fig. 10.3 Visitors are made welcome on private property, here on agricultural land converted into woodland as part of the Mersey Forest (Photo Sylvie Nail)

The *Countryside and Rights of Way Act* (CroW) 2000 went further to give access to the countryside. Yet, it went against the (strong) grain of private property, which may explain why the decision was taken during the legislative process not to impose a statutory right of access to woodland in England: access remains a voluntary issue. On the other hand, the principle of land dedication was introduced, providing public access in perpetuity, but without liability. This applies to the Forestry Commission, local authorities as well as private landowners. Forestry Commissioners decided to dedicate the whole of the Forestry Commission's freehold estate to ensure permanent access on foot. The Government provided the funds and the dedication was launched by the Minister for Rural Affairs, Landscape and Biodiversity in August 2005.

Stronger protection of open spaces has also been encouraged in the planning system, for instance through the *Policy Planning Guidance* (PPG) *Notes* issued by the Department of the Environment. PPG 17, devoted to 'Planning for Sport, Open Space and Recreation', was revised in 2002. It takes into account the research mentioned above, and in particular the need for different types of open space underlined in the English Nature report. Thus, it turns its back on the NPFA's criteria, which focused on population/open space ratios, and encourages the provision of open spaces fulfilling five criteria: accessibility, quality, multi-functionality, a hierarchy of purposes, and only last, quantity [ODPM 2002: 9]. The *PPG Note* stresses the need to involve local communities to assess their needs, before auditing local provision. It also stresses the impracticability of fixing national thresholds of provision, as English Nature recommended in its *Accessible Natural Greenspace Standard*, preferring to let local authorities assess their own assets and needs, in terms of provision to be protected, enhanced, relocated and finally, created. Through the creation of Social Regeneration Priority Areas (SRPAs), the Forestry Commission has identified areas with a population of 'at least 10,000 people living in, or within 4 kilometres of, the 40% most deprived wards combined with the neighbourhood renewal areas'. In these areas, support is targeted 'for bringing woodland closer to people of greatest need' [Forestry Commission 2006: 18]. The Forestry Commission is confident that its approach 'strikes the proper balance between the "three R's" – rights, responsibilities and restrictions' [Small Woods Association 2001: 7].

In the woodlands open to the public, various types of facilities are present, corresponding to three types of needs. Infrastructures like toilets, tearooms and play areas are situated close to entry points; short/health walks, too. Signage often consists of boards pointing to the main services and itineraries, or interpretation boards signalling species.

Fig. 10.4 Simple boards at Risley Moss, headquarters of the Mersey Forest, help the visitor to find his/her way round and choose between various walks (Photo Sylvie Nail)

Further away, sculpture trails, footpaths, bridle and cycle paths are provided. Finally, further away still from access points, low-level facilities are provided, discreetly marked, for 'crowd fugitives' [Price and Chambers 2000: 71]. To provide for the special needs of visitors, access for handicapped users as well as for pushchairs is being improved. In the areas covered by the Capital Modernisation Fund projects, over 70 kilometres of wheelchair accessible paths have been produced. A typical issue is that of motorbikes, the presence of which can be both dangerous and intrusive due to the noise. In a number of Community Forests and Forestry Commission sites, motorbikers' habits have been studied so as to cater for their needs: special areas have been set aside for them, while the other users are protected thanks to simple devices like A-shaped gates that permit access to pedestrians only.

10.3.3 Provision of Access Versus Land Use Structure

Yet, as was said before, private property dominates in the new woodlands as in the old ones and it is not possible to impose permissive access, even though landowners increasingly depend upon grants and subsidies. Thus, access policies can only be implemented in so far as they do not encroach upon the landowners' independence, and they therefore rely on goodwill. This was the case with the 1949 Act on the right to wander freely over upland areas, which gave immunity to somebody who entered without doing any damage if an access agreement or order had been made: as Wendy Darby notes, there is some distance between not being considered as a trespasser and having a right of access [Darby 2000: 190]. It is even more the case today, in the present context of distrust between much of rural England and the New Labour government.

The experience of the Community Forests testifies to constraints and opportunities in that respect. The case of the Mersey Forest was explained previously, with most farmers unwilling to open their woodlands for access. In the Forest of Mercia, before focusing on providing new access, the stress was put on linking the various new sites of the Forest with the existing public rights of way network [Hunt 2001: 22]. But securing access to private land has proved more difficult, with the notable exception of the creation of a permissive walkway around Wyrley N°3, a former spoil mound on the Little Wyrley Estate. In order to ensure greater access with high quality provisions, the Forest of Mercia took part in a partnership project co-funded by the Directorate General for Environmental Affairs of the European Commission (DGXI) between 1997 and 2000. The programme, entitled TOURFOR (European Award for Tourism in Forest Area Project), was based on the implementation of an environmental management system (EMS) to promote well-managed tourism in forests as part of multifunctional forest management. The sites complying with this EMS get an award similar to the Blue Flag Award for well-managed beaches, or the more recent Green Flag Award for well-managed public parks.

A study of landowners' attitudes to public access throws light on the reasons for landowners' reluctance and gives hints on how to improve the situation. The authors distinguish between three types of landowners/managers: the 'dutyists', the 'reluctants' and the 'marketeers' [Church et al. 2005: 6]. The first group accept access as a counterpart to receiving public aid. The second one oscillates between giving access, where and when they see fit, and preventing access altogether. The third group responds more readily to incentives and see access as a market good, therefore prefers incentives to be short-term in order to possibly adapt to new market opportunities. Since the last two groups are mostly composed of private landowners, these attitudes need to be taken into consideration when devising aid packages to facilitate access. Access is not a priority in itself for landowners, it is seen as part of a wider management picture, in which grants and subsidies can contribute to the maintenance of the woodlands. Thus, in order to attract interest from more landowners, aid for access should be related to broader issues related to management.

With the new standard for greenspace in mind, the Woodland Trust devised an inventory of accessible woodland, *Woods for People*, in order to assess the adequacy of provision to needs. The Trust's definition of woodland is that used by the Forestry Commission in the *National Inventory of Woodland and Trees*, i.e. 'land under stands of trees with, or the potential to achieve, tree crown cover of more than 20 per cent', and the definition of accessible is 'any site that is permissively accessible to the general public for recreational purposes' [Woodland Trust 2004: 6]. Data were gathered from 735 public or private British organisations providing accessible woodland according to the definitions given above. They established that in England, out of the 13,295,236 hectares of land, the woodland area represents 1,059,771 hectares, a little below 8%. Out of these, 46% are accessible to the public, with wide regional variations. Only around 10.2% of the population have access to a wood of a minimum of 2 hectares within 500 metres of where they live, and slightly over 55% have access to a wood of at least 20 hectares within 4 kilometres of their dwellings.

These figures testify to the difficult task facing the Forestry Commission and the Community Forests teams. They justify both the importance of the Commission's efforts to convince more landowners to open their lands to the public in exchange for grants, and that of the dedication scheme under the CRoW Act. As the Woodland Trust emphasises, permissive access to private woods has steadily increased as a result of the Forestry Commission's grant schemes. Supposing the existing woodlands not presently accessible were opened to the public, this would mean that an extra 26% of English people would benefit from woodlands of 2 hectares and more within 500 metres of their homes, and an extra 26.7% could have access to woodlands of more than 20 hectares within 4 kilometres.

To make up for the absence of relevant standards specific to woodland in planning policy, the Trust advocates a Woodland Access Standard very similar in its requirements to those concerning natural greenspace provision. According to the Standard, nobody should live more than 500 metres from an area of accessible woodland of at least 2 hectares, and there should be an area of woodland of at least 20 hectares within 4 kilometres of people's homes. This would entail creating 48,683 hectares of new 2-hectare woodlands within 500 metres of where people live, and 15,392 hectares of 20 hectares and more within 4 kilometres of people's dwellings.

The English Woodland Grant Scheme has integrated this demand in woodlands close to people's homes in its Woodland Management Grant (WMG), but twinned with the more traditional standard concerning density of population, since it considers that

> for WMG purposes, a demand for public access will be deemed to exist where at the time of application there is less than 1 hectare of free public access woodland for every 500 residents within 8 km [Forestry Commission England 2005: 2].

This means a more restricted provision than the one advocated by the Woodland Trust, although in exceptional cases where the criteria are not fulfilled, special needs and public support may justify the allocation of a grant for access projects.

The population is encouraged to take part in operations of cleaning, planting and protecting the local woodlands, as was mentioned before. Yet, because of the importance of the private property principle, this does not necessarily entail that they will have the right of access to these spaces that they have contributed to enhance. Similarly, public money can be spent on plantations in which the public will not be allowed to walk. Such is the case in parts of the National Forest and of the Community Forests. The battle over the 'right to roam', articulated by groups such as the Socialist Walking and Climbing Club, Red Rope, or the Ramblers' Association through their Access Days, has brought to the daylight the conflicting issues of access and private property.

As noted by the Community Forest Partnership itself, although the area of woodlands brought under management has increased year on year, woodland with new access has actually declined [Community Forest Partnership 2005: 10]. There is a crying contradiction between the promotion of regular contact with local woodlands and a significant proportion of them being inaccessible to the public. This raises the question of the relevance of expenditure on sites where there is no public access, whereas access is recognised as a priority by the local populations (90% of the respondents at Telford consider that one of the benefits of greatest importance is woodland being open to everyone, for example).

The answer given to this author by a Community Forest Officer was nicknamed by him the 'blue whale theory', a variation on Michael Jacob's 'existence value' [Jacobs 1995: 19]. This Officer makes a yearly contribution to the salvation of the blue whale, yet does not expect to see one in his lifetime; similarly, he deems environmental benefits sufficient for the quality of life of the neighbouring populations to justify public expenditure on woodland creation, irrespective of access. This abstract – should we say vicarious – appreciation of woodlands relies on their intrinsic value, for ecological purposes for instance, but that argument does not convince all critics. There is no doubt that access has become a key issue and that the same Officer makes a lot of effort to convince participants to the Community Forest to open their lands to the public. Access is encouraged, in particular by the granting of subsidies that take into account the proximity of existing woodlands and the existence of paths allowing walkers to roam through the new plantations. Yet, resistance sometimes comes from the owners themselves, as was shown before.

10.3.4 Access as the Best Enemy of the Woodlands

Access and the provision of facilities on site are not only limited by the legal status of land use, however. There are also physical limits to access, in so far as excessive frequentation can result in wear-and-tear likely to damage the resource. The belief that the carrying capacity of woodlands is more or less infinite and that the purposes of conservation can easily be reconciled with those of recreation has now reached its limits. Over-use can entail a number of negative effects for the woodlands, and the damage threshold is much lower than was hitherto believed: footpaths become eroded, tree roots are exposed, young shoots are trampled underfoot and bulbs crushed, all of which results in a reduced viability of these 'loved-to-death' woodlands. The social and the ecological objectives cannot always be reconciled: 'recreation can be a major contributor in facilitating unwanted ecological change in woods of conservation value' [Littlemore 2001: 30]. Taking this risk into account, a code of conduct was elaborated to balance the needs of conservation on Natura 2000 sites and recreation, in one of the most frequented English forests, the New Forest. The stakeholder forum of the PROGRESS (PROmotion and Guidance for Recreation on Ecologically Sensitive Sites) Project, led by the Forestry Commission and co-financed by Interreg IIIB North West Europe, thus produced four illustrated guides in 2005: a general *Out and About* guide, as well as three specialised guides for dog walkers (estimated at 27,000 households within the National Park), horse riders and cyclists.

Vandalism is often pointed out as a cause of damage to the woodlands, and thus likely to increase with greater access. Among examples of anti-social behaviour affecting trees, recurring ones in literature and in interviews include uprooting or burning newly planted trees, breaking branches or driving in 4-wheel drive cars through woodlands. According to a European survey, no less than 30% of newly planted street trees are affected by vandalism in some urban areas [Pauleit et al. 2002]. A means of combating vandalism can be to leave the woodland unmanaged: brambles and fern seem to have so far saved the core of the medieval woodland of Mill and Alder Wood (Merseyside). The downside of this protective measure is that a woodland needs to be maintained in order to thrive, so that protecting it by negligence, so to speak, jeopardises the health of the woodland and may only be a short-term alternative to tackling vandalism.

However, whatever the media may convey, most researchers agree to say that the share of vandalism is the tip of the iceberg as far as tree death is concerned. Out of an estimated 50% rate of mortality within the first 5 years, a minority can be laid at the door of anti-social behaviour, while the majority is due to neglect or inadequate maintenance, as was explained before. In the London Borough of Tower Hamlets, the death rate due to vandalism only accounted for 9% of tree deaths, while lack of maintenance represented 50%. Bradshaw and Gilbertson came to similar conclusions in the north of England [Chambers 1987: 5; Bradshaw and Gilbertson 1985: 131].

Fig. 10.5 Stolen and burnt cars are among the most visible forms of vandalism in woodlands, as here in Mill Wood, an ancient wood straddling Speke, Halton and Knowsley (Merseyside) (Photo Sylvie Nail)

Due to the thrust of public benefits in forestry, criteria for the allocation of resources and the very definition of these resources have changed. The Sustainable Development Research Network, supported by DEFRA, thus undertook in 2006 a review of *Emerging Methods for Sustainability Valuation and Appraisal*. It lists six methods which go far beyond strict economic analysis, since they include sustainability valuation.

Access has become much more central to the discourse; moreover, access deficit has been identified, permitting remediation. Yet there are undeniable gaps between discourse and practice. Discourse may be holistic, but on the ground, there are limits to how values and policies can be conciliated, or to how access and ecology can cohabit. Whether the same remark applies to a central tenet of community forestry, namely participation, will be the object of the next chapter.

References

Bengston, D. 1994. Changing forest values and ecosystem management. *Social and Natural Resources* 7: 515–533.

Bishop, K., L. Kitchen, T. Marsden and P. Molbourne. 2002. Forestry, community and land in South Wales Valleys. In *Trees Are Company. Social Science Research into Woodlands and the Natural Environment*, eds. L. O'Brien and J. Claridge, 29–41. Edinburgh: Forestry Commission.

Box, J. and C. Harrison. 1993. Natural spaces in urban places. *Town and Country Planning*: 231–235.

Bradshaw, A. and P. Gilbertson. 1985. Tree survival in cities. The extent and nature of the problem. *Arboricultural Journal* 9 (2): 131–142.

Burgess, J., C. Harrison and M. Limb. 1988a. People, parks and the urban green: a study of popular meanings and values for open spaces in the City. *Urban Studies* 25: 455–473.

Burgess, J., M. Limb and C. Harrison. 1988b. Exploring environmental values through the medium of small groups. 1. Theory and practice. *Environment and Planning A* 20: 309–326.

Chambers, K. 1987. Urban forestry in the London borough of Tower Hamlets – an account of the European urban forestry project sponsored by the EEC. *Arboricultural Journal* 11 (1): 1–14.

Church, A., N. Ravenscroft and G. Rodgers. 2005. *Woodland Owners' Attitudes to Public Access Provision in South-East England*. Edinburgh: Forestry Commission Information Note.

Coles, R. and S. Bussey. 2000. Urban forestry landscapes – urban lifestyles and urban living. In *Urban Lifestyles: Spaces. Places. People*, eds. J. Benson and M. Roe, 185–192. Rotterdam: Balkerna.

Coles, R. and M. Caserio. 2003. *Understanding and Facilitating the Social Outputs of Urban Green Spaces*. Paper read at the International Conference 'Open Space, People Space', Edinburgh. http://www.openspace.eca.ac.uk/conference/proceedings/PDF/Coles1.pdf. Accessed 16 May 2006.

Community Forest Partnership. 2005. *Community Forests Monitoring Report 2004–2005. Final Report*. Bury Saint Edmunds: Projects Partners Research. http://www.communityforest.org.uk/resources/monitoringreport_2004_05.pdf. Accessed 7 February 2007.

Countryside Agency. 2004. *Leisure Day Visits GB. Report of the 2002–2003 Great Britain Day Visits Survey*. London: Countryside Agency.

Darby, W. 2000. *Landscape and Identity: Geographies of Nation and Class in England*. Oxford: Berg.

Davies, C. and J. Vaughan. 2002. Bridging the gap: the experience of England's Community Forests in engaging excluded communities with urban woodland and green spaces. Congress *Urban Woods for People*, 25–27 November 2002, Paris. http://www.onf.fr/europe/life-urbanwoods/pdf/davies_GB.pdf. Accessed 3 April 2006.

Forestry Commission. 2005. *England Forestry Strategy Review. Topic Paper: Community Forestry*. http://www.defra.gov.uk/corporate/consult/forestry-strategy/community.pdf. Accessed 12 April 2007.

Forestry Commission. 2006. *England Forestry Strategy. Implementation Progress Report. 1998–2006*. Cambridge: Forestry Commission.

Forestry Commission England. 2005. *English Woodland Grant Scheme 6*. 'Woodland Management Grant Guide'. Cambridge: Forestry Commission England.

Forestry Commission England. 2006. *English Woodland Grant Scheme 1*. 'General Guide to EWGS'. Cambridge: Forestry Commission England. http://www.forestry.gov.uk/pdf/ewgs1-guide.pdf/$FILE/ewgs1-guide.pdf. Accessed 3 April 2006.

Forestry Commission England. 2007. *English Woodland Grant Scheme*. http://www.forestry.gov.uk/ewgs. Accessed 13 November 2007.

Forestry Commission England. DEFRA. 2003. *The English Woodland Grant Scheme*. Edinburgh: Forestry Commission.

Frumkin, H. 2002. Urban sprawl and public health. *Public Health Reports* 117: 201–217.

Gobster, P. 1995. Aldo Leopold's ecological esthetic: integrating esthetic and biodiversity values. *Journal of Forestry* 93 (2): 6–10.

Handley, R. et al. 2003. *Accessible Natural Green Space Standards in Towns and Cities: A Review and Toolkit for their Implementation*. English Nature Research Reports, n°526. Peterborough: English Nature.

Harrison, C. 1981. A playground fro whom? Informal recreation in London's green belt. *Area* 13: 109–114.

Harrison, C., M. Limb and J. Burgess.1986. Recreation 2000: views of the country from the city. *Landscape Research* 11: 19–24.

Harrison, C., J. Burgess, A. Millward and W. Dawe. 1995. *Accessible Natural Greenspace in Towns and Cities. A Review of Appropriate Size and Distance Criteria*. English Nature report N° 153. Peterborough: English Nature.

Henwood, K. and N. Pidgeon. 1998. *The Place of Forestry in Modern Welsh Culture and Life*. Report to the Forestry Commission. Bangor: University of Wales.

Hill-Tout, P. 1999. Costed options for managing state forests in Britain. In *Non market Benefits of Forestry. International Symposium held in Edinburgh*, eds. C. Roper and A. Park, 165–170. London: Stationery Office.

Hull, B. and A. Harvey. 1989. Explaining the emotion people experience in suburban parks. *Environment and Behavior* 21 (3): 323–345.

Hunt, G. 2001. Woodlands and public access. In *A Way into Woodland. Making Access Work Well for All Concerned.* Conference report of the National Small Woods Association, Small Woods Association, 21–24. Cannock: Small Woods Association, Forestry Commission.

Jacobs, M. 1995. *Sustainability and Socialism.* London: Socialist Environment Resources Association.

Littlemore, J. 2001. Managing Britain's woodlands for conservation and recreation. In *A Way into Woodland. Making Access Work Well for All Concerned.* Conference report of the National Small Woods Association, Small Woods Association, 30–32. Cannock: Small Woods Association, Forestry Commission.

National Playing Fields Association. 2001. *The Six Acre Standard. Minimum Standards for Outdoor Playing Space*. London: National Playing Fields Association.

O'Brien, E. 2003. Human values and their importance to the development of forestry policy in Britain: a literature review. *Forestry* 76 (1): 3–17.

O'Brien, M. and Y. Guerrier. 1995. Values and the environment: an introduction. In *Values and the Environment: A Social Science Perspective*, eds. Y. Guerrier, N. Alexander, J. Chase and M. O'Brien. Chichester: Wiley.

O'Callaghan, D. 2003. The challenges of planning for sustainable development in the UK. *Arboricultural Journal* 27 (2): 93–116.

Ode, Å. and G. Fry. 2002. Visual aspects in urban woodland management. *Urban Forestry and Urban Greening* 1 (1): 15–24.

ODPM (Office of the Deputy Prime Minister). 2002. *Assessing Needs and Opportunities: A Companion Guide to PPG 17*. London: Her majesty's Stationery Office.

Office for National Statistics. 2006. *Social Trends N°36*. London: The Stationery Office.

Pauleit, S., N. Jones, G. Garcia-Martin et al. 2002. Tree establishment practice in towns and cities – results from a European survey. *Urban Forestry and Urban Greening* 1 (2): 83–96.

Penning-Rowsell, E. 1981. Fluctuating fortunes in gauging landscape value. *Progress in Human Geography* 5: 25–41.

Penning-Rowsell, E., G. Gullett, G. Searle and S. Witham. 1977. *Public Evaluation of Landscape Quality*. Polytechnic Planning Research Group Report N°13, Middlesex University. London: Middlesex.

Price, C. 1997. Twenty-five years of forestry cost-benefit analysis in Britain. *Forestry* 70 (3): 171–189.

Price, C. 2003. Quantifying the aesthetic benefits of urban forestry. *Urban Forestry and Urban Greening* 1 (3): 123–133.

Price, C. 2007. Do cultural landscapes have a sustainable future? In *Social and Cultural Values of Forests – Benefit for Today's Society. 12th IUFRO World Congress – Proceedings. News Of Forest History* 4 (38): 12–24.

Price, C. and M. Chambers. 2000. Hypotheses about recreational congestion: tests in the Forest of Dean (England) and wider management implications. In *Forest Tourism and Recreation: Case Studies in Environmental Management*, eds. X. Font and J. Tribe, 55–74. Wallingford: CABI.

Rhode, C. and A. Kendle. 1994. *Human Well-Being, Natural Landscapes and Wildlife in Urban Areas. A Review*. English Nature Science Report N°22. Peterborough: English Nature.

Russell, D. 2000. Planning for the compatibility of recreation and forestry: recent developments in woodland management within the National Trust. In *Forest Tourism and Recreation: Case Studies in Environmental Management*, eds. X. Font and J. Tribe, 249–258. Wallingford: CABI.

Sheppard, S. and H. Harshaw (eds.). 2000. Landscape aesthetics and sustainability: an introduction. In *Forests and Landscapes. Linking Ecology, Sustainability and Aesthetics*. IUFRO Research Series N°6, 1–12. New York: CABI, in association with IUFRO.

Shroeder, H. 1989. Environment, behavior, and design research on urban forests. In *Advances in Environment, Behavior and Design*, eds. E. Zube and G. Moore, 87–117. New York: Plenum.

Simson, A. and J. Ryan. 2003. *NeighbourWoods Case Study Report*. Leeds: Design Task Force/Leeds Met University. http://www.sl.kvl.dk/euforic/docs/NeighbourWoods/D10d%20Leeds.pdf. Accessed 17 April 2007.

Slee, B., D. Jones and N. Curry. 2001. *Social Exclusion in Countryside Leisure in the United Kingdom*. Cardiff: Countryside Recreation Network.

Small Woods Association. 2001. *A Way into Woodland. Making Access Work Well for All Concerned*. Conference report of the National Small Woods Association. 2001. Small Woods Association, Forestry Commission.

Sustainable Development Research Network. 2007. *Emerging Methods for Sustainability Valuation and Appraisal*. SDRN Briefing Four. http://admin.sd-research.org.uk/wp-content/uploads/2007/04/briefingprooffinal.pdf. Accessed 1 May 2007.

Tsouvalis, J. 2000. *A Critical Geography of Britain's State Forests*. Oxford: Oxford University Press.

Woodland Trust. 2004. *Space for People. Targeting Action for Woodland Access*. Grantham: The Woodland Trust.

Chapter 11
Woodland Participation and Community Building

The three main components of a sustainable urban forest are a healthy tree resource, comprehensive management and community-wide support

[Konijnendijk et al. 2004: 3–5]

Abstract The discourse on the benefits of urban and community forestry stresses people's participation as part of sustainable community building. The eleventh chapter explores the theoretical framework behind participation and how it is interpreted in the Community Forests' agenda. This entails defining such ubiquitous terms as 'community' and 'stakeholders', as well as showing the wide array of participatory techniques devised and implemented. Whether these actually contribute to favouring social inclusion is another matter. Giving people a measure of control over their local spaces and reconnecting them to their local environment, is bound to be positive in terms of self-confidence and social capital building. Yet, fieldwork and other studies show, first, that woodland represents a minority interest and, secondly, that there remain obstacles to social inclusion, let alone empowerment in English society.

Beyond the individual benefits related to quality of life, part of the social forestry agenda relies on the assumption that woodlands can contribute to something wider, that we might call social health, often termed 'sustainable communities' in governments' publications. The *Sustainable Communities Plan*, published by the Office of the Deputy Prime Minister in 2003, aims at injecting funds into some 'growth areas' specifically to tackle issues related to housing and local infrastructure as well as the development of skills and jobs and the improvement to town centres and green spaces. Beyond this particular programme, the development of sustainable communities has specific implications in terms of woodlands, with a strong emphasis laid on participation in woodland projects. This chapter will explore what Clive Davies (Director of North East Community Forests) and John Vaughan (Group Director of England's National Community Forest Partnership) term the 'tree-led solution' [Konijnendijk and Simson 2001: 6] in relation to this very recent agenda.

This entails briefly summing up the particularly voluminous literature on the subject in search of the underlying principles to public participation and the participatory methods applied in England, and also bringing to the surface the beliefs and purposes, in relation to the cultural substract presented before, which underlie public participation for social inclusion, so as to show the achievements and obstacles to a more complete participation.

11.1 Sustainability and Social Inclusion

Over the past 15 years, environmental decision-making has integrated a positive vision of a participative approach, thus shaping the contours of a new system of natural resource management. Contrary to the hitherto dominant interpretation of resource management as a question best left to the professionals, the belief in the benefits of a participative approach to environmental decision-making has gone from strength to strength.

11.1.1 The Agenda

Although participation was advocated and practised before the 1992 Rio conference, studies on the meanings, stakes and methods of participation have multiplied since then. At Rio, participation was defined as a key instrument in the sustainability agenda, particularly through *Local Agenda 21*. This found an echo in The European Commission's 5th Environmental Action Programme in 1992, called *Towards Sustainability*, which recognised public participation as the *sine qua non* condition for success in sustainable development strategies.

The *Aarhus Convention* of 1998 provides an advanced treaty on public participation, with means for implementing the rules of international environmental law within the national institutions [UNECE 1998]. It closely relates environmental rights to human rights, and requires public authorities to provide rights of access to information, public participation in decision-making and access to justice in environmental matters.

The rationale for public participation as defined at the international level is threefold. First, involving non-state actors may provide valuable knowledge and perceptions. Secondly, public participation is an indispensable tool to combat the democratic deficit and promote human rights. Thirdly, public participation is seen as a prerequisite for the legitimation of environmental policies [Appelstrand 2002: 282–283]. The last aspect is to be related to public trust in institutions and scientists being very low in many countries, including Britain. Put differently, participatory mechanisms are necessary in the name of efficiency and effectiveness, of ethics and of public demand [Warburton 1997: 25]. The Ministerial Conferences on the Protection of Forests in Europe have put forward social inclusion as part of the objectives of sustainable forestry, as the title of the declaration made at the Vienna conference in 2003 shows: *European Forests—Common Benefits, Shared Responsibilities*. Among the objectives of National Forest Programmes, multifunctional management, including moral and scientific criteria, ranks high.

In Britain, the Government's Strategy for sustainable development (*Sustainable Development. The UK Strategy*) and the Biodiversity Action Plan (*Biodiversity. The UK Action Plan*), both published in 1994, endorse Rio's commitment to participation and stress the need for capacity building. However, a literature review of participation methods in environmental matters noted in 2000 that 'legislative requirements [...] for public participation are limited currently to passive consultation activities, placing the onus on the public to identify that an issue may be of interest and to take steps to comment' [Petts and Leach 2000: iii]. At the heart of New Labour's approach since 1997 lie the three concepts of performance management, democratic renewal and joined-up partnerships, which have repercussions on all areas of public policy. Government is often seen as shifting from an 'executive' to an 'enabling' role, which has implications in terms of community involvement [Curry 2000: 21]. In the 1998 White Paper *Modern Local Government: In Touch with the People*, citizen participation is thus presented as one of four pillars of democratic renewal [DETR 1998a]. The same year, the *Guidance on Enhancing Public Participation in Local Government* asserted that 'engaging with the public has become a touchstone for the general effectiveness of local authorities' [DETR 1998b: 3]. The 'Best Value' regime, meant to replace Compulsory Competitive Tendering, also puts forward the participatory approach through the four Cs: challenge, compare, consult, compete [Petts and Leach 2000: 11].

In parallel, but no stranger to this first trend, the former Prime Minister Tony Blair's speech in Croydon in April 2001 is credited with triggering off a review of policies affecting the local environment:

> We need stronger local communities and an improved local quality of life. Streets where parents feel safe to let their children walk to school. Where people want to use the parks. Where graffiti, vandalism, litter and dereliction is not tolerated. Where the environment in which we live fosters rather than alienates a sense of local community and mutual responsibility [Blair 2001].

The resulting report, also drawing on the final report of the Urban Green Spaces Taskforce, *Green Spaces, Better Places*, states the Government's objectives. These include commitments to improve the environment, but citizens are clearly expected to do their share, too: 'ultimately, each of us as individuals has a role to play in caring for our neighbourhoods and respecting the communities in which we live, work and play' [Prescott 2002]. Among the 'components for success', community involvement looms large:

> Where communities have been effectively engaged in projects the outcomes are better – and stay that way for longer. That has been shown time and time again. Moreover, there are important benefits for local people themselves from being involved in community activity – the opportunities for meeting and working closely with other local people, for developing new skills and for building confidence that can lead to greater community cohesion. Increasingly these are being recognised and efforts are being made to build community involvement into all stages of policy and action [ODPM 2002: 14].

So, from the macro (international) level down to the micro (local) one, 'multi-level' governance has become the new order of the day, with institutions perceived as facilitators for people empowerment. In forestry, the ethos of multi-level governance twinned with the objectives of multi-purpose forestry have led to top-down approaches being rejected. As a report by the Social Forestry Unit puts it, 'approaches that are more effective are needed, i.e. consultative, democratic, collaborative, deliberative and interactive' [O'Brien 2001: 24]. This has entailed a dramatic change of attitude on the part of the delivering bodies, the Forestry Commission in particular. Instead of delivering forests for the public good, the latter being assessed by expert opinion, the Commission has started to envisage itself as a facilitator, responding to the values, needs and priorities of the local populations rather than imposing decisions derived from scientific assessment.

The inclusion of values, on top of needs and priorities, is important to provide bridges between different value systems. In order to take this pluralism on board, humility is a prerequisite: instead of 'educating' the public, the attitude has become one of responsiveness, learning, aiding, sharing, developing 'a common vocabulary', encouraging and demonstrating [Hislop 2001: 12], so that people feel part of the process. On the official website of England's Community Forests, the objectives are presented along three lines, the first and last ones of which have already been discussed: 'Supporting regeneration and growth; Building sustainable communities; Creating better places'. The concept of sustainability present in the second objective is social as much as environmental, and urban and peri-urban forestry are conceived as tools for social outcomes. Part of the work of the officers in Community Forests consists in going out to meet people to find out about their needs and their views, to build them into the decision-making process, not only to improve their quality of life, but also to increase their sense of ownership and their participation to community-building. By the by, as was implied in the quotation taken from *Living Places: Cleaner, Safer, Greener*, this participation reduces their possible animosity towards landscape change and the chances of vandalism to the trees. How this is done conditions the results; developing appropriate tools is therefore crucial to the efficient inclusion of public participation.

11.1.2 Building Up the Toolbox for Implementing and Assessing Participation

In order to be effective and representative, participation has to follow certain principles and to lend itself to assessment.

In its programmes, the Forestry Commission and the Community Forests essentially follow the theoretical framework of Sherry Arnstein's 'ladder of participation' [Arnstein 1969], elaborated in the aftermath of the 1960s' social unrest in Europe and the United States. The ladder is made up of eight rungs, going from manipulation, at the bottom, to citizen control, at the top.

Sherry Arnstein makes it very clear that the two bottom rungs, ('manipulation' and 'therapy') do not partake of participation: they correspond to citizens rubberstamping decisions already made, and to participation being reduced to information, education and persuasion. The next three rungs ('informing', 'consultation' and 'placation') partake of 'degrees of tokenism': the possibility to give one's opinion and to negotiate marks the passage from non participation to participation, but attitude surveys, neighbourhood meetings and public hearings come under 'window-dressing' in her eyes. Only when negotiation leads to real partnerships, in which power is shared between citizens and powerholders, is the participation process fully implemented and power achieved. Decisions cannot be unilaterally changed, citizens are included in decision-making authorities, and conflict-solving procedures are anticipated. The upper three rungs ('partnership', delegated power' and 'citizen control') correspond to genuine participation and empowerment.

A more recent model, developed by Peter Oakley, identifies three levels of participation: contribution, organisation and empowerment. While the first one does not entail any delegation of power, the last two do to a certain extent, since they involve a transfer of control [Oakley et al. 1991: 6]. In all cases, argues Diane Warburton in her review of participation literature, a number of key issues must be addressed by institutions if they want to implement public participation and restore trust. These are

> willingness to change; commitment to participation; balancing participation, statutory duties and public accountability; clarity of purpose and stance; recognising the importance of context; following principles of good practice; recognising that participation requires resources; choosing the appropriate techniques; assessing whether participation has been successful [Warburton 1997: 35].

In the realm of forestry as in that of social policies such as housing, the desire of decision-makers to take into account the beliefs and opinions of the stakeholders, representing cross-sectoral interests, is crucial and it implies being prepared to change if the latter's interests do not match those of the institutions. Effective collaboration between all partners, whether from the public, private, voluntary and/or community sectors is the other crucial ingredient, since it should permit to move away from a competition setting with winners and losers, to win-win situations, i.e. 'creative policy-making that fulfils all the criteria: socially inclusive, economically viable, resource-conserving and aesthetically pleasing' [Barton 2000: 7–8]. In a nutshell, in order to achieve maximum efficiency of the participatory process, participation must take place from the start of the project, the public must be allowed to make comments and ask questions, the outcome of public participation must be taken into account in the final decision, and all decisions should be susceptible to review by a court, as articles 6.1, 6.2, 6.4, 6.8 and 9 of the *Aarhus Convention* state. There is no doubt that, as has been pointed out in various studies, there are difficulties and downsides to participation: it is a time-consuming process, which can entail extra costs (but also savings if consensus improves the outputs and prevents conflicts later on); it can generate extra conflicts and may not result in better decision-making and; landowners may feel threatened to lose their free will on the decisions concerning their lands. So it is a phenomenon which requires careful handling to lead to a 'win-win' outcome.

The Community Forests teams operate along Arnstein's lines, but the ladder is more often than not converted into a pyramid. At the bottom of the pyramid of participation are the people who turn up on the day of an event (e.g. tree-planting, taking part in woodland walks, helping to clean a wood or attending a meeting). Further up are the people who get involved on a more regular basis (through consultation, answering questionnaires, by becoming members of a 'Friends of' group, for example), while the top represents the category where real stewardship takes place. Thus, the claim by Community Forests that they follow Arnstein's ladder as a proof of their commitment to a participatory approach in social forestry, is excessive and even ironical since, as the above description shows, the large majority of the possibilities to participate come under non-participation or tokenism according to Arnstein (incidentally, the pyramid shape admits few people at the top). Furthermore, the rationale seems to be that each rung of the ladder needs to be experienced before the next one up can be reached, in a kind of meritocracy alien to Arnstein's thoughts:

> usually, it is not possible to progress up the ladder, unless the previous level has been achieved. Consultation is the first level of participation. [...] Without a mechanism for consultation, the progression from working for, to working with local people cannot be achieved [Forestry Commission 1996: 2, 12–14].

This leads one to wonder whether this implies that non-participation techniques have to be fully explored before the right to true participation can be gained. This ambiguity may be partly due to lack of experience and lack of tools: admittedly, 10 years ago, there was no such thing as a participatory agenda in Community Forests, and tools take time to develop. It may also be linked to the difficulty of implementing a fully-fledged participatory approach with the overwhelming majority of lands being in private ownership.

To try and make the process clearer and more effective, a *Framework for Public Involvement in Forest Planning* was developed jointly by Forest Research, Forest Enterprise and the USDA Forest Service. It gave birth in 2004 to a *Toolbox*, put together by the Forestry Commission, designed to help woodland managers decide which tools they could use in the woodland planning process, by answering four basic questions:

- Who should be involved in decisions about the forest or woodland?
- When should they be involved?
- In what ways could they be involved?
- What resources will I need? [Hislop et al. 2004: 1]

The introduction starts by recalling the bases of citizen participation derived from Sherry Arnstein's work, summarising it in four levels: information, consultation, involvement and partnership. The explanation of each one of these is a positive presentation of citizen contribution at every level, much more so than the equivalent rungs were for Arnstein. Reference is not made either to Oakley's transfer-of-control grid or to the *Aarhus Convention* and its demanding agenda concerning a commitment to integrate the public's decisions into the decision-making process, and access to justice. The *Toolbox* is accompanied by 50 practical *Toolsheets*, each one of which revolves around one tool (from 'advertisements' to 'workshops') and is graded according to the degree of participation implied, following the four categories mentioned above.

Although the *Toolbox* lists an impressive array of recently developed tools, only 14 of these come under 'partnership', while 22 correspond to 'involving', 40 to 'consulting' and 25 to 'informing' (one technique can correspond to various levels of implication). There is no strict correspondence between the number of techniques and the number of initiatives using each one of them on the ground, so in order to form a better idea of the participatory approach on the ground, in its scale and scope, one needs to assess the type of events that actually take place, for instance in the Community Forests, which represent the widest-ranging programme for community involvement. Before doing this, a rapid look at assessment and monitoring methods is necessary to complete the theoretical background to public participation.

While the development of participation theories and methods has given birth to numerous publications, both academic and institutional, the evaluation of the effectiveness of the methods has been much slower. Evaluating methods require clarity of objectives and the existence of criteria of effectiveness, both of which are far from being present in all the initiatives undertaken in the name of participation. Assessing participation means having a clear idea of what is being promoted – a product, a process, or both. Indeed, participation can be used as a method to improve projects by using the resources of the public, but it can also be seen as part of the purpose of the project, in other words, participation can be 'instrumental' or 'transformative'. Assessing the transformative nature of the participatory process implies more than seeing a project through: participation has to produce benefits to the participants. These can range from the development of skills, confidence and/or self-reliance, increased trust and understanding between individuals/groups and institutions, or other personal direct benefits like cash, improved access to services, social status or satisfaction [Warburton 1997: 26]. This makes clarity of purpose as well as assessment vital to any long-term participation, and the lack of them result in a number of 'knowledge gaps' [Petts and Leach 2000: 56] which hinder social inclusion. Evaluation has to cover inputs (resources, policies), processes (the way inputs are applied), outputs (products) and outcomes (consequences of outputs) and success will be assessed according to efficiency ('ration of inputs out outputs'), effectiveness ('measuring outputs against targets') and equity, separating between short-term alterations and long-term change. Flexibility is all-important to make sure that unanticipated outcomes are welcomed and acted upon if needs be.

Against this demanding framework various assessment methods have been tested, essentially since 2000. Criteria drawn from recent studies not directly related to environmental projects can be applied, for instance 'inclusivity, timeliness, focus, openness, resourcing, responsiveness and appropriateness' [Petts and Leach 2000: 46]. Through the Local Environment Agency Plan (LEAP) of the New Forest, meant to involve all the interested parties in planning for the future of the area, agree to a vision and establish a strategy for 5-year periods, Jacquelin Burgess has tested a methodology to assess stakeholder participation entitled 'multi-criteria analysis'. It focuses on the discursive process as much as on the product and lists a number of requisites for successful participation. First, it stresses the need to 'acknowledge explicitly, and to value, the different perspectives that individuals brought to the process'. Secondly, 'to make explicit, and to value, the different knowledges stakeholders possessed'. And finally, 'to encourage networking between the participants to increase local support and access to expertise and resources for local Agency staff' [Burgess 2000: 277–279].

The example was deemed successful and all the Stakeholder decision analysis has been tested successfully since on other LEAPs. Another method used in waste strategy development revolves around ten key questions to ensure that true participation has taken place: they deal with representativeness, the degree of initiative the participants are permitted within the agenda, the promotion of dialogue for mutual understanding of values and concerns as well as for the resolution of conflicts, access to scientific knowledge, consensus-building and transparency [Petts and Leach 2000: 47].

Among the seven priority targets of the Community Forests agreed by the National Partners and DEFRA in 1994, listed in a previous chapter, First Order Priority number four revolves around 'Securing involvement in the Community Forest by a wide range of local communities, especially those socially excluded from their environment' [Land Use Consultants/SQW 2005: 19]. This includes 'any activity in which local people are actively considered as an integral and valued part of the woodland management process'. Foresters are warned not to try and force participation, which might 'alienate the community and appear patronising', but on the contrary to 'stress equal opportunities and make a special effort to include under-represented groups in the community, using multi-language where necessary' [Forestry Commission 1996: 2, 10].

This Priority has only been assessed in its present, more extensive form, since 2000. Assessment is made through structured interviews of key stakeholders, meaning representatives of national, regional and local organisations, as well as the Directors of the Community Forests Teams. The following table shows the quantitative evolution in the number of events held between 2000 and 2003:

Table 11.1 Targets and number of events held in the Community Forests [Land Use Consultants/SQW Ltd, table 6.1, 71]

2000/01		2001/02		2002/03			
N° events held	Target	N° events held	Target	N° events held	Target	Total to date	2003–05 target
9,466	12,210	12,955	12,210	16,029	12,210	38,450	24,420

What these figures indicate is that, after a slow start, events are now well above target in numerical terms. What they do not indicate is the nature of these events, which matters at least as much as their number. The following table indicates the breakdown of events for 2004–2005.

Table 11.2 Breakdown of events in the Community Forests for 2004–2005 [Community Forest Partnership, 2005: table 1, 14]

Type of event	Mean number of events per forest area
Involvement/action	678
Information/interpretation/publicity	303
Schools	416
Training	70
Total	**1,467**

The eight rungs of Sherry Arnstein's ladder, the three stages of Oakley's grid and the requirements of the *Aarhus Convention* are replaced by four headings, the logic of which is difficult to perceive in participatory terms. They provide a mixture of broad objectives and publics. The category 'involvement/action' is far too vague to provide a clear idea of the degree of participation implied: participation in a planting event, consultation during a public meeting, membership of a group or stewardship of a woodland could equally come under this heading, and they correspond to very different degrees of involvement and decision-making. As far as 'schools' are concerned, their relevance as a category is debatable: all sorts of different activities may take place in schools, and the mention of the category as such is not telling of any particular degree of involvement. Training may be the only category where empowerment may actually take place, in so far as it foments capacity-building and skills transfer. Yet, this represents only 4.7% of all events (6% in 2000–2001, 5% in 2001–2002, 7% in 2002–2003, 10% in 2003–2004).

All in all, it is difficult to see exactly what degree of participation has been achieved, and there is a distinct feeling that quantity matters more than quality. The stress on schools as a distinct category corroborates this: a lot of initiatives take place within schools, for reasons mentioned in a previous chapter, and they involve a high number of participants, so figures paint a rosy picture. Although they are often of good quality, the degree of real participation they imply is a moot point. Most of the time, children do not choose to take part, they happen to be at school and to participate in 'one-of' events planned for them, but they may not perceive the benefit or gain any form of empowerment from them. What they often see is the immediate benefit they get from these activities that take them out of the classroom: a little girl planting trees in a primary school told this author radiantly, 'and we managed not to work all day!', while another one struggling to dig in hard soil during a National Tree Week event handed her fork to the organiser and was heard asking wearily, 'can I have my badge now?'.

The emphasis on figures is logical for three reasons: first, community participation has often become a prerequisite of funding, so organisers have to be seen to be involving the local population; secondly, it is easier to count the number of days or the number of volunteers than the degree of satisfaction of the participants; finally, it corresponds to the Forestry Commission's instructions in its Guide on involving communities:

> if your principal aim is achieving a 'product', then the output of volunteers on the ground can be measured. If your principal aim is to get volunteers participating in the planting and managing of woodland then your success can be measured by the number of people or

> person-hours involved. If your principal interest is in 'process', then measuring success is more difficult because community involvement in forestry is primarily for community and individual development [Forestry Commission 1996: 23].

At the local level, quantitative assessment dominates, too. An exhibition held in Huyton (Merseyside) public library in the winter of 2003, entitled *Creating Knowsley's Forest*, assessed 10 years of work as part of the Mersey Forest, and the emphasis there was on figures, too: one million trees and 9 kilometres of hedgerows planted in Knowsley in 10 years; 45 kilometres of paths created; 45,000 people involved in 6,000 events and campaigns. As part of the exhibition, a photograph competition provided a qualitative element, as the exhibitors conveyed their reactions to trees in the captions to the photographs, but this interesting element did not relate to empowerment.

One thus needs to be very cautious as regards the assessment of participation, as the rhetoric used to describe the benefits to the community does not always match the reality. One example among many others will show this. As part of the Great Western Community Forest, a community woodland, Shaw Forest Park, has been established on an old rubbish tip in West Swindon. The presentation on the website of the Community Forest starts by claiming that, more than the trees planted, 'the most important element in the success of this site has been the community ownership and involvement'. But the next sentence puts into question the definition of both terms 'ownership' and 'involvement': 'the Forest has held three community tree planting events, and over 10,000 local people have taken part in these and other activities' [Great Western Community Forest 2005]. Tree-planting activities, by any of the standards listed above, can hardly count as 'ownership', not even as 'participation' according to Arnstein. The leaflet displays the same ambiguity as far as 'consultation' is concerned. It mentions the organisation of a Consultation Day 10 years after the launch of the Great Western Community Forest:

> a Shaw Forest Park Consultation Day was held in early December 2004 to give the people living in West Swindon the opportunity to meet the Forest team and officers from the Borough's landscape architects. As well as finding out more about the changes that had taken place over the years, local people were able to learn about the technicalities of turning a tip into a park and to discuss their thoughts and ideas on how they might get involved in the Park's future [Great Western Community Forest 2005].

Admittedly, it is better to consult the local population late than never, but consultation here should more appropriately be called information, and consultation in any form represents a rather narrow interpretation of community involvement, which stops short of responsibility or control since it does not guarantee an input in decision-making.

The preliminary report of *Public Opinion of Forestry 2007: England*, established at 7% of the people questioned (that is, about three million adults) the number of people who have been involved in the previous 12 months in tree planting activities (4%) woodland voluntary work (3%) or are members of a community-based woodland group (2%) [Forestry Commission 2007: 17].

The assessment is not entirely positive. This probably reflects what Judith Petts and Barbara Leach see as cultural factors preventing the increase in public involvement, i.e. 'a culture of professionalism and expertise; a culture of minimal compliance;

and a culture of confidentiality' [Petts and Leach 2000: 23]. An essential part of the participation equation thus revolves around fitting content and assessment to objectives and correct timing; another equally important issue is that of the definition of the participants to promote social inclusion.

11.2 From No Man's Lands to Thriving Communities

If democratic renewal is at stake in community forestry programmes, then targeting and involving as broad a range of citizens as possible and involving them in creative ways is the key to success.

11.2.1 Desperately Seeking Stakeholders

Britain has a long tradition of voluntary participation: 'getting involved' is a deeply held idea and charitable organisations like the Royal Society for the Protection of Birds (RSPB), the Council for the Protection of Rural England or the British Trust for Conservation Volunteers are the precursors of many recently created groups. The term 'community' which underlies the scope of these organisations undeniably has a positive ring in British culture. Thus, the choice of the term 'Community Forests' for the main programme of peri-urban forestry is culturally rooted.

The Forestry Commission guidelines to involving the population, published in 1996, aptly noted that 'the first step in the process of community participation is to define the relevant communities. This is a difficult task', and went on to distinguish between two types of communities: 'communities of neighbourhood' and 'communities of interest' [Forestry Commission 1996: 4]. The new guidelines published in 2004 broaden the range of populations targeted for participation in forestry programmes, ranging from the 'public' ('the community or people in general'), to 'the stakeholders' ('those who have an interest in a particular decision, either as individuals or representatives of a group') and lastly, the 'community', described as 'a subset of all the stakeholders' [Hislop, Twery and Vihemäki. 2004: 4]. Very often, the term 'community' alludes to one or all of these categories, as it seems to evoke best the public for whom and with whom most social forestry programmes are undertaken. As such, it is worth exploring, both in its explicit and its implicit dimensions, as a key to understanding whose participation is sought.

In many publications, the term 'community' refers to the public, woodland users in general, for whom the State or various organisations act as mediators to make their visits as safe and pleasant as possible by providing facilities such as car parks, toilets, footpaths, signs, etc. Behind the provision of facilities for these users lies another stake already mentioned before, that of the protection of the countryside by providing 'honeypots' to urban dwellers.

More than this loose conception of the 'community' of users who have in common nothing more than an occasional sharing of time and space, the term frequently takes on a spatial meaning, referring to 'communities of neighbourhood', as in the example below:

> The term 'community' implies all those who live and work in an area and includes organisations such as local authorities, town councils, parishes, businesses, clubs, societies, schools and community groups, farmers and landowners and the millions of individuals who live within reach of the Mersey Forest area [Mersey Forest 2001: 40].

This meaning of the term has given rise to a lot of debate. This 'community' refers to the people of a district or to the district itself, and is often synonymous with 'neighbourhood' or 'locality'. This spatial construct is closely connected with social networks of people knowing each other and possibly relying on each other for relationships and support. The traditional 'community of neighbourhood' which comes to mind in England, at times unconsciously, is a rural one, usually a village, perceived as an organic unit in which every individual person is known and has an allocated place in the overall functioning of the whole. The following vision, expressed by local 'communities' as part of the implementation of LA21, reflects this:

> The locality has a role in maintaining the 'social capital' of community networks based on local activities and propinquity. Certain attributes of the shared visions are surprisingly persistent: an attractive and green neighbourhood which is safe, pollution-free and uncongested; a sense of local community and excellent access to friends and facilities both locally and regionally [Barton 2000: 10].

This sustainable neighbourhood/community, with its physical and social characteristics, is reminiscent of William Morris's utopia in *News from Nowhere* or the Garden City ideal. This definition is not unproblematic however. There is a mythical dimension to this 'community' which has all but disappeared with rural exodus and social changes, but also because of the critiques made of it [Whitt and Slack 1994: 12]. First, this organic view, now as in the past, stresses the solidarities but denies the existence of social stratification. Moreover, the term 'community' in its sense of 'communities of neighbourhood' is a fluctuating notion whose contours are difficult to elicit. Even if more than half of British adults spend their life within 5 miles of where they were born [Gray 1997], up to 10% of a community may change in a year depending on the areas [Alison 2000: 77], which makes it difficult to engage full community support on a long-term basis. The definition itself is not stable, as an anecdote testifies: in a High Court ruling in 1996, the judge decided to favour the opening of a pharmacy in a commercial centre by using the argument that commercial centres can be considered as 'neighbourhoods' [Anon. 1996]. The notion of commercial centres as neighbourhoods challenges traditional meanings and values and raises questions in terms of the communities these neighbourhoods might stand for.

All the same, there is no doubt that for many English people, 'communities of neighbourhood' evoke local connections and attachment to the local environment: several of the author's interviewees involved in charitable organisations evoke their desire to pay a debt to the neighbourhood where they grew up by 'putting something back into the community'. 'Community', beyond its geographical associations with 'neighbourhood' and 'locality', evokes people's aspiration for society to exist and to provide both a background to individual development and a commitment to shared values [Etzioni 2000: 188].

Fig. 11.1 Commercial outlets, a new kind of neighbourhood? Cheshire Oaks, Chester (Cheshire) (Photo Sylvie Nail)

As Jacquelin Burgess puts it,

> 'imagined location' is intimately connected with an 'imagined community' who might be available in time of need [Burgess 1995: 28].

Community Forests and other peri-urban community woodlands conjure up this physical and sociological image of the 'I and we' society [Etzioni 2000: 189], possibly as a post-modern response to the disappearance of traditional communities, and as an invitation to take rural values into the urban world for social regeneration purposes. The literature on the Thames Gateway, a 40-mile corridor East of London hailed as the largest regeneration scheme in Western Europe, is enlightening in this respect. Below is an extract from the Local and Communities website on the new Jeskyns Community Woodland (between Gravesend and Cobham, Kent), created as part of the Sustainable Communities Plan launched in February 2003. It exemplifies how derelict agricultural land is rebranded as heir to local heritage landscapes and how native woodlands are commodified as the perfect setting for the new concept of urban life.

> The site will bring together some of the key landscape features in North Kent such as traditional orchards of apples, cherries and cob nuts, open parkland, grass and wildflower meadows and native woodlands. A network of trails and paths across the site will make Jeskyns accessible to pedestrians, cyclists and horse riders [LGC 2007].

Lastly, 'community' is used to evoke the entire population of England: in 2003, the Community Forests' page on the Forestry Commission's website welcomed visitors with the following assertion: 'Forestry Commission – working with England's community'. The national identity dimension of this assertion will be explored in the next chapter, but there is no doubt that the Community Forests programme tries to provide visual continuity between the present landscapes and real or supposed past social and cultural landscapes.

It is striking that the term 'community' evokes all levels of social and spatial connections, from the local to the national. Yet all communities exclude, precisely because they are based on bonds of affinity. Paradoxically, one of the challenges of Community Forests is to go against the grain by including within the 'community' specific groups suffering from what sociologists term 'social exclusion'. When the *England Forestry Strategy* states that

> the presence of damaged and disturbed land is also frequently linked to areas of social deprivation where unemployment runs at high levels, environments are poor and the quality of life is often perceived as very low. The Government has begun to tackle these challenges. At the heart of our approach is the need to give these communities a new chance [Forestry Commission 1998: 14],

it is clear that the communities targeted include the deprived minorities living in inner cities, a lot of whom may feel disengaged from the British context. Community Forests, by providing equal access to all, whether car-owning or not, aim both at improving quality of life and making social mix possible, thus creating connections and hopefully leading to what Amitai Etzioni calls 'megalogues', that is to say 'society-wide dialogues, that link many community dialogues into one' to redefine their values and moral commitments [Etzioni 2000: 193].

'Communities', apart from groups characterised by their local attachment or by common social conditions, also mean 'communities of interest' and in that sense overlap with the ubiquitous 'stakeholders'. This meaning differs from the previous ones of 'community' in that they are defined in relation to a specific issue, independently of a place [Glicken 2000: 307]. Such communities, in connection with woodlands, might include charitable organisations like the Woodland Trust or people forming part of a national network like the Tree Wardens, among many other possibilities. The stakeholders of the twelve Community Forests include a wide range of actors, e.g. local authorities and other institutions, businesses (whether local or national), organised groups (whether environmental or other), landowners or land managers. Their participation is often actively courted by the teams responsible for the implementation of urban/community forestry projects. Indeed, on their active involvement, and especially that of the landowners, relies the implementation of the programmes. It is therefore important to make them feel that by being involved, they do not relinquish their rights associated with private property, as the term 'Community Forest' might imply, but that on the contrary they gain in the process. An example of discourse indirectly addressed to landowners can be found in the Forestry Commission's guide to public involvement:

> Of key importance in achieving these benefits is the need to develop a feeling of community ownership. This does not imply land ownership or a veto over the landowner's rights as an owner. It does imply allowing people sufficient participation to develop a bond with community woodland. This bond increases the responsibility that people feel towards a woodland which can manifest itself in reduced antisocial behaviour, residents wardening and an increased willingness by local people to invest time and effort in voluntary work in the woodland [Forestry Commission 1996: 9].

The benefits of community involvement to the landowners are stressed in unequivocal terms: increased responsibility from the public, reduced vandalism, as well as free manpower. As a counterpart, the 'community' of users is given a new sense of ownership, however symbolic. Although nobody would dare today to put the advantages of community participation as bluntly as 20 years ago, when citizen involvement was assessed in terms of significant cost savings in planting, maintenance and surveillance [McPherson and Johnson 1988: 185], this advantage of community participation still looms large for many. As far as private companies are concerned, their financial support is essential for match-funding. Large companies gladly associate their names with urban forestry programmes, especially if their activities involve environmental damage.

This use of the word 'community' raises the question of the representativeness of the stakeholders for two main reasons. The first reason has to do with the respective weight of the stakeholders. In one woodland of a Community Forest, the leader of the Friends group interviewed by this author complained of the decision taken at a stakeholders' meeting in response to the need to improve the footpath in the wood (to permit access to wheelchairs and pushchairs). The decision taken, against the Friends' group's opinion, was to strew sawdust over the footpath, as the local council had a large supply of sawdust that had burnt and that was useless for any other purpose. The project was seen through, and the new surface proved inadequate for its intended purpose. The bottom line of this anecdote is that several of the stakeholders supporting the decision also happened to belong to a famous motor company which has a local factory, and that these always support the local council in all its decisions, possibly in gratitude for the State's financial support of the company. In this case, the interests at stake may have nothing to do directly with the site and the interests of the 'community' of users, especially those in wheelchairs, but everything to do with power relations. Some stakeholders are definitely 'more equal than others', and the rules of the game are fraught.

The second reason why one should question to what extent the 'community' – or communities – targeted in these programmes do represent the whole community and its aspirations, is linked with the recruitment of stakeholders. Indeed, when looking at the reports of 'community' forestry partners (when they are listed), one is struck by the absence of people not belonging to established groups. Understandably, it is more difficult to contact and enlist the support of individual people who do not belong to established groups than that of people who are already members of an organisation, as the leaders of various programmes know very well. The REACT assessment report explains that the reason why success has been achieved mostly with ethnic minorities of three out of the five wards targeted has to do with the 'support of the PCT's bi-lingual Health Link Workers and the fact that many people from these groups are organised into Community organisations and are therefore much more contactable' [Central Liverpool Primary Care Trust 2004: n.p.]. As an assessment of the 'Woods on your Doorstep' programme concludes, 'social exclusion is such a buzz word at present but acting in a socially inclusive way is a real challenge to us all' [Alison 2000: 77].

The Audit Commission has drawn attention to this difficulty, particularly acute in the case of people with little spare time or unsocial working hours, as well as that of people who do not speak English well, are suspicious of institutions, or are physically isolated. Yet, it also pointed out the importance of finding strategies to overcome these barriers so that the participatory process should be inclusive [Audit Commission 1999]. Indeed, can 'community' objectives pretend to be all inclusive if the 'community' in question boils down to institutions and already organised groups of the population who are more likely to include 'mainstream' English citizens than disabled people, single mothers or ethnic minorities, to take just a few examples of groups that tend to suffer from isolation and under-representation? Not only does this phenomenon go against the inclusion, not to mention empowerment, of new types of people, but it may end up reinforcing the power already held unofficially by some groups. Some field observations make one aware of the risk involved in the process: as one interviewed regional officer of the Woodland Trust put it, 'there are professional join-a-group people', who before joining a woodland initiative, often belonged to other groups. Here, participation entails giving more voice locally to those who already had one, rather than opening participation to include those voiceless members of the community. Forestry Commission's statistics confirm the lack of extensive consultation: according to the 2005 *Public Opinion of Forestry*, only 6% of the respondents had been consulted about plans for creating, managing or using woodlands in their area [Forestry Commission 2005a: 13].

The concept of 'community' is complex and maybe more ambiguous in a Western urban world than in a developing country where the concept was first forged. The choice of the word clearly tries to bridge the gaps and make everyone feel they have something to give, something to take, and more than anything, something to share. Yet, in spite of its widespread, all-embracing, and far from explicit use, some 'communities' have very little in common with others, and some individuals and groups clearly remain outside the range of the 'community'. Inclusion of the wider community through the participation of its isolated members, remains wishful thinking. This is not to suggest that attempts have not been, and are not being made, to overcome this obstacle. There is no doubt that a lot of funds, energy and creativity have been invested to enlist the support and participation of as wide a section of the public as possible, with the purpose that stakeholders should constitute part of the community, rather than the contrary, which the Forestry Commission's 2004 guidelines advocated.

11.2.2 Looking Outside the Box to Build Social Capital

One way of encouraging participation of those not normally involved and often cut off from the participatory process has consisted in changing the way in which the consultation/participation takes place. Benchmarks have been developed at the European level through an interdisciplinary research project, URGE, which attempts to review urban green space systems in order to develop a set of social criteria to assess green space performance. Similarly, the Forestry Commission has increasingly tried to take public demand into account, although sometimes providing quality means more providing services of good quality rather than taking into account qualitative analyses of people's needs and values. Recognising that this is a brand new area for which they have no tradition or skills, they have begun to buy in new expertise to allow them to pass from hard-nosed to people-based policies.

Forest Design Plans have tried to take on board the public dimension. These documents underlie the management of one million hectares by Forest Enterprise, and are meant to chart the next 40–50 years of management in matters of recreation, conservation, timber production and visual amenity. They involve setting objectives, surveying the forest, analysing the information in terms of constraints, opportunities and landscape character, sketching felling and restocking, as well as monitoring [Broadhurst and Harrop 1999: 190]. So they are *par excellence* the type of technical document produced by experts. Yet, the process now often includes information and/or consultation at various stages.

Among other examples, in the New Forest, the adoption of the Plan was preceded by a Forest Design Plan Forum and public meetings under the title 'New Forest – New Future'. The degree of response of the public to such types of participation is debatable, among other things because of the technical aspect of such documents. Map reading, or even reading, can be a problem to some participants. Several innovative methods have been recently added to the range of techniques available to officers so that, once people have been convinced to take part in a consultation process, it is ensured that they understand them and can participate more, especially in the design phase. The Red Rose Community Forest, for example, has developed a 'Participation package', which includes funds to help projects get off the ground, but also support from a dedicated worker and a toolkit of resources [Whitaker 2000: 59].

Technology has contributed to creating new possibilities of enrolling the participation of local populations. The new discipline of ecological engineering has permitted the visualisation of projects with databases, computer graphics and simulation, among others. These new technologies can contribute to the inclusion of public preferences in the designing of new woodlands, in so far as they do not require previous knowledge, only the presence of a technician capable of translating people's wishes into pictures. The application of Geographic Information Systems (GIS) is a favourite one in many community woodland projects, with interactive GIS-based web pages used to enable members of the public to try out scenarios and make recommendations.

An example of such a website, ambiguously called *Virtual Decision Making Environment* (one can muse on whether it is the environment which is virtual, or the decision-making process), has been developed by Leeds University's School of Geography, and previously tested on three case studies. It allows users to suggest changes of uses for buildings shown on a map or identify zones for new developments. Obviously, as any tool, visualisation techniques raise the question of the objectivity of the inputs. There is no doubt that these techniques make the process and the project more attractive, but in return this depends very much on the parameters that have been used (size, colour). It is easy to be convinced by a beautiful 4-D (3-D plus movement) presentation of a future recreation area, and to forget that the result depends on the inputs. Visualisation techniques can certainly be a help, if they are reliable. Otherwise they can act as a filter between the layman and the specialist, to get the go-ahead of the local population with a semblance of consultation.

The authors of the Exploratory Workshop held in Vancouver in 1999 also conclude on the need to make people question the visualisation tools as a key to win their acceptance of landscapes based on sustainable aesthetics [Sheppard and Harshaw 2000: 276–279]. It is to be noted that a lot of recent research has been devoted to the improvement of the validity of visualisation techniques, by including movement, realistic tree shapes and species through virtual landscape simulators for instance.

Participation issues have become so central to community development that the Neighbourhood Initiatives Foundation, a charity based in Telford, has developed, specialising in community participation, training and development. They have created 'Planning for Real', a participation technique to foster consensus-building which has become widely used in England. It involves community participation through group discussions, but also 3-D models of areas that people contribute to building so as to visualise what might happen to their area, and take part in it. After the model has been elaborated, workshops are held in the various places where the model is exhibited, during which people are asked to write their concerns, opinions and desires on cards which are placed on the model. Cards can thus be rearranged, issues negotiated and action plans developed.

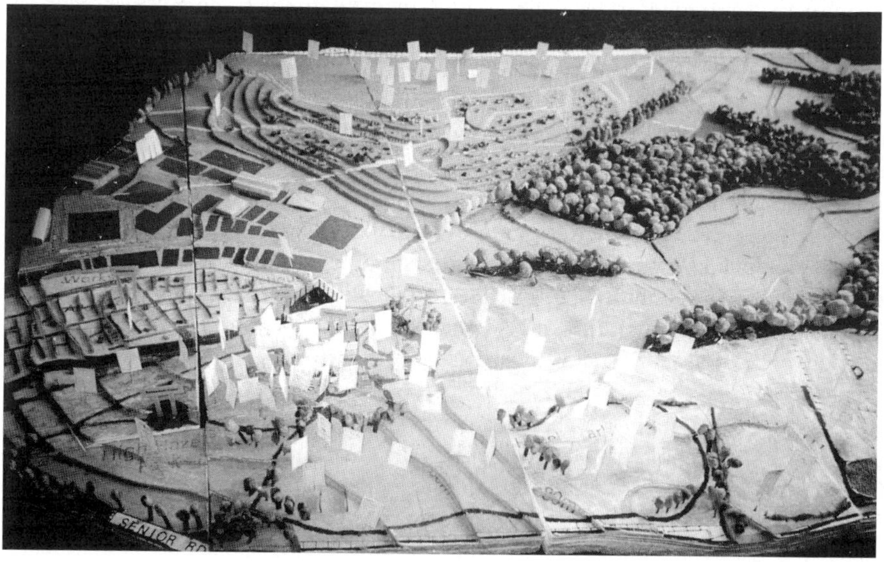

Fig. 11.2 Model resulting from a Planning for Real exercise in High Hazels, Sheffield (Photograph with kind permission of Planning for Real, a registered trademark of Neighbourhood Initiatives Foundation)

Apart from new tools, new methods of getting to people outside organised groups are being experimented. The consultation process can be made more attractive by providing refreshments, prizes or childcare facilities, and by consulting with people at venues which they usually frequent, rather than try to invite them to alien places [Audit Commission 1999: 31]. In Liverpool, the officer in charge of the REACT project thus visits venues where the ethnic minorities of the wards concerned usually meet.

More than any other tool, however, participatory strategies hinge on bringing people back into the woods from which they have become alienated and which they do not understand any more. Rangers in local nature reserves organise several 'green days' per year. The 'Local Woods Campaign' set up by the Small Woods Association in the West Midlands, aims at restoring the links between local communities and their woodlands and therefore organises a lot of outings in England and Wales, with activities (charcoal-burning, compost-making, tree-hugging, etc.). In the spring, the 'Bluebell Express' takes coach parties from various urban areas to discover the carpets of the native bluebells (*Hyacintoides non-scripta)* in bloom in the Mersey Forest. During the summer holidays, the 'Forest Fever' provides for more walks through the woods.

Fig. 11.3 The UK has 30% of the world population of bluebells. Joining the Bluebell Express offers communities the chance to see them – and to reconnect with the local countryside (Photo courtesy of the Mersey Forest)

The Tree Council organises every May a festival entitled 'Walks in the Woods', to discover woodlands through a wide array of activities ('teddy bears' picnics, bat and badger watches, bluebell walks, dawn chorus events, learning games and story telling, art and craft activities, treasure hunts, tree trails around towns, and woodland open days' [Tree Council 2007]). The same charity organises seed-gathering Sundays each year at the end of September. The scheme was so popular that it was turned into a seed-gathering month in 2006. Charitable groups vie with each other for creativity: bat or badger observation, circus, kites, orienteering, bird-feeding, craft workshops, fruit gathering for jam-making, among many others. All these actions exemplify the principle according to which the more you frequent a place, the less frightened of it you are. People may go into a wood to do something specific unrelated to the woodland setting, but the chances are that they will reconnect with the woodland and come back to visit it at some later point.

One favourite method used to bring people into the woods, already mentioned in relation to schoolchildren, is tree-planting projects, which have gone from strength to strength since the first 'Plant a Tree in' 73' campaign. This was so successful that it was renewed the following year under the catchphrase 'Plant More in' 74', and became National Tree Week in 1975. This has become one of the main occupations of the Tree Council, a charitable umbrella body for around 150 organisations around the UK which promote the planting and conservation of trees both in urban and rural areas. Every year in the last week of November, hundreds of opportunities to plant trees, visit woodlands, hear about woods, visit exhibitions, take part in activities, etc. gather numerous city dwellers in National Tree Week up and down the country. Similarly, between 1996 and 2001, the 'Woods on Your Doorstep' project, coordinated by the Woodland Trust, gathered over 250,000 local volunteers to plant a million and a half trees in 250 new woods in the UK to celebrate the new millennium. In many areas suffering from economic deprivation, tree planting is seen as a means of contributing to environmental justice, providing a better environment with public access in areas often sadly lacking in it.

A clue to the success of such initiatives is their being practical, easy, entertaining and rewarding, and a large measure of the reward is the creation of social networks. The aptly named 'Walk and Talk' Project in the Forest of Mercia revolves around social interaction and physical activity, while the 'Green Streets' Project of the Red Rose Forest consults people so as to reactivate links between people who 'don't talk to each other' [Blackman and Thackray n.d]. The presence of the others, the company, are both an aim in itself and a means to make people feel comfortable. The link between the visitors and the world of the woods is ensured by a growing number of intermediary, often volunteers, who have swelled the ranks of charitable organisations. It is because they are accompanied by rangers, Community Forest officers or community liaison officers that people unfamiliar with the woodlands feel secure enough to discover and enjoy them: 'people need people' [Burgess 1995: 7].

One of the ways in which this need is being fulfilled is through such initiatives as the Tree Wardens, created in 1990 by the Tree Council in association with an electricity provider. There were over 7,000 of these volunteers by the end of 2000, appointed by parish councils or local organisations. Their mission, in the words of one of them interviewed in the Wirral, is 'as much or as little as you wish, really'. More than anything else, they act as mediators, between the local community and the trees on the one hand, and between the local community and institutions on the other hand. This may mean gathering information and attending training courses in order to undertake surveys of local trees and then take people on guided tours of local remarkable trees; measuring trees or monitoring the changes in hedges with schoolchildren; negotiating with local landowners who cut their hedges with machines and thus destroy their character; answering local people's queries about trees; and keeping an eye open for vandalism against trees in order to warn the relevant authorities.

Another intermediary between local organisations and institutions has been created in the Mersey Forest, under the Community Contracting Initiative (CCI). This aims at coordinating 12 groups in charge of a wood, helping them with paperwork, especially for fund-raising and organising events. The dynamic leader of the CCI is aware that some people blossom from taking on new responsibilities, while others, too keen to achieve results, burn themselves out and can even get into trouble with the community, so a regulator, or 'social worker' as the CCI leader calls herself, is very useful to prevent the conflicts, channel the energy and lend a sympathetic ear.

Fig. 11.4 Planting trees during National Tree Week, a multi-purpose, inter-generational pursuit (Photo courtesy of the Mersey Forest)

Participatory techniques are numerous and varied. Whether they actually contribute to combating the democratic deficit, which in British terms means engaging the population in the political process again and reestablishing trust in institutions and politicians, is another matter. Mike King's assertion that 'planting trees is what you do as a citizen, as important, for instance, as voting' [King 2000: 64], emphasises this connection. Giving people a measure of control over their local spaces, 'adding eyes to the street' [Burgess 1995: 35] and reconnecting people to their local environment, including the various communities which form part of that environment, has to be positive in terms of self-confidence and social capital building.

However, there is no room for complacency, as Community Forests officers are well aware, and key issues need to be addressed for the process to really address democratic deficit, otherwise Mike King's sentence will become a very ironical reflection of people's lack of belief in the electoral process rather than a proof of citizens' commitment to planting trees.

First of all, very few of the initiatives described above actually grant a significant level of empowerment. Moreover, internal conflicts between stakeholders, as in the case of Telford [Janse and Konijnendijk 2007: 34], may make the implementation of the participatory process more difficult. Another limitation is that the key functions in the multi-level governance are occupied by stakeholders who are not necessarily representative of the public's choices: this is the case of the Government offices, which occupy key functions in the Regional Development Assemblies without being elected, or of the private companies whose interests may be at variance with those of the local population. Finally, empowerment is also limited by some form of control, if only because the targets must match governmental strategies or the criteria listed to obtain the necessary funds. In the last resort, the tree-led solution remains one solution among many others to build social capital, but it can also act as an alibi to save on public expenses thanks to cheap manpower without letting go of the real power when it comes to decision-making [Murdoch and Abram 1998: 42]: 'hired hands rather than local voice' [Burgess 2000: 285]. It can also mask the reality of communities' different agenda.

11.3 Ladders of Participation and Scales of Interest

In order to make the social agenda of woodland progress, defining the public and devising innovative tools is not enough. A last aspect, often overlooked even now, has to be taken on board for the equation to be complete. It has to do with perceptions, and this is where the cultural background takes on its full meaning. Relationships between the English and woodland throughout history have been marked by ambivalence, as has been shown. The 21st century interpretation of woodlands is decidedly positive. According to present fashion/political discourse, nothing bad can come out of the woods; they have the potential to solve virtually every problem, individual or collective, related to the economy, the landscape, not to forget health, education and social relationships.

The class-related conflicts of the past over the uses of woodlands have been smoothed out. The dominant imagery related to woodlands is that of a consensual landscape, where Robin Hood is a cartoon character gently rebuking the king with a view to improving social harmony, not a rebel threatening to overthrow the political order for the benefit of an oppressed minority. The myth lies in conjuring up the past history of the woods as supposed spaces of social harmony from which inspiration can be drawn to plant new woodlands. This historical reading leads to a paradoxical interpretation of past exclusion as basis and ferment for present inclusion. This vision is far from being shared by everybody. The memory of the love-hate relationship between humans and woodlands since the beginning of humankind, and in specific forms throughout the history of England, remains vivid and takes special nuances in today's society.

Two major aspects which play against the full implementation of the community forestry agenda will be addressed here: first, the part played by fears associated with woodland, and secondly the low priority that woods occupy in many people's lives.

11.3.1 Love-Hate Relationship

The reasons why people do not go into the woods or get involved in charities dealing with woodlands are not easy to elicit, probably because it is difficult to strike a discordant note in the concert of approval of their beauty and multiple uses. With time and an atmosphere of trust, people begin to explore their real feelings, which may well include reservations and fears.

People who do not get involved in community forestry initiatives often have assumptions about the kind of people who do. It is assumed – not entirely wrongly – that people strongly involved belong to the category of 'do-gooders', that is people belonging to the higher, educated social classes, who get involved in all the local projects [Cooke 2000: 47]. Some organisations, if only because of their homogeneity, can make outsiders feel unwelcome. Groups develop their own identity, and it is striking to see how they seem to attract like magnets more of the same: executive men in their 50s, retired women, etc., which can be quite daunting to those who do not share this profile. Conversely, the feeling that some projects are aimed at disadvantaged groups can also act as deterrents for the others, thus marginalising the participatory process as well as the minorities [Warburton 1997: 14].

A major cause for not visiting woodlands is fear in relation to safety and security, which has two facets to it. Woodland, like any other natural environment, has inherent dangers: those most exposed are forestry workers, with a rate of accidents standing at 1% [Forestry Commission 2004–2005 and 2005–2006: 41]. As far as visitors are concerned, accidents are quite rare: according to statistics concerning 2002–2003, the total number of visitors to woodland amounted to 252 million, out of whom 40% in the Forestry Commission estate [Forestry Commission Economics 2004: 37]. Accidents to members of the public on the Forestry Commission estate for that period were 106 [O'Brien and Tabbush 2005: 25]. The *National Tree Fatality Database* puts the average annual number of tree-related deaths between 1998 and 2003 at six, that is one in ten million averaged over the national population [Forbes-Laird 2003]. In order words, there is very little connection between actual and perceived risk. However, the 'risk-adverse' British society presented in the report *Risk, Responsibility, Regulation. Whose risk is it anyway ?*, published by the Better Regulation Commission in October 2006, is not necessarily willing to take these minimal risks. In many people's mentalities, safety can and should be guaranteed through norms, insurance schemes or court action.

Accidents, when they do occur, get wide media coverage and bring to the surface the ancestral fears of the wild. In England in the summer of 2002, a man died of rabies after being bitten by a bat. Even though all people involved with bats are vaccinated against rabies, it is extremely difficult to eliminate this kind of danger for the public at large, as bats can easily be carried on air currents from Europe or even America. Around the same time, a woman was killed by the fall of a dead branch as she was walking through a wood. Some landowners or land managers are concerned about the risk involved though the presence of veteran trees and urban trees. Even though they present biodiversity value, they are at times felled to reduce the risk of claims for injury or damage, thus of liability [Barkham 2007]. A lot of landowners opening their woodlands to the public are worried that the risks lie at their door, hence the importance of the removal of liability to the landowner or manager in the dedication scheme mentioned before.

The Visitor Safety in the Countryside Group (VSCG) was formed in 1997, made up of various institutions and charitable organisations, to promote best practice in visitor safety. The guidelines published by the group focus on awareness, partnership, responsibility and risk control, that is to say they attempt not to reduce people's sense of freedom, but to warn visitors, so that they can exercise responsibility for themselves. Similarly, plans are under way through the Forestry Commission's 'Visitor Safety Project' to limit risks in the mountain bike trails of the Commission, which have seen the number of casualties rise in proportion with the rise in the number of participants. Here, too, the challenge is to cut the risk of accident through safe infrastructures and information to the users, while keeping the sense of fun and adventure.

More than safety due to natural risks, security in relation to other people, the second cause of fear, is a key preoccupation. This is a topic which comes up quickly in conversations with interviewees, women and parents of young children in particular, and yet it has taken a long time to reach the social forestry research agenda. Apart from a few ground-breaking studies on the relationships between open spaces and fear of attack [Millward and Mostyn 1989; Burgess 1995], most studies on the topic have been made since 2002. Let us first review the facts. The *British Crime Survey* 1991 established that the incidence of crime in open spaces was far less than in other environments. Moreover, all assaults reported were against men, while half of the murdered women were killed in their homes. According to the latest *British Crime Survey*, based on the interviews of 48,000 people in England and Wales, those most at risk from violent crime are young men between 16 and 24, with 12.6% of those interviewed having experienced violent crime of some form in the year prior to the interview. For women the risk was much lower: 7% for women between 16 and 24, 2.8% for women aged 25–34 and down to 0.4% for women aged 75 and over [Walker et al. 2006: 74]. Perceptions vary markedly from these data. According to various surveys, those who feel more at risk in public places, the woods among them, are women, children and ethnic minority members, while men mostly worry about getting lost [Burgess 1995: 10; ODPM 2002: 14]. Figures are not sufficient to reassure, and it is essential to understand where perceptions, those of women in particular but not exclusively, stem from in order to build self-confidence.

In interviews, the first element which constitutes a barrier to the enjoyment of woodlands is the perception of lack of social control which is conveyed visually in public places through a poorly maintained environment (litter, dumped cars, damaged benches, etc.). This anxiety is readily converted into a fear of crime. In one of the woods which is part of the CCI in the Mersey Forest, 57 cars were dumped in the wood over 1 year and rubbish is regularly thrown into the woods and can be set on fire if it is not removed.

Fig. 11.5 A 'Vandal-proof' bench, also devised to evoke the area's past industrial activity. Freeman's Copse, Ellesmere Port (Photo Sylvie Nail)

Newly planted trees can be a favourite target: although the emphasis put on vandalism as a cause of tree mortality is exaggerated, as was shown above, there is no doubt that in some areas, the incidence of vandalism can be quite high. This may be linked with the present image of trees in society: emblems of care and community spirit can easily turn into targets for antisocial behaviour. Tree mortality seems to be highest in areas of low socio-economic status. As a consequence, rangers try to be as discrete as possible when they plant new trees, so as not to attract attention to them.

Fig. 11.6 Newly-planted trees may act as scapegoats to vent frustration and anger (Photo courtesy of the Mersey Forest)

Another cause of fear is that, although men are more frequently victims of attack than women, the likelihood of sexual attack is greater for women. Jacquelin Burgess's interviewees, and our own, testify to how encounters with 'flashers' (men who expose themselves), even though they may not involve any risk of physical attack, undermine women's confidence to use the woods alone and constrain them to more open spaces, possibly where wardens are present:

> I don't walk around these areas because it is not safe for a woman to walk on her own in isolated places anymore, but it is alright for groups to walk around them [Simson and Ryan 2003: 35].

Decline in wardening and/or policing has not helped to alleviate the fears. The case of Liverpool City Council Calderstones Park is typical of many areas: not only has the number of rangers been cut drastically over the last 40 years, but their functions have changed, with the patrolling part of their functions being contracted out to a private company. On top of the fact that rangers and charitable organisations' members cannot be expected to fulfil the same functions as the police, the police presence itself has changed: in Speke (Merseyside), out of the 24 police officers who used to be present, 12 have been moved to Liverpool and have been replaced by 24 security officers. In terms of presence, this is a gain, but these security officers have more limited powers than police officers, and cannot arrest someone guilty of vandalism for instance.

Another crucial element in public perception of risk is the media coverage of incidents. As the *British Crime Survey* found out,

> Readers of national 'tabloids' were around twice as likely as those who read national 'broadsheets' to think the national crime rate has increased 'a lot' in the previous two years (39% and 19% respectively) [Walker et al. 2006: 35].

Since the sensationalising of very rare violent crime and the coverage of sex crimes have increased since the 1980s, the impression created in readers is that things have actually worsened. In response to these fears, strategies have been implemented in order to ensure a safe environment which, for 52% of the respondents to the *Public Opinion of Forestry*, is an essential criterion for the choice of a woodland to visit [Forestry Commission 2005a: 51]. The Woodland Trust has published guides to help managers involved in creating or managing woodlands tackle 'damage and misuse' and 'litter and fly-tipping'.

The Forestry Commission's Social Forestry Unit dedicated a seminar to the topic of accessibility and crime and safety issues in 2004, a sign in itself of the institution's growing awareness of the importance of not dealing only with technical aspects of participation but with the underlying perceptions and hindrances to community-building through a participatory approach. The agenda of the seminar included practical issues like managing antisocial behaviour and ensuring good design, but also dealing with people's perceptions in relation with actual risk and reassuring landowners in relation to their concerns on liability versus public access [O'Brien and Tabbush 2005: 5].

The first strategy consists in removing the elements which create an impression of disorder, in order to avoid a spiral of decline and improve the public's perception of the place. Part of the rangers', wardens' or group leaders' task consists in incessantly waging war on fly-tipping and on all forms of litter which spell disorder. Part of the routine of the rangers interviewed, whether in a local nature reserve in Cheshire or in various sites of the Mersey Forest, consists in spending their Monday morning collecting the rubbish strewn during the weekend, replanting uprooted trees, while several of the 'Friends of' groups meet once a month to clean up their woods. Some of the groups interviewed even originate from the members' despair faced with fly-tipping. Rapid intervention is essential to erase marks of unsociable behaviour: replacing burnt furniture, collecting rubbish, disposing of dumped cars, among others. Local authorities or land managers also devise systems to prevent unsociable behaviour and make woodlands more user-friendly: putting chains across paths to prevent cars from entering, putting car parks and toilets at the disposal of visitors. The battle is permanent and sometimes lost, as night-time in some areas is the realm of gangs with a lot of imagination.

Attention to design can also help to alleviate fears in relation to possible assailants, for instance through paying attention to the height of trees, to paths being clear of overgrowing vegetation and having clear sight lines, to good lighting and clear waymarking, to reducing the number of places where people can hide, possibly installing a CCTV system. However, as CABE Space (part of the Commission for Architecture and the Built Environment) pointed out,

> a space denuded of all risk will be a soulless place indeed as this would reduce opportunities for young people to gain an understanding of limits and boundaries, which they need to learn as part of their growth and development. This takes us back to the reduction in scale of negative perception associated with woodlands and natural spaces, from the landscape itself, right down to the micro-level and the individual components that are essential in making a good space [Fox 2005: 20].

More than anything else, however, the presence of a ranger or warden is what makes people feel secure. In the case of ethnic minority visitors, the presence of a non-white warden may counter the impression that the woods are white-dominated. Making sure that the media show the positive and not only the negative sides of the woodlands and reduce the incidence of crime to its real level would certainly improve perceptions, but this may be asking too much. In a speech entitled 'Common sense culture, not compensation culture', delivered at the Institute of Public Policy Research on 26 May 2005, former Prime Minister Tony Blair declared

> we need to involve the media in a better dialogue about risk. To that end, I have asked John Hutton [the Secretary of State for Work and Pensions] to invite newspaper and broadcast editors to discuss with the Chief Medical Officer and the Government's Chief Scientist the best and most appropriate forum for ensuring that risk is communicated effectively so that the maximum information can be put into the public domain with the minimum of unnecessary alarm [Blair 2005].

Undoubtedly, beyond the reality of unsociable behaviour in the woods, the feeling of insecurity which is often mentioned has to do with something else. First, it is linked with the alienation of urban dwellers from the world of the woods. Then it is related to the social and symbolic meanings of forests inherited from the past: those associated with landscapes of power, power covering issues of race, gender and wealth, and those associated with landscapes of exclusion in the two senses of the term: landscapes that exclude those who do not correspond to the definition above, and landscapes that harbour those who are excluded from society (outlaws and criminals).

11.3.2 Woodlands as a Minority Interest

The Progress Report of the *England Forestry Strategy*, published in 2006, lists an impressive array of programmes, initiatives, projects and reports. The energy, funds, manpower and imagination inverted over the past 10 years into making woodlands more amenable to the wide public in England are impressive by all accounts. These efforts are all the more admirable as they go against the flow: contrary to what one might believe when reading the innumerable studies and reports devoted to the issue, woodlands have little or no relevance to many people's lives and they represent a minority interest in the English population, which is another obstacle to the social agenda of forestry.

It is of course impossible to make a typology of people to whom woodlands mean nothing, first because individual tastes defy categorisation, and secondly because not going into the woods may be linked to a series of internal (likes and dislikes, knowledge, attitudes, time available, etc.) and external (controversy about a woodland, etc.) factors. The attempt to classify people who are not attracted to the woods must remain an attempt, but a worthwhile one to try and address questions related to motivation.

Children are probably among the keenest visitors of the woodlands, possibly because they are less touched by prejudice and more attracted by adventure. Within the group of children, however, a proportion of them is reluctant to get into the woods for different reasons [Bixler et al. 1994; Bixler and Floyd 1997]. Urban children, for instance, tend 'to be more fearful than rural children who possess greater ease in pastoral settings' [Rhode and Kendle 1994: 33]. This is related to what was said before, and in particular to their parents' fears being passed on to the children: according to a MORI telephone survey of 1,008 women in London, 67% of them never allowed their young children to play in local parks or open spaces unsupervised [O'Brien and Tabbush 2005: 11]. Even within rural children, not all of them relate positively to the woods, as the observation of schoolchildren revealed: in a rural class of 10-year old children, only a minority answered positively to the question 'do you like walking in the woods?'. As a study of children and woodlands found out, 'some children (in 3 out of the 6 schools) said they found the countryside "boring" and "too quiet"' [Bell et al. 2003: 93]. This is related to age, too: with adolescence, boys and girls lose interest in the countryside and woodlands, especially the 15–17 year-old group. Furthermore, to repeat a point made above, boys from 16 to 24 are the most exposed to crime, which is probably a deterrent, unless they want to indulge in activities away from the prying eyes of their parents or of the police.

Woods are not 'cool', so attracting children and teenagers to woodlands has been the object of a lot of imaginative strategies, some of which have already been mentioned in relation to schoolchildren. In order to allow teenagers to enjoy the woodlands, Community Forest teams take them to other sites, away from peer pressure, for instance with the Bluebell Express. It has been suggested that one of the ways in which the use of natural areas could be increased is by publicising them more. A lot of effort has gone into advertising woodlands and diversifying the range of activities offered in them to include a sense of adventure and excitement, like the 'Go Ape!' adventure in 11 English forests, with rope bridges and 'Tarzan swings' 12 metres above ground level.

Fig. 11.7 Rediscovering the woods and a sense of adventure high up in the trees (Photo Sylvie Nail)

The vocabulary used on community woodland websites is not the one usually associated with the woods, but it is in-keeping with the need to sound attractive to new audiences, teenagers in particular: woods are 'exciting', 'fun', 'cool', 'magic' with a lot of 'secrets' and 'legends' to discover through 'scavenger', or 'treasure hunts'. Involving celebrities has become classic since Richard Branson climbed to the top of the tallest tree in London to advertise the Forest of London project [Johnston 1997: 120].

Rather than information meetings, which do not allow for real feedback from the public and may only attract the people already interested in the first place, effort has been put into setting up public events, which combine the advantages of being fun and providing a setting for informing people. A recent addition is that of the Forestry Commission's Forest Tour, created in 2000 to bring more and different people into woodlands, and to make woodlands part of people's lives, by changing the conservative image of the woods and their association with mature visitors. It involves classical music concerts, but more than anything else pop bands playing in various of the Commission's sites. In 2006, the Tour attracted over 100,000 people in seven venues: Bedgebury, Cannock Chase, Dalby, Delamere, Sherwood, Thetford and Westonbirt Arboretum. The repercussions of these events on woodland visiting and woodland perceptions are difficult to establish, but one can suppose that the same principle applies to concerts as to crafts workshops and teddy bears' picnics, and that these 'passive' visitors may (re)connect with woodlands through pleasant activities and later become 'casual visitors' and maybe finally 'forest specific visitors' [Roberts et al. 2000: 8–9].

Vic Cooke has found out that an obstacle to people getting into woodlands or taking part in community programmes is related to socio-economic difficulties. Clearly, their priorities lie elsewhere and they fail to see the connection between woodland and an improvement to their quality of life. So even though they may find trees pleasant, they do not see them as a priority. An officer in one Community Forest confessed to the author that he sometimes had doubts as to the scope of community forestry, especially in occasions like the following. Going round a neighbourhood in Liverpool in search of people's views on what they would like to see on a piece of school land which had been sold close to their homes, the answer he got was 'a police-station', not 'an urban woodland'. The same gap between local people's expectations and the woodland agenda was found in the National Forest, where people preferred the prospect of a new supermarket to that of a new forest [Kitchen et al. 2006].

For a lot of people living in difficult economic and social circumstances, there is very little connection between landscape regeneration and social regeneration, and maybe it is not so surprising if, as the Forestry Commission notes, 'disadvantaged communities and some social groups are under-engaged in both shaping and using their local environment' [Forestry Commission 2005b: 3]. Still, the order of the day is urban woodlands, and funds are available there.

Finally, although they may represent a significant proportion of urban populations, ethnic minorities are also characterised by low attendance in woodlands, both as visitors and as participants in woodland projects. Part of what has just been said on deprived communities applies to some ethnic minorities even though it would be daring to classify ethnic minorities in one homogeneous group. On top of suffering from deprivation and social exclusion which makes their involvement in decision-making difficult, they may feel disengaged from the British culture, its institutions and its social issues, all the more so in the cases where language constitutes a barrier to communication.

Several methodologies have been proposed specifically aimed at involving members of ethnic minorities into the social forestry agenda. These include involving community organisations representing them, making sure that the relevant material is translated, and that there are translators in public meetings and for interviews and written questionnaires, meeting them in the places where they normally go, and training staff in cultural awareness. Nationwide organisations such as the Countryside Agency, the National Trust, the Black Environment Network, Urban Wildlife Trusts and Groundwork Trusts have attempted to involve ethnic minority communities in environmental matters.

Ethnic minority interest in environmental issues appears to develop when connections are made with non-environmental interests such as culture, food and access to green spaces [Mackenzie and Paget 1999]. Indeed, the importance of the public's cultural background in relation to landscape preferences has been shown in a study of three community preferences in Toronto [Fraser and Kenney 2000]. The differences they encountered between British, Chinese and Mediterranean preferences in terms of trees have direct bearing on participation: the Chinese community, for instance, may not willingly participate in tree-planting activities because this is not their typical idea of a beautiful landscape. On the other hand, they might want to participate in community gardens to grow vegetables or orchards to grow fruit. Gardens may be more appropriate in some under-privileged areas, because on top of sociability spaces they produce food. Awareness of the community's perception is essential before trying to involve its members, and the methodology elaborated by Jacquelin Burgess et al., detailed in a previous chapter, could usefully be implemented to elicit these at times unconscious traits.

11.3.3 The Wider Picture: Community Development, Empowerment and Social Forestry

Journalist Duncan Campbell commented in 1987 that society could 'look forward with confidence to the creation of more well-designed forests' [Campbell 1987: 36]. This certainly has happened, but whether their relevance to all social groups has been made tangible, that is, whether forestry can contribute to the objectives of the *National Strategy for Neighbourhood Renewal* (1998), is another matter. If woodlands cannot provide means to improve all communities' quality of life, and not only the aesthetic setting (however valuable this might be in itself), then there is considerable room for scepticism about the assertion according to which 'nature can apparently provide a setting for the establishment and maintenance of both close ties and relations with the wider community' [Rhode and Kendle 1994: 140].

Failing these connections, the social forestry agenda may be a means of disguising other issues, such as unemployment, housing or insecurity, instead of dealing with them. A 'sense of pride and ownership in the local community' [Cooke 2000: 46] can hardly be achieved by planting trees alone, or even by taking part in community action around forestry. There are obvious advantages to involving the local population in the designing and maintenance and their environment. Vandalism is much less frequent when people actively take part, if only because it is not the people who design/maintain a place who vandalise it, as numerous examples show. Part of the reason why public participation in Community Forests is implemented right from the start of projects is precisely in order to create a feeling of ownership, and it is striking to see the higher incidence of vandalism in places where design and implementation has been contracted out to private enterprises so as to save time.

Taking care of nature cannot go without taking care of people, which is the sense of Mike King's provocative proposal to

> stop planting trees and devote the limited resources we have at our disposal to the development of Social Capital. Social capital can best be described as 'the capacity of the community to act together to improve their quality of life' [King 2000: 65].

The basis of this capacity is trust. Woodlands, like public parks, public swimming-pools or public libraries, rely on the belief that social bonds exist that permit people to use them with confidence, and confidence is precisely what seems to be lacking in contemporary cities [Worpole 1997: 3]. This may explain the attraction of such private places as commercial centres (the new 'neighbourhoods'?) or theme parks, because they give the appearance of being public while offering guarantees of private security. Restoring trust entails re-engaging the public in democratic processes, through transparent decision-making and through accountability procedures.

Participation has to be an end at least as much as a means to attain the goal of participatory democracy. This implies taking into account a triangle made up of the knowledges people possess, scientific knowledge and policies and may be a way of countering the current scepticism as to what people can do in front of the global environmental challenges [Cooke 2000: 46]. There is a marked difference in motivation, within the Community Contracting Initiative groups, between those 'cause groups' who mobilised themselves in response to the threat of development locally, and those who were set up intentionally as part of the CCI: the latter type of group struggles because they feel they are doing it because they have been asked to do it, not because it is *their* cause.

Similarly, discussions around woodlands should not replace, but rather foster, debates on violence, drug abuse or inequalities. This process requires time, for group dynamics to take its full course, lead to decisions that have not been predetermined and permit transfer of skills. Measures of success include concrete results, in terms of products and policies, but also of processes, like interests, relationships or responsibilities. Implementation of the participatory process around urban/community forestry issues may foster the habit of participation around other issues relevant to people. A sign that prospects may be improving for the participatory democracy lies in the legislative framework, which is stronger today than it was a few years ago. In an assessment of the *Humans Rights Act 1998* published in 2006, the authors note that the Act has had a positive effect on the relationship between the citizen and the State,

> by providing a framework for policy formulation which leads to better outcomes, and ensuring that the needs of all members of the UK's increasingly diverse population are appropriately considered both by those formulating policy and by those putting it into effect. The Act therefore directly contributes to greater personalisation and better public services [Department for Constitutional Affairs 2006: 35].

In 2005, two other important tools for public participation came into force. First, the *Freedom of Information Act 2000* was implemented, making information more widely available from Departments of State and other institutions. Secondly, the *Aarhus Convention* was ratified by the UK. These tools should contribute to improving scrutiny and trust and making participation more effective. Already, in spite of the fact that funding has become more complex and that match-funding doubles the amount of paperwork without doubling the outputs, officers in Community Forests claim that the private sector is definitely more involved than it was, that partnerships function and that community participation is a real tool for empowerment. This is not to say that effective community participation is a straightforward process, far from it: reconciling stakeholders' agenda, scientific knowledge with popular knowledges and values, with the biases and assumptions of each of these, is far from obvious. Moreover, public involvement cannot follow recipe-book approaches, it has to be permanently reinvented according to time, place and issue. Yet, it is at that cost that decisions concerning woodlands, and possibly much wider issues, can truly be legitimised, and that the participatory approach can enter the practice of as wide a community as possible. On the other hand, if participation remains at a low level and does not allow a voice in the decision-making process, the odds are that the expectations raised and left unfulfilled will result in higher scepticism, more frustration and a greater feeling of exclusion: 'the cumulative effect of ill-fated participation is a demoralised citizenry' [Bradley 1995: 70].

References

Alison, H. 2000. Learning the lessons from Woods on your doorstep – where next? In *Community Forestry – A Change for the Better. Conference Proceedings, 7–8 December 1999, The Guildhall, London*, Forestry Commission, n.p. Forestry Commission, Countryside Agency.

Anon. 1996. Tesco planning targets 'neighbourhood' ruling. *Chemist and Druggist Niewsweekly* 6 July. http://www.dotpharmacy.co.uk/jul6.html. Accessed 6 August 2007.

Appelstrand, M. 2002. Participation and societal values: the challenge for lawmakers and policy practitioners. *Forest Policy and Economics* 4: 281–290.

Arnstein, S. 1969. A ladder of citizen participation. *Journal of American Institute Planners* 35: 216–224.

Audit Commission. 1999. *Listen Up! Effective Community Consultation*. London: The Audit Commission.

Barkham, P. 2007. Chainsaw massacre. *The Guardian* 3 May.

Barton, H. 2000. Conflicting visions of neighbourhood. In *Sustainable Communities – The Potential for Eco-Neighbourhoods*, ed. H. Barton, 3–18. London: Earthscan.

Bell, S., C.W. Thompson and P. Travlou. 2003. Contested views of freedom and control: children, teenagers and urban fringe woodlands in Central Scotland. *Urban Forestry and Urban Greening* 2: 87–100.

Bixler, R. and M. Floyd. 1997. Nature is scary, disgusting and uncomfortable. *Environment and Behavior* 29: 443–456.

Bixler, R., C. Carlisle, W. Hammitt and M. Floyd. 1994. Observed fears and discomforts among urban students on school field trips to wildland areas. *Journal of Environmental Education* 26: 24–35.

Blackman, D. and R. Thackray. n.d. *The Green Infrastructure of Sustainable Communities*. Community Forests Partnership. http://www.communityforest.org.uk/resources/ECF_GI_Report.pdf. Accessed 6 August 2007.

Blair, T. 2001. Improving your local environment. Speech at Fairfield Halls, Croydon, April. http://www.number-10.gov.uk/output/page1588.asp. Accessed 6 August 2007.

Blair, T. 2005. Common sense culture not compensation culture. Speech delivered at the Institute of Public Policy Research, 26 May. http://www.number-10.gov.uk/output/Page7562.asp. Last accessed 6 May 2007.

Bradley, G. (ed.). 1995. *Urban Forest Landscapes: Integrating Multidisciplinary Perspectives*. Seattle: University of Washington Press.

Broadhurst, R. and P. Harrop. 1999. Forest tourism. Putting policy into practice in the Forestry Commission. In *Forest Tourism and Recreation: Case Studies n Environmental Management*, eds. X. Font and J. Tribe, 183–200. Wallingford: CABI.

Burgess, J. 1995. *Growing in Confidence. Understanding People's Perceptions of Urban Fringe Woodlands*. Technical Report CCP457. Cheltenham: Countryside Commission.

Burgess, J. 2000. Situating knowledges, sharing values and reaching collective decisions: the cultural turn in environmental decision making. In *Cultural Turns, Geographical Turns: Perspectives on Cultural Geography*, eds. I. Cook, D. Crouch., S. Naylor and J. Ryan. Harlow: Prentice Hall.

Campbell, D. 1987. Landscape design in forestry. *Landscape Design* 166: 31–36.

Central Liverpool Primary Care Trust. 2004. *The REACT Project. A Celebration of Achievements to Date*. Liverpool: Central Liverpool Primary Care Trust.

Community Forest Partnership. 2005. Community *Forests Monitoring Report 2004–2005. Final Report*. Bury St Edmunds: Project Partners Research.

Cooke, V. 2000. Social issues in community forestry. In *Community Forestry – A Change for the Better. Conference Proceedings, 7–8 December 1999, The Guildhall, London*, Forestry Commission, n.p. Forestry Commission, Countryside Agency.

Curry, N. 2000. Community participation in outdoor recreation and the development of millenium greens in England. *Leisure Studies* 19 (1): 17–35.

Department for Constitutional Affairs. 2006. *Review of the Implementation of the Human Rights Act*. London: DCA.

DETR (Department of the Environment, Transport and the Regions). 1998a. *Modern Local Government: In Touch with the People*. London: HMSO.

DETR (Department of the Environment, Transport and the Regions). 1998b. *Guidance on Enhancing Public Participation in Local Government*. London: DETR.

Etzioni, A. 2000. Creating good communities and good societies. *Contemporary Sociology* 29: 188–195.

Forbes-Laird, J. 2003. National tree fatality database. *Arboricultural Newsletter*, No 123, December 2003.

Forestry Commission. 1996. *Involving Communities in Forestry... Through Community Participation*. Forestry Practice Guide 10. Edinburgh: The Forestry Authority and the Forestry Commission.

Forestry Commission. 1998. *England Forestry Strategy. A New Focus for England's Woodlands*. Cambridge: Forestry Commission.

Forestry Commission. 2005a. *GB Public Opinion of Forestry Statistics 2005*. Edinburgh: Economics and Statistics Unit, Forestry Commission.

Forestry Commission. 2005b. *England Forestry Strategy Review. Topic Paper: Community Forestry*. http://www.defra.gov.uk/corporate/consult/forestry-strategy/community.pdf. Accessed 12 April 2007.

Forestry Commission. 2005 and 2006. *Corporate Plan for England and Great Britain Activities. 2004/2005 and 2005–2006*. http://www.forestresearch.gov.uk/pdf/200406FCGBEngCorpPlan.pdf/$FILE/200406FCGBEngCorpPlan.pdf. Accessed 5 May 2007.

Forestry Commission. 2007. *Public Opinion of Forestry 2007: England (preliminary report)*. Edinburgh: Economics and Statistics.

Forestry Commission Economics. 2004. *Forestry Statistics 2004*. Edinburgh: Forestry Commission Economics and Statistics.

Fox, Deborah. 2005. Woodlands, natural spaces and the media. In *Accessibility of Woodlands and Natural Places. Addressing Crime and Safety Issues*, ed. L. O'Brien and P. Tabbush, 19–21. Famham: Forest Research.

Fraser, E. and A. Kenney. 2000. Cultural background and landscape history as factors affecting perceptions of the urban forest. *Journal of Arboriculture* 26 (2): 106–113.

Glicken, J. 2000. Getting stakeholder participation 'right': a discussion of participatory processes and possible pitfalls. *Environmental Science and Policy* 3 (6): 305–310.

Gray, J. 1997. Do we really want more US decadence? *The Guardian* 21 July.

Great Western Community Forest. 2005. Forest and football: Shaw Forest Park, West Swindon. http://www.communityforest.org.uk/resources/forest_football.pdf. Accessed 7 February 2007.

Hislop, M. 2001. The social forestry programme of forest research. Past, present and future. In *Social Forestry: Questions and Issues*, ed. E. O'Brien, 11–13. Farnham: Forestry Commission.

Hislop, M., M. Twery and H. Vihemäki. 2004. *Involving People in Forestry. Introduction*. Edinburgh: Forestry Commission.

Janse, G. and C. Konijnendijk. 2007. Communication between science, policy and citizens in public participation in urban forestry—Experiences from the Neighbourwoods project. *Urban Forestry and Urban Greening* 6 (1): 23–40.

Johnston, M. 1997. The early development of urban forestry in Britain, Part I. *Arboricultural Journal* 21: 107–126.

King, M. 2000. Community involvement and community development – a review of lessons for the future. In *Community Forestry – A Change for the Better. Conference Proceedings, 7–8 December 1999, The Guildhall, London*, Forestry Commission, n.p. Forestry Commission, Countryside Agency.

Kitchen, L., T. Marsden and P. Milbourne. 2006. Community forests and regeneration in post-industrial landscapes. *Geoforum* 37: 831–843.

Konijnendijk, C. and A. Simson. 2001. *Partnership-Led Regeneration*. Papers of the 4th European Forum on Urban Forestry held in Durham 15–19 May 2001. http://www.vbv.be/efuf/_docs/EFUF4_Papers.PDF. Accessed 7 June 2007.

Konijnendijk, C., M. Gauthier and R. van Veenhuizen. 2004. Editorial. *Urban Agriculture Magazine*, special issue 'Trees and Cities: Growing Together' 13: 3–5.

Land Use Consultants/SQW Ltd. 2005. *Evaluation of the Community Forest Programme. Final Report*. Cheltenham: Countryside Agency.

LCG (Local Communities and Government). 2007. Case Study: Jeskyns Community Woodland., http://www.communities.gov.uk/index.asp?id = 1509469. Accessed 25 August 2007.

McPherson, E. and C. Johnson. 1988. A community forest planning progress: case study of citizen participation. *Landscape and Urban Planning* 15: 1–2.

Mackenzie, A. and G. Paget. 1999. *Maximising Ethnic Minority Involvement in Environmental Activities*. Blackburn: Groundwork.

Mersey Forest. 2001. *Mersey Forest Plan*. Birchwood: Mersey Forest.

Millward, A. and B. Mostyn. 1989. *People and Nature in Cities. The Social Aspects of Planning and Managing Natural Parks in Urban Areas*. Peterborough: Nature Conservancy Council.

Murdoch, J. and S. Abram. 1998. Defining the limits of community governance', *Journal of Rural Studies* 14 (1): 41–50.

O'Brien, E. (ed.). 2001. *Social Forestry: Questions and Issues*. Farnham: Forestry Commission.

O'Brien, L. and P. Tabbush (eds.). 2005. *Accessibility of Woodlands and Natural Places. Addressing Crime and Safety Issues*. Alice Holt: Forest Research.

Oakley, P. et al. 1991. *Projects with People. The Practice of Participation in Rural Development*. London: International Labour Office (via Intermediate Technology Publishing).

ODPM (Office of the Deputy Prime Minister). 2002. *Living Places: Cleaner, Safer, Greener*. London: ODPM.

Petts, J. and B. Leach. 2000. *Evaluating Methods for Public Participation: Literature Review*. R&D Technical Report: E135. Bristol: Environment Agency.

Prescott, J. 2002. Ministerial foreword. In *Living Places: Cleaner, Safer, Greener*, Office of the Deputy Prime Minister, n.p. London: ODPM.

Rhode, C. and A. Kendle. 1994. *Human Well-Being, Natural Landscapes and Wildlife in Urban Areas. A Review*. English Nature Science Report N°22. Peterborough: English Nature.

Roberts, D., D. Eiser, B. Crabtree and G. Broom. 2000. *Forests' Role in Tourism: Phase 1*. Aberdeen: Macaulay Institute.

Sheppard, S. and H. Harshaw. (eds.). 2000. Priorities for reconciling sustainability and aesthetics in forest landscape management. In *Forests and Landscapes. Linking Ecology, Sustainability and Aesthetics. IUFRO Research Series n°6*, 3–12. New York: CABI, in association with IUFRO.

Simson, A. and J. Ryan. 2003. *NeighbourWoods Case Study Report*. Design Task Force/Leeds Met University. http://www.sl.kvl.dk/euforic/docs/NeighbourWoods/D10d%20Leeds.pdf. Last accessed 17 April 2007.

Tree Council. 2007. *Head for the trees — Enjoy a Healthy 'Walk in the Woods' Throughout May 2007*. Tree Council New Release, March. http://www.treecouncil.org.uk/. Accessed 3 May 2007.

UNECE (United Nations Economic Commission for Europe). 1998. *Convention On Access To Information, Public Participation in Decision-Making And Access To Justice In Environmental Matters*. Aarhus: UNECE.

Walker, A., C. Kershaw and S. Nicholas. 2006. *Crime in England and Wales 2005–2006*. Home Office Statistical Bulletin 12/06. London: Home Office.

Warburton, D. 1997. *Participatory Action in the Countryside. A Literature Review*. Cheltenham: Countryside Commission.

Whitaker, J. 2000. Get involved – working effectively with local people. In *Community Forestry – A Change for the Better. Conference Proceedings, 7–8 December 1999, The Guildhall, London*, Forestry Commission, n.p. Forestry Commission, Countryside Agency.

Whitt, L. and J. Slack. 1994. Communities, Environments and cultural studies. *Cultural Studies*, 8 (1): 5–31.

Worpole, K. 1997. *The Richness of Cities: Working paper 2: Nothing to Fear? Trust and Respect in Urban Communities*. London: Comedia and Demos.

Chapter 12
Grafting the Past onto the Present: The Heritage of Woodlands in the 21st Century

To understand our past helps us to come to terms with the present and provides the foundations for the future. Our heritage plays an important educational role, but, even more importantly, a vital social role

[English Heritage 1992: 2].

Abstract The purpose of the final chapter is to explore the 'social nature' of ancient woodlands, in order to undo the bundle of meanings and values attached to the heritage of trees and woodlands, and to look for resonances between these perceptions and the conservation policies implemented over the last 20 years. The species of trees protected, the forms taken by protection and preservation programmes, as well as the discourse on these, even though they may appear as motivated solely by issues of biodiversity/landscape, are informed by a wealth of representations and identity markers. In return, conservation practices contribute to forming people's perceptions of woodlands in the 21st century.

No less than 60% of the people questioned by the Forestry Commission in 2005 felt that, over the past decades, the woodland area in Britain had decreased [Forestry Commission 2005: 3, 12], whereas in actual fact the extent covered by woodlands has steadily increased. Similarly, emotions run high when trees are at risk of destruction (e.g. the Newbury Bypass protests in 1996), while the strategies devised over the last 3 decades ensure that trees have probably never been as well protected as they are today. These reactions are very much linked to the role that it is felt humans should play in the defence of the environment, and so to ethics. The purpose of this final chapter is to explore this topic, that is, the 'social nature' of ancient woodlands, in order to undo the bundle of meanings and values (as defined in the introduction) attached to the heritage of trees and woodlands, and to look for resonances between these perceptions and the conservation policies implemented over the last 20 years.

12.1 'Trees of Time and Place'

The conservation of ancient trees and woodlands has recently given birth to copious research and numerous strategies, not least to protect their rich biodiversity, which can spread into new, neighbouring woodlands. Veteran trees, snags, old stumps, as well as dead and decaying wood provide ecological niches for a wealth of fauna and flora. But ancient woods have 'dual existences' [Muir 2005: 229] and, on top of their objective value as part of the natural heritage, they also possess subjective meanings, which may be shared by a whole group of people or be individual-related. Ancient woodlands provide a visual and imaginary link with the wildwood, a sort of time capsule unaffected by agricultural and woodland management practices or by urbanisation, which have all progressively caused habitat loss and degradation. As such, they constitute an important component of the national natural heritage. In this first part, we shall focus on these 'features of the mindscape' [Muir 2005: 229], mental images connected with identity, the persistence of sylvan myths and religions.

12.1.1 Bastions of Englishness

The appeal to past landscapes, and woodlands in particular, is all the more noticeable in times of rapid change, causing concern over the future, with trees conjuring up stability and continuity in such circumstances [Corvol 1997]. Ray Lankester, writing during the agonising years of the First Wold War, expressed this vividly:

> we shall... be blessed by future generations of men for having saved something of Britain's ancient nature, when all else, which is not city, will have become manure, shooting greens, and pleasure gardens [Lankester 1915].

Similarly, the 'Neophiliacs' [Booker 1969] of the Swinging Sixties, despite their craving for progress and throwaway furnishings, collected *objets trouvés*, old objects of any description given a new lease of life by being removed from their original context and turned into decorative items. Nevertheless, the 1960s were not conducive to woodland conservation, with the productivist ethos of the Forestry Commission in full swing. Old trees were considered as geriatric, ailing specimens waiting to be removed and for many, old hardwood woodlands smacked of the declining aristocracy, its flagging energy and outmoded field sports. Many ancient woodlands disappeared under development, defined by Section 55 of the *Town and Country Planning Act 1990* as 'building, engineering, mining or other operations in, on, over or under land, or the making of any material change in the use of any building or other land'. This was all the more tempting as a high proportion of them was to be found in the South-East of England, where land is at a premium and pressure is high. Alternatively, they were turned into conifer plantations. Things were soon to change.

In the revised edition of his famous book, *Trees and woodland in the British landscape*, published in 1990, Oliver Rackham marvelled at this rapid reversal. He noted that half of the books included in his 1990 bibliography had been published since 1976 and concluded in the following terms on how different the context was from that of the first edition, published in that year:

> The first edition of this book ended with a depressing prognosis. If the changes between 1950 and 1975 had continued, there would by the turn of the century be almost no ancient woodland left except in nature reserves. This will not happen. Little ground has been lost in the last fourteen years, and much ground thought lost has been regained. The vast, intractable problems of conserving woodland have passed into history, and a new set of issues has come to the fore [Rackham 2002: 197].

This change may be linked with the identity crisis England has faced over the last decades, with the demise of the Empire, the entry into Europe, as well as devolution to Scotland and Wales. If one adds to these the advent of globalisation, it is easy to understand the sense of loss and the need to redefine national identity which have fostered a mental context favourable to an idealisation of the past. Landscape being a major component of Englishness, it has attracted a lot of attention in this attempt.

> Like imperialism itself, landscape is an object of nostalgia in a postcolonial and postmodern era, reflecting a time when metropolitan cultures could imagine their destiny in an unbounded 'prospect' of endless appropriation and conquest [Mitchell 1994: 20].

The past which the contemporary imagination takes as a reference point in its nostalgic quest is significant in terms of associated values. In his celebration of the old oaks of Sherwood Forest in the 1790s, Major Rooke stressed among other features their historical associations, which included the ancient Britons and the Druids. The same ancient oaks became medieval icons in the decades following the publication of *Ivanhoe* in 1820 [Watkins 1998]. Incidentally, then as now, the commercial prospects of these iconic landscapes (Sherwood Forest, the Cotswolds, the Lake District or the Peak District), where consumption is displaced onto the landscape itself, were not lost on the owners. As early as 1844 at Clipstone (Nottinghamshire), the Duke of Portland placed statues of Robin Hood, his merry men, King Richard and Maid Marian, together with a new lodge with views of the venerable oaks, to cater for the numerous visitors [Watkins 1998: 110]. The Robin Hood Festival which has taken place in the Forest of Sherwood every summer since 1986, is but a recent variation on this theme. It combines the appeal for ancient woodland, rural recreation and consumption of the national heritage through 'arts', a 'medieval market' and 'medieval merriment' [Nottinghamshire Country Council 2007].

For David Lowenthal, it is the Pre-Raphaelites' influence which contributed significantly to associate Englishness with the medieval landscape [Lowenthal 1994]. If one associates the Middle Ages with the Norman Conquest, the reference to that period in the case of ancient woodlands may seem paradoxical. More than the political context, it is an imagined lifestyle and national virtues which are extolled through this association:

> the hierarchical certainty of the old England, the amalgam of faith, diligence, loyalty, independence and authority [Morris 1981: 146].

Fig. 12.1 Robin Hood takes on the Sheriff of Nottingham close to the Major Oak in Sherwood Forest during the Robin Hood Festival 2006 (Photo courtesy of the Nottingham County Council)

In other words, what has been described in the previous chapter as the community, idealised in its medieval attire. Words shape expectations and the expectations conveyed in the medieval associations relate to a moral order, or 'moral message', as Deborah Fox also found in her analysis of medieval landscapes in stories [Fox 2005: 19]. Thus, in the first years of their existence, the Community Forests literature constantly referred to the Middle Ages as point of reference and proof that the new programme was rooted in national history and national values. This appeal to the past supposedly warranted authenticity and continuity (a particularly well tried and tested process throughout history). On the website of the English Community Forests in the early 1990s, one could read that they drew their origin, not from Third World Community Forests, but from the medieval definition of a forest, namely a mosaic of woodland, cultivated fields and villages, with the implicit association of the close-knit medieval community. Interestingly, a similar trend appears in the 'current discussions of bioregionalism and sustainability', with the medieval city supposedly mirroring nature and being tied to the land and its cycles [Keil and Graham 1998: 194]. The author has found in several interviewees the same yearning for the local (whether local apples, local beef or local trees), with its familiar bearings, all the more so since various agricultural disasters have made the alienation from the local a corollary of environmental and agricultural hazards. Farmers' markets, a recent trend which is going from strength to strength, offer a reassuring, humane counterpoint to anonymous global exchanges.

The precise reference to the Middle Ages has disappeared from Community Forests literature since then, possibly as a consequence of the wide acceptance of the Community Forests programme which has rendered unnecessary a claim to continuity. In ancient woodlands, as in landscape in general, history reads like a palimpsest, embracing with the same benevolence (almost) all the layers constructive of national identity – the Celts, the Romans, the Saxons, the Danes and the Normans, whose combined virtues contributed to shaping the national character – and all the rural landscapes, from openfield to enclosures and from the Royal Forests to the Lake District.

Protecting these landscapes, loci *par excellence* of this heritage, means contributing to their stability, hence to that of the nation, which is the unique result of this palimpsest. But it is also an indirect statement against foreign landscapes. The geography of insularity and the history of hybridity can sometimes result in landscapes of xenophobia. Indeed, 'identities are contingent, constructed in relation to a social or a territorial other' [Darby 2000: 75]. Witness Gilpin's declaration at the end of the 18th century:

> perhaps of all species of landscape, there is none, which so universally captivates mankind, as forest-scenery: and our prepossession in favour of it appears in nothing more than in this; that the inhabitants of bleak countries, totally destitute of wood, are generally considered, from the natural feelings of mankind, as the objects of pity [Gilpin 1791: 269].

It echoes the belief in greenwood liberties mentioned before and displays a belief in the superiority of the national wooded landscape, just as Wordsworth's praises of the Lake District as 'the perfect metaphor for the English national character' smack of 'ecoxenophobia'[Hazucha 2002: 69]. When English Nature asserts that 'Britain has the largest and *best* array of ancient trees in north-west Europe so we have a special responsibility for their conservation' [English Nature 2000: 1. italics added], it is making a statement about both the superiority of the English landscape and the species of trees that are part of the national identity, which has consequences in policies, as we shall see. The conservation of local trees, or local tree species, is thus clearly an instrument of continuity, and many initiatives revolve around this, out of which are the following two examples.

'Trees of Time and Place', a millennium campaign to encourage people to gather local indigenous tree seeds or saplings to plant them locally, partakes of this desire, based more on emotional bonds than on genetic merit. Indeed, part of the ancient trees which remain standing may have been discarded because they were less than perfect for timber, hence their gene pool may not be the most appropriate one for replanting. An original approach to cultivating bonds is the project 'Fuelling a Revolution', funded by the Heritage Lottery Fund, which took place in South Yorkshire between 2000 and 2005. It had the double purpose of protecting and managing 35 heritage woodlands, but also, more unusually, of celebrating their role as the source of energy which permitted the development of the iron and steel industry, that is, of vindicating the part they took in the identity formation process of the region. The links between wooded landscapes and identity are so strong and their connotations so positive in England today that, through the use of woodlands, identity changes can even be negotiated. Such has been the case in the National Forest and in Consett (County Durham). In both cases, far from celebrating the local or regional industrial past as in South Yorkshire, it was estimated that environmental greening was an indispensable part of the process of changing the identity of the area, to adapt to the postindustrial era (and attract inward investment). In the words of a member of staff at the Department of Town and Country Planning at the university of Newcastle,

> in conjunction with the economic redevelopment programme, a more profound and meaningful regeneration of the community's sense of place and identity is to be brought about by the adoption of the environmental strategy and the extensive planting of trees [Hughes 1991: 2].

Planting trees to hide the scars and provide new local roots to the population – the idea was simple. The assumption was that the regeneration of the natural environment would contribute to 'the psychological stabilisation of the community's self-image and visitor impressions concurrently, while improving the general character and atmosphere of the area' [Hughes 1991: 5]. Yet, it proved less straightforward than it was meant to be: reintroducing trees from where they had disappeared is a far from neutral act and threatens the identity of the local populations, who are not deceived or necessarily delighted by the landscape change, and do not necessarily want their image 'stabilised' [Cloke et al. 1996a, b]. Moreover, different communities had different views: local farmers and residents did not wish to see the residents of nearby cities (Derby, Nottingham, Leicester, Birmingham or Sheffield) walk over their lands or change the identity of their village into a leisure park, while others were sceptical over the public benefits of the mostly private 'National' forest. Additionally, in the case of Consett, the population was not consulted, which may be explained by the fact that the project was one of the early regeneration ones and consultation was not yet the dogma it has come to be. However, the study notes the demand for 'action and involvement' from the community [Hughes 1991: 5].

These examples show the variety of responses, from a positive assessment of the past via landscape enhancement, through to makeover initiatives bordering on landscape manipulation, which can be elicited from the close association between wooded landscapes and local, regional or national identity. Acting as the stewards of heritage woodlands and ancient trees does not only entail interpreting the landscape for its consumers, but also making decisions when it comes to management practices.

12.1.2 Perpetuating Practices

Even though not all veteran trees are pollards, research and intervention have often hinged on these recently, probably because of the aesthetic appeal and of the dramatic presence in the landscape of their gnarled branches and imposing trunks. Veteran trees are part of those trees that English Nature calls 'working trees': the avenue of pollarded Spanish chestnuts at Croft Castle and the beech pollards in Selbourne Common (both in Herefordshire), the beech and oak pollards in Burnham Beeches (Berkshire), the oak and beech pollards in Naphill Common, Bradenham (Buckinghamshire), the willow pollards in Wicken Fen and Woodwalton Fen (both in Cambridgeshire), the oak pollards in Kingston Lacey (Dorset), Aspal Close, Mildenhall (Suffolk) and Ashstead Common (Surrey), not to mention the numerous species of pollards in Borrowdale (Cumbria), Lullingstone Park (Kent) and in Epping Forest (Essex). These are just a few examples of the heritage of managed trees. Many of them have been saved precisely because they stood in historic parklands and wood pastures, where coppicing and pollarding were part of woodland management [Rackham 2004]. Yet, as was explained in the second chapter, pollarding and coppicing disappeared because they were outdated in the rural economy, so what justification can there be in restoring them today?

One answer is that, for lack of management over many decades, they are vulnerable to crown collapse [Lonsdale 1999]. But the opposite is also true, and returning trees to coppice after decades of neglect may harm or kill them. Another argument is that they provide niches for some specific species. But these niches can to some extent be artificially reproduced: the Tree Council thus promotes, in sites of special biodiversity interest, the idea of inducing decay by injuring trees with chainsaws, inoculating heart rot-inducing fungi, using explosives or damaging tree bark so as to provoke habitats for various species. Beyond debatable utilitarian reasons, one could argue that the focus put on reintroducing the practices of pollarding and coppicing as forms of management of these ancient trees has a deep resonance in cultural terms. They spell continuity, not with primeval nature, but with culture in the form of the working practices of generations of forest workers who managed those trees previously. As much as providing a connection with landscapes of the past, these trees act as intermediaries between people past and present.

Fig. 12.2 An old pollarded hornbeam in two halves in Hatfield Park (Photo courtesy of Mark Johnston)

The case of Hatfield Forest (Essex), property of the National Trust, is telling in this respect. Until the 1970s, the National Trust, like the rest of society, did not value pollarded or coppiced trees, which means that in management terms, there was a break of 100–150 years in their maintenance. Yet, in the 1970s–1980s, the staff decided to 'get coppices back on cycle' – meaning cutting back the trees after 12–18 years of growth – and to plant new trees to train them as pollards [Garner 2004: 91, 94].

An appreciation of the biodiversity value of these trees and of their contribution to the historical landscape around the National Trust properties possibly lies behind this decision. However, the staff's comments during the interviews are highly revealing of something deeper. They evoke power, stability and responsibility. Indeed, a pollarded or a coppiced tree is the material representation of authority, of dominance over the landscape. This may explain why, on top of its irrelevance to modern economy, it was abandoned in the first half of the 20th century. Indeed, if we follow Keith Thomas [Thomas 1984] and see in the treatment of trees, horses and children a reflection of the current ideas on education, then this break in a practice which can be seen as a mutilation of the trees and a limitation of their natural potential, makes sense at a time of moral liberalisation. Similarly, their reintroduction at the end of the 1970s makes sense too, with the Victorian value of discipline coming back into favour at the time.

With the feeling of power conferred by shaping the ancient trees comes a strong sense of responsibility, too, another Victorian value. Technical expertise is indispensable to perpetuate the shape and the health of trees which have been standing there for centuries, developing a life of their own beyond the reach of a human's lifetime, but which can be damaged by the wrong move – or cut. Perpetuation and continuity are therefore a key motif in the staff's reactions: far from thinking that they are maiming the trees, they choose to see themselves as 'renewing a deeply patterned cycle, of echoing (…) the movements and practices of the past' [Garner 2004: 91], as perpetuating a form of human-environmental interaction and contributing to the stability that these trees stand for in a rapidly-changing world. As Andrew Garner expresses it, these trees 'shape human responses in patterned ways': just as pollarding and coppicing give the trees continued meaning into the present, the process of cutting them 'makes the past' [Garner 2004: 96–98].

For all these reasons, which have very little to do with niche markets or job creation, pollards, after decades of neglect, are now very much cared for by the National Trust and the Forestry Commission. At Burnham Beeches, re-pollarding of the oaks and beeches and starting new pollards was first undertaken in the mid-1990s, while at Epping Forest, crown restoration of the oak and beech pollards, pollarding of the hornbeams and the creation of new pollards started in 2002.

The planting of new pollards (the so-called 'maiden pollards') raises even more questions as to the underlying motives than the maintenance of the old ones. The arguments put forward underline the need to replace the old pollards when they eventually die, so as to hand down the same landscape to the future generation. The question of why it should be so, why the landscape given as an inheritance to the next generation should be yesterday's, rather than today's, landscape is not tackled, apart from reference to the habitat niches they represent. It looks as though preparing the veterans of the future was sufficient reason in itself, without the need being felt to connect these 'working trees' with today's agricultural practices.

Yet, the maintenance of the old pollards is based on the assumption that they prolong past history into the present. What is the rationale of creating a past for the future with anachronistic practices? The production of a landscape that David Lowenthal calls a 'vast museumised ruin' [Lowenthal 1994: 24] questions the choice of the landmarks that define the limits of past and present, and of the versions of the past one chooses to preserve for the future. These issues are clearly political, in the sense that they correspond to choices and hierarchies that in their turn reflect readings of history. Today's cherished pollards were yesterday's abandoned ones, with hardly more justification in one case than in the other, except for the social construction of reality which makes one version of the past more amenable to contemporary interpretation than another one. The Tree Council, aware of this ambiguity, adopts a flexible attitude towards pollarding:

> rather than automatically follow historic practices such as pollarding, which were developed for particular purposes, the specific nature conservation objectives should be considered before deciding on the height of pollarding, the frequency of re-pollarding, and the proportion of poles to be cut [Green et al. 1999].

Another controversial revival of an extinct practice throws additional light on attitudes to past working practices. It relates to the renewed use, by some organisations (the British Trust for Conservation Volunteers, for example), of horses instead of machines, to pull logs out of woodlands. In the Spring-Summer 2007 issue of *Tree News*, the magazine of the Tree Council, an article entitled 'Horsepower evolution' presents the case of a forester who works with horses.

> 'Most woodland owners don't realise the full potential of a horse.' Jon West strokes his beard, pulls on the stub or a hand-rolled cigarette and clicks his teeth as he flicks the long reins. 'That means we only work in the most inaccessible places.' Five yards ahead Cokey, a shaggy bay gelding, lumbers forward and begins to plod up the steep bank, his flank muscles rippling. 'And because we only ever work in difficult places, it's not surprising the figures don't look that good.' In reality, however, he feels horses can compete on equal footing with machinery and in some conditions can even be more efficient than a skidder when working on level ground close to the road. At the moment Jon, Cokey and his partner Tor, are patiently pulling logs out of a neglected wood in Mid-Wales. Like most good foresters, he grumbles quietly as he works. In this case the grouse is that the trees were felled before his team was called in: 'They've cut them all at once thinking there would be heavy machinery to haul them out from below,' he says. 'It would have been much better for the horses to work bit by bit' [Butler 2007: 11].

The picture conjured up in this passage is the epitome of nostalgia. It harks back to the pre-industrial rural imagery of an unsophisticated forester, complete with beard, hand-rolled cigarette and appropriate grumble, finding satisfaction in a challenging job ('inaccessible', 'difficult places') following a natural rhythm conducive to self-fulfillment ('patiently', 'bit by bit') rather than the rules of productivity ('figures'). Yet, in the harmonious relationship with nature presented here, today's world is just round the corner, and the forester is at pains to demonstrate that his work makes

Fig. 12.3 'Horsepower' to remove logs from woodland (Photo *Tree News*, Spring-Summer 2007: 11, with kind permission of Daniel Butler)

sense in a world ruled by productivity ('compete on an equal footing', 'can even be more efficient').

The same applies to many conservation or restoration projects. Thus, the rural development project based on the Wyre Forest in the West Midlands, the third largest area of contiguous ancient woodland in England with its 300 hectares, combines objectives related to biodiversity, tourism, education and economic outlets for the products of the Forest, demonstrating that biodiversity objectives are not separated from the rest of the woodland agenda. There is clearly an imperative need to demonstrate that resuscitating past practices and landscapes can make good economic sense.

The vision of landscape as repository of ancestral working practices is made explicit in the same article of *Tree News*. It refers to the forester as an immutable figure who, apart from the fluorescent jacket and the helmet, 'would have been instantly familiar to a Celtic peasant in the Dark Ages or even before that in the days of Roman rule' [Butler 2007: 13]. A similar connection can be found in relation to Community Forests:

> Seed Gathering on Seed Gathering Sunday brings the community together, building social contact and networks, as a significant event to the cultural life of that community as was once the Harvest Festival to previous generations [King 2000: 63].

The comparison of those 'rituals' past and present is again meant to give validity to the present one by comparing it with a very old tradition, but it neglects two differences, and essential ones at that. First, the Harvest Festival was part and parcel of the community's culture and professional rhythm, while Seed Gathering Sunday is

In the author's experience, this new type of funeral rite has suffered from a certain lack of professionalism, as regards information given to the family, tree planting and management of the site. This might be due to the novelty of 'green burials'. However, this alternative to traditional burials certainly corresponds to the philosophy of a growing number of people (there are now around 200 woodland burial sites in the UK) and to their symbolical relationships to the natural world. The juxtaposition of trees (symbols of life and eternity) and death is a vivid statement about lifecycles.

Woodlands symbolically are at the heart of the search for identity, not only collectively, but also individually. In folklore and fairy tales, the hero or heroin, willingly or by a whim of destiny, embarks on a journey leading to the discovery of self, through the trials of a long and solitary wandering through the forest. These journeys, among their characteristics, comprise the isolation of the characters in the forest, the absence of social hierarchies, and an unusual solidarity between humans and nature, whether animals or trees. One may wonder whether the modern walks in the woods may not be distant echoes of these heroes' and heroines' journeys of initiation, with the discovery of self as the prize after confrontation with the wildwood. This would throw a different light on the numerous projects and initiatives to encourage people to rediscover the woodlands.

In view of the all the layers of meanings of ancient trees and woodlands, how have public policies responded to this heritage and its multiple interpretations? The second part of this chapter will shed light on the official interpretation of woodland heritage and the analysis will show how policies embrace perceptions or shape them.

12.2 Heritage, a Modern Word

In parallel to urban foresters developing their own methods and tools, strategies have been devised over the last 3 decades to protect and conserve old trees and woodlands. Ancient trees and woodland conservation has given birth to a wealth of publications and strategies.

12.2.1 Context and Definitions

Without going back into the distant origin of the preservation/conservation movement at large, let us focus on the last of the four periods identified by Philip Lowe in the evolution of the conservation values and institutions, starting in the 1960s, and which he entitles 'the popular/political period' [Lowe 1983: 344]. The disappearance of many buildings, objects and landscapes felt to belong to the national heritage (among them, some 595 country houses demolished between 1945 and 1980) spurred the birth of numerous charitable organisations: the Georgian Group, the Victorian Society, the Ancient Monuments Society, the Garden History Society, the National Piers Society and the Railway Heritage Trust, among many others. The media played an important part in interpreting and advertising nature conservation issues for the wider public thanks to, for example, the BBC Natural History Unit or press advertising campaigns like that of the Royal Society for the Protection of Birds (RSPB) in 1958. Political discourse began to reflect these new preoccupations in the late 1970s, when heritage moved up the political agenda with Prime Minister Margaret Thatcher's appeal to these values in 1979, the mention that same year of the word 'heritage' in a speech made by the Queen, and the vote of the *National Heritage Act* in 1983 which created English Heritage.

At the same time as heritage and green issues were rising on the political agenda, the academic world contributed to defining new approaches to the environment, with the new disciplines of landscape history, environmental history and environmental arboriculture, as well as the development of applied ecology with the creation of the Nature Conservancy in 1949. The study of biodiversity developed and contributed to a better understanding of the life and needs of the woodland, focusing in part on a discussion of the respective values of native and exotic trees for the development of flora and fauna, a lot of it based on the work of T. Southwood [Southwood 1961], from the Department of Zoology of Imperial College, London. Thanks to the indefatigable work of some scientists, among whom Keith Kirby, Alan Mitchell, George Peterken and Oliver Rackham, ancient woods began to regain a status that had been lost in the Forestry Commission's productivist period, and to enter into their own as areas of intervention for public policies, not only for landowners.

The last decades of the 20th century thus witnessed the categorisation by science and the endorsement by the State of the heritage of trees: the concept of 'ancient woodlands' was born, negotiated, categorised and progressively integrated into public policies [Tsouvalis-Gerber 1996], becoming so internalised that no one today would deny either their existence or their importance. As the statement of policy concerning ancient woodlands puts it, 'our ancient and semi-natural woodlands are the jewels in the crown of English forestry, and protecting and enhancing them will be a high priority [DEFRA/Forestry Commission 2005: 17]. It was agreed that continuity rather than naturalness would constitute the dominant criterion to categorise 'ancient' woodlands, defined as the areas which have been continually wooded or under woodland management since at least 1600 A.D. To this criterion was then added the natural, or native, variable, resulting in a complex definition. Thus, within the category of ancient woodland, 'primary' (also termed 'native' or 'ancient semi-natural' woodlands) woodlands descend directly from the former natural forests and have either regenerated naturally or have grown up from coppice stools or pollards, they do not obviously originate from planting, while 'secondary woodlands' (or 'planted ancient woodland sites') are ancient woodlands with plantations of conifers or non-native broadleaved species [Spencer and Kirby 1992: 78; Land Use Consultants 2001: 4–5]. According to that definition, it was estimated in the early 1980s that 574,000 hectares of British woodland was ancient, that is about 28% of the total woodland area and 2.6% of the land surface of England [Spencer and Kirby 1992: 78, 84]. By the early 1990s, 7% of the area under ancient woodland in 1945 had been grubbed out, while a lot of it had been replanted, often with conifers: plantations occupied around 40% of the total area of surviving ancient woodland. Most of these had occurred between 1945 and 1985 and had been supported by grant-aid to private owners, in order to increase timber production [Spencer and Kirby 1992: 90].

To assist monitoring of ancient woodlands over 2 hectares, the *Ancient Woodland Inventory* project was elaborated by the Nature Conservancy Council in 1981 and continued by its successor English Nature in 1991, before being integrated into the National Digital Archive of Datasets in 2001–2002. With all the data gathered county by county and summarised in reports and databases, an assessment could be made, serving as a baseline against which further change could be measured and as a source of information for policy-making. Not only are ancient semi-natural woodlands rare[1] (the 22,000 ancient woodland sites identified in England by the *Inventory* represent only around a quarter of the total woodland area); they are also fragmented and distant from each other. Scientific knowledge greatly assisted the elaboration of policies, and the *Inventory* played an important part in supporting the *Broadleaves Policy* launched in 1985, mentioned in a previous chapter, and English Nature's *Position Statement on Environmentally Sustainable Forestry and Woodland Management*, issued in 1994. The characteristic features of ancient woodlands, in terms of the communities of animals, plants (including the ubiquitous bluebells) and fungi they support [Kirby 1996: 35], are important as part of the Natura 2000 agenda.

The terms 'old' and 'ancient' are often used interchangeably (the shades of meanings of 'veteran' have already been explained.), although there are some differences between them and disagreement on the definitions. 'Old' tends to focus on the chronological age of the tree, whereas 'ancient' refers to the stage of development of a tree (subdivided into 'early ancient' with the onset of crown retrenchment, 'late ancient' where annual rings become discontinuous and 'senescent' with the onset of terminal decline' [Fay 2002]). Paradoxical as it may seem at first sight, in relation to the image one has of woodland, and also in relation with other objectives of woodlands like carbon absorption, a last category has attracted attention from environmental arboriculturists and tree ecologists, that of dead wood. Due to the habitat it provides for some specific species (e.g. saproxylic, or dead wood, insects), dead wood is an essential part of woodland ecosystems, and it therefore requires management so as to enhance its sustainability. Experiments have been carried out at Windsor Great Park, an outstanding European site for dead wood fauna, to improve and increase the quantity of dead wood.

As in other areas of forestry and of heritage protection, external circumstances have accelerated a process engaged in the scientific arena.

---

[1] 'Semi-natural' woodland, a term introduced by Tansley in 1939, is slightly different from 'ancient' woodland in that it refers to stands that have not been planted, and therefore consist mostly of native trees and shrubs, as opposed to plantations, where the structure, age and species are mostly dependent on human influence. The *Ancient Woodland Inventory* makes a distinction between Ancient Semi Natural Woodland, Ancient Replanted Woodland, also known as Plantation on Ancient Woodland Sites (PAWs), and Ancient Woodland Sites which have been grubbed out.

12.2.2 Policies and Incentives

With regards to ancient woodlands, one of the main factors behind the change in perceptions and policies is certainly attributable to the shock caused by the great storms of 1987 and 1990. Millions of trees were blown down across Southern England, provoking the emotional background against which a new, coherent policy for ancient woodlands could take place. Another impulse to policy came from the *UN Convention on Biological Diversity*, which spurred the publication of the *UK Biodiversity Action Plan* (UKBAP), with particular stress laid on the broad-leaved, mixed and yew woodland habitat. The *England Forestry Strategy Review* estimates that the most important woodland for biodiversity is ancient woodland and woodland comprising native species. The objectives of *Sustainable Forestry: the UK Programme* revolve around the sustainable management of the existing woodland and its steady expansion for multiple benefits. Cases like that of Oxleas Wood, a site of special scientific interest (SSSI) in South-East London, provide an opportunity for ecologists to demonstrate the wealth of species present in an ancient woodland as well as their fragility. No replacement wood can hope to give shelter to the same species, contrary to the arguments put forward by the Department of Transport when a plan was made in the 1980s to build a dual carriageway which would have passed through Oxleas Wood [Goldsmith 1988]. Tree conservation is more effective for biodiversity than new plantings [Gale and Cordray 1991], and this has certainly contributed to protective measures for ancient woodlands to be enshrined in the *UK Forestry Standard:*

> The United Kingdom has no natural forest, but has about 650,000 hectares of semi-natural woodland of which 288,000 hectares are classed as ancient and semi-natural (1.2% of land area). (…) Semi-natural woodlands are especially significant for wildlife conservation because they support a high proportion of rare and threatened species. They are also important for landscape and cultural heritage. (…) Ancient semi-natural woodlands are especially valuable as some are remnants of the original post-glacial forest. Conservation of natural habitat is of prime importance [Forestry Commission 2004: 2].

Part of the radical change in Forestry Commission's policies has thus consisted in putting 'ancient woodland at the heart of forestry policy in England', to 'guide much of [the Forestry Commission's] development work in the coming years' [Hill-Tout 2005: 32], that is, in-keeping with the national mood, putting the past at the core of the future. The launch, in June 2005, by the Minister for Rural Affairs, Landscape and Biodiversity, of a new statement of policy entitled 'Keepers of Time', made the protection of ancient and native woodland a top priority of the Government's *England Biodiversity Strategy*. In a swing of the pendulum typical of the last 20 years in forestry, grants to protect ancient woodlands have replaced grants to put conifer plantations in them. In the name of ecological continuity, the creation of new native woodland, as well as natural regeneration, are encouraged, so as to enlarge ancient woodlands and create buffer zones to protect them from surrounding land uses (whether pesticide drift or cattle incursions). Moreover, by bringing the neglected ancient woodlands into management, their ecological viability is improved. Conservation purposes combine with aesthetic requirements in forestry as enunciated by Sylvia Crowe, which also helped to make the agenda of conservation progress.

Several measures have permitted to translate the value English society puts on ancient/native woodland into financial terms. Vice versa, grants and regulations have helped to convey this value to landowners and managers. Felling regulations constitute one way of protecting ancient trees and woodlands. New *Tree Preservation Orders* were voted in 1999 and came into force in June 2006 [DCLG 1999]. They apply to individual trees, groups of trees or whole woodlands, but their enforcement can prove difficult [Eden 2006], and ancient trees are still being cut, for leisure and housing for instance, as the report *Development Threats to Ancient Woodland* shows [Land Use Consultants 2001]. Another Government instrument which complements the previous one is *Planning Policy Statement 9*, devoted to *Biodiversity and Geological Conservation*, published in August 2005. It includes a section on the value and need to protect ancient woodland from development. Local authorities are urged not to grant permission for development that would result in the destruction or deterioration of ancient woodland or veteran trees. However, and it is an interesting restriction, this applies unless 'the need for, and benefits of, the development in that location outweigh the loss of the woodland habitat' [ODPM 2005: 6]. This reveals an attempt to reach a balance between the conflicting demands of an ethnocentric ethos and those of a biocentric one.

To repeat a point made above, the species of the trees contribute markedly to the character and distinctiveness of a place, as the Forestry Strategy recognises, and therefore to the feeling of a national landscape. Removing woodlands from the scope of income and corporation tax has helped to protect and promote native woodland, as it has ruled out assistance to softwood plantations. The higher grants for deciduous trees than for conifers in the Woodland Grant Scheme certainly take into account the higher cost of management of broadleaf woodland, including the fight against rodents which do a lot of damage to new hardwood plantations. But they also reflect the priority of native broadleaf planting over plantations of exotic conifers, in particular in Plantations in Ancient Woodland Sites (PAWS). Three grants are more specifically targeted at conservation and biodiversity in ancient woodlands. The first one is the Woodland Improvement Grant (WIG), already mentioned, which funds capital investment for management operations aimed at increasing public benefits. It includes three types of funds, the first of which is the Woodland Biodiversity Action Plan. This involves putting a stop to changes in woodland due to 'undesirable' species: in one of the turnarounds of perceptions of landscapes, yesterday's pets have become today's pests, and both the deer so close to William I's heart and the rhododendrons which form the backbone of the English garden rank among the undesirables, as does the grey squirrel. The other two funds of the WIG cover woodland SSSIs and woodland access.

The second grant is the Woodland Regeneration Grant, meant to support change in woodland composition, the analysis of which throws light on how policies can contribute to changing landscapes by putting monetary value on some and denying it to others. Under this grant scheme, restocking an ancient or a semi-natural woodland with conifers would not (logically enough) entitle owners/managers to grant aid; restocking a conifer plantation with conifers would be eligible for £360 per hectare; and restocking a conifer plantation with broadleaved woodland would entitle owners to £950 per hectare. But restocking an ancient/semi-natural woodland with native woodland would be eligible to a grant of £1,100 per hectare, while restocking a conifer plantation on an ancient woodland site with native woodland would give access to the maximum grant, reaching £1,760 per hectare [Forestry Commission England 2005].

The third grant, already mentioned previously in relation to rural regeneration, is the JIGSAW Scheme, which ran from 2000 to 2005 in some targeted areas of the eight regions of England. This favoured the establishment of new native woodland alongside ancient one, and/or permitted to piece together fragments of ancient woodland by creating woodland links. Both measures were meant to encourage biodiversity, in particular endangered species (the red squirrel and the dormouse in the Isle of Wight, for instance). Thus, PAWS should progressively be replaced by native woodland. This is timely, as three-quarters of these plantations are already mature or will reach economic maturity by 2015, which should allow for the regeneration of the corresponding areas with native broadleaves.

The Forestry Commission leads delivery of the woodland sector of the England Biodiversity Strategy and chairs the England Woodland Biodiversity Group. The Biodiversity Strategy launched in 2002, entitled *Working With the Grain of Nature*, illustrates the principle of policy integration, according to which environment must play a part in decision making within all sectors of public policy, and includes one chapter on woodlands and forestry. It was revised in 2006, which allows us to see how progress was – or was not – achieved. In 2002, it acknowledged gaps in knowledge and stressed the need for further research to establish an information base on the situation and the dynamics of habitats and species. This is probably why the text remained vague in its proposals, mostly echoing the *England Forestry Strategy*. Research and incentives were the main priorities in order to develop adequate strategies. Seven indicators were developed, concerning woodland birds, butterflies in woodland, conditions of woodland SSSIs (woodland represents 7% of all SSSIs, that is 75,500 hectares), status of woodland Biodiversity Action Plan priority species and habitats, woodland plant diversity, area of ancient woodland and public enjoyment of woodland [DEFRA 2006: 78].

Against these objectives, the assessment made in 2006 noted that two indicators pointed to positive change, i.e. the percentage of SSSIs supporting woodland habitats that are in favorable condition, which had increased to 79% but still fell short of the 95% objective to be reached by 2010, and the proportion of woodland priority species and habitats showing a positive trend (i.e. no longer declining, or even increasing). On the other hand, the long-term trend was negative concerning woodland bird species and butterflies. Two trends were not assessed, concerning plant diversity in woodland and the maintenance of the area of ancient woodland. But another source, pitting today's situation against surveys made in the mid-1970s, reveals important long-term changes in terms of woodland structure and composition, in particular a 36% decline in woodland plant diversity, a reduced number of young trees and shrubs as well as an increasing impact of deer browsing [England Forestry Forum 2005: 3–4]. As far as the last criterion of BAP priorities are concerned, namely public enjoyment of woodland, the figures, based on *UK Day Visits Surveys*, are slightly erratic and it is difficult to establish a trend [DEFRA 2006: 78–95].

These results are important, and confirmed by other sources. They demonstrate that quantity, that which the public notices and which allows public bodies to show their credentials, is not sufficient when it comes to conservation value. More woodland does not mean the same woodland: broadleaf, mixed and yew woodlands may have increased in size, but this is partly offset by evidence of a decline in habitat quality. The conditions of growth of the newly planted trees are very different from those of their elders, especially in the case of trees planted on former agricultural land enriched with nutrients. This has foreseeable consequences in terms of biodiversity, so that the new woodlands are unlikely to ever replace the ancient ones in the future in ecological terms [Haines-Young et al. 2000: 4].

12.2.3 On Feeding Retrophilia

Although heritage focused mostly on buildings or industrial sites to begin with, as well as historic gardens and parks, ancient woodlands have since then been caught up in the whirlwind of what some call the museumisation of the country. Old tree specimens in particular have caught the imagination of both researchers, the government and the media, and have reached the status of icons.

The Forestry Commission contributed to making them known by listing trees of exceptional measurements, before progressively handing this mission over to the Tree Register of the British Isles (TROBI), a charitable association set up in 1988 by Alan Mitchell after he retired from the Forestry Commission, where he had worked for several decades seeking exceptional trees for seed collections. *Champion Trees in the British Isles*, published by the Forestry Commission in 1994, thus lists 2,118 exceptional trees belonging to 1,065 species, cultivars and varieties. More initiatives have recently been undertaken to improve the knowledge of these heritage trees. This is the case of the Ancient Tree Forum (ATF), founded in 1993 by scientists coming from various disciplines, which also aimed to convey the value and importance of the existing population of ancient trees. It inspired, and became a partner of, English Nature's ‘Veteran Trees Initiative’ between 1996 and 2000, which was meant to raise awareness, harmonise survey methodology and provide training in relation to veteran tree protection.

Other means have been devised to familiarise the wider public with ancient trees, in the media or through direct involvement. Old trees have also become the favourite topics of books and/or TV programmes such as *Meetings with Remarkable Trees* [Pakenham 1996] and *Spirit of the Trees*. These have undoubtedly contributed to a better knowledge and appreciation of these rarities as part of the national heritage. Yet, they seem to be of much greater interest to scientists than to the general public. The 'Ancient Tree Hunt', set up in 2004 by the Ancient Tree Forum, together with the Woodland Trust and TROBI, enlists volunteers' help to map all the old trees across the UK, to feed a database and foster protection. In common with others presented earlier, the initiative is marketed thanks to appealing terms like 'treasure hunt' and the use of a non-scientific vocabulary (trees are not 'ancient' or 'veteran', but 'fat, old trees'); they elicit 'wow!' as a response, and their diameter can be measured in 'hugs' rather than in metres [Ancient Tree Hunt 2007]. The programme obtained over £500,000,00 in grants from the Heritage Lottery Fund for the year ending 31 March 2006 [NHMF 2006: 46]. Yet, enticing vocabulary does not seem to be sufficient to attract help: the objective of recording at least 100,000 trees between 2004 and 2011 is unlikely to be achieved, with only 6,000 recorded by mid-2007.

Defined in DEFRA's policy statement *Keepers of Time* as 'very old trees of cultural and/or biological interest [DEFRA/Forestry Commission 2005: 7], and by English Nature as trees that are 'of interest biologically, culturally or aesthetically because of [their] age, size or condition' [Read 2000: 13], many of these 'veterans' are notable trees associated with national events, from the Ankerwycke Yew at Runnymede, said to have witnessed the sealing of *Magna Carta* in 1215, to Isaac Newton's Apple Tree at Woolsthorpe Manor. The term veteran can also refer to a tree that has been through hard times and has survived an accelerated passage through the ageing process (abiotically induced, physiological stress or wounding). It is a term borrowed from the human experience of war, where soldiers mature rapidly under duress and suffering, and it therefore calls for clarification when applied to trees. What have these veteran trees triumphed over? Could it be over the Nazi-like – or alternatively, of the Communist-like – platoons of conifers marching across the land, taken as metaphors of totalitarian states and landscapes [Wright 1992: 7]? Today's conifers are no more Nazi or Communist than the oaks were Royalist during the Commonwealth, but the connections between nation, nature and native are anything but neutral.

Heritage has been the object of a lot of commentaries and analyses, both in relation to context (e.g. the loss of the Empire, Britain's identity and place in Europe), places (e.g. country houses, landscapes) specific issues (e.g. tourism), not to forget functions, since it came into its own as signifier for the social imaginary of the United Kingdom in the 1970s. It has benefited from the rise of environmental concerns nationally and internationally and of the competence and tenacity of dedicated scientists, who have permitted the translation of emotions and aspirations into policies, in a reactive approach typical in British public policy-making. Success has probably gone beyond their expectations: ancient woodlands have become a top priority both for the Forestry Commission and for the government within the framework of its biodiversity policy.

Conversely, the appropriation by the State of the issue of woodland heritage ensures that a centralised version of heritage is conveyed. This does not go without its drawbacks, and one may muse over the landscape legacy, and its version of Englishness, that the early 21st century will leave to future generations. Heritage holds an ambiguous position, poised like a two-faced Janus between the national glorious past and the economically-driven present, with the resulting synthesis of heritage providing an impetus for commercial outlets to feed the national 'retrophilia'. Courtesy of the National Trust, Past Times and many other ventures, the national past can be purchased in the form of Celtic sundials, 18th century topiary shrubs, Victorian lace parasols, teddy bears or pot-pourris. Reflecting this, the *National Heritage Act 2002*, amending the *National Heritage Act 1983*, makes the marketing of nostalgia official by allowing the Historic Buildings and Monuments Commission to 'produce souvenirs relating to ancient monuments or historic buildings situated in England and sell souvenirs' [OPSI 2002].

David Lowenthal concludes that the English landscape is the country's 'prime anachronism' and a 'vast, museumised ruin'. That may be the case, but only when anachronism and ruin serve a specific purpose, dictated by economy or ecology. Ancient woodlands, like other honeypots in the English landscape, attract attention, and channel visits, to specific sites. On the other hand, vibrancy in the forestry sector is expressed elsewhere: forestry business continues as usual in more discrete places and in less picturesque ways.

References

Ancient Tree Hunt. 2007. http://www.ancient-tree-hunt.org.uk/. Accessed 19 August 2007.

Booker, C. 1969 . *The Neophiliacs. A Study of the Revolution in English Life in the Fifties and Sixties*. London: Collins.

Brosse, J. 1989. *Mythologie des arbres*. Paris: Plon.

Butler, D. 2007. The Horsepower evolution. *Tree News* Spring-Summer 2007: 11–13.

Cloke, P., P. Milbourne and C. Thomas. 1996a. From wasteland to wonderland: opencast mining, regeneration and the English National Forest. *Geoforum* 27 (2): 159–174.

Cloke, P., P. Milbourne and C. Thomas. 1996b. The English National Forest: local reactions to plans for renegotiated nature-society relations in the countryside. *Transactions of the Institute of British Geographers* NS21: 552–571.

Corvol, A. 1997. La Forêt. In *Les lieux de mémoire*, ed. P. Nora, quarto 2, 2765–2816. Paris: Gallimard.

Daniels, S. and D. Cosgrove (eds.). 1988. Introduction. Iconography and landscape. In *The Iconography of Landscape*, 1–10. Cambridge: Cambridge University Press.

Darby, W. 2000. *Landscape and Identity: Geographies of Nation and Class in England*. Oxford: Berg.

DCLG (Department for Communities and Local Government). 1999. *Tree Preservation Orders: New Regulations,* 14p. London: DCLG.http://www.communities.gov.uk/pub/70/TreePreservation Ordersnewregulations_id1507070.pdf. Accessed 8 May 2007.

DEFRA (Department for Environment, Food and Rural Affairs). 2006. *Working with the Grain of Nature. Taking it Forward. Vol. II*. London: DEFRA.

DEFRA/Forestry Commission. 2005. *Keepers of Time. A Statement of Policy for England's Ancient and Native Woodland*. London: DEFRA/Forestry Commission.

Eden, N. 2006. Councils must publicise tree crimes. *Horticulture Week*, 6 April. http://www.hortweek.com/news_story.cfm?ID = 1838. Accessed 14 November 2007.

England Forestry Forum. 2005. *England Forestry Strategy Review. Topic Paper: Natural Resource protection*. http://www.defra.gov.uk/corporate/consult/forestry-strategy/interactions.pdf. Accessed 12 April 2007.

English Heritage. 1992. *Managing England's Heritage: Setting our Priorities for the 1990s*. London: English Heritage.

English Nature. 2000. *The Future for Veteran Trees*. Peterborough: English Nature.

Fay, N. 2002. Environmental arboriculture, tree ecology and veteran tree management. *Arboricultural Journal* 26 (3): 213–238.

Forestry Commission. 2004. *The UK Forestry Standard. The Government's Approach to Sustainable Forestry* (1st edition 1998). Edinburgh: Forestry Commission.

Forestry Commission. 2005. *GB Public Opinion of Forestry Statistics 2005*. Edinburgh: Economics and Statistics Unit, Forestry Commission.

Forestry Commission England. 2005. *Funding for Farm Woodlands in England*. Leaflet published in October 2005, n.p.

Fox, D. 2005. Woodlands, natural spaces and the media. In *Accessibility of Woodlands and Natural Places. Addressing Crime and Safety Issues*, eds. L. O'brien and P. Tabbush, 19–21. Alice Holt: Forest Research.

Gale, R. and S. Cordray. 1991. What should forests sustain? Eight answers. *Journal of Forestry* 89 (5): 31–36.

Garner, A 2004. Living history: trees and metaphors of identity in an English forest. *Journal of Material Culture* 9 (1): 87–100.

Gilpin, W. 1791. *Remarks on Forest Scenery and Other Woodland Views*. London.

Goldsmith, F. 1988. Threats to woodland in an urban landscape: a case study in Greater London. *Landscape and Urban Planning* 16: 221–228.

Green, T., K. Alexander and R. Key. 1999. *Managing Our Ancient Trees,* § Creating decay. Injuring existing trees. http://www.treecouncil.org.uk/. Accessed 19 May 2007.

Haines-Young, R. et al. 2000. *Countryside Survey 2000. Accounting for Nature. Assessing Habitats in the UK Countryside*. London: Department Of The Environment, Transport and the Regions.

Harrison, R. 1992. *Forests: the Shadow of Civilization*. Chicago: University of Chicago Press.

Hazucha, A. 2002. Neither deep nor shallow but national. Eco-nationalism in Wordsworth's *Guide to the Lakes. Interdisciplinary Studies in Literature and Environment* 9 (2): 61–73.

Hill-Tout, P. 2005. The Forestry Commission in England. In *Great Britain and England Annual Report and Accounts 2003–2004*. Edinburgh: Forestry Commission.

Hughes, R. 1991. The role of urban forestry in environmental strategies and the economic regeneration of the post industrial town. *European Environment* 1 (2): 1–5.

Keil, R. and J. Graham. 1998. Reassserting nature. Constructing urban environments after Fordism. In *Remaking Reality. Nature at the Millenium*, eds. B. Braun and N. Castree, 100–125. London: Routledge.

King, M. 2000. Community involvement and community development – a review of lessons for the future. In *Community Forestry – A Change for the Better. Conference Proceedings, 7–8 December 1999, The Guildhall, London*, Forestry Commission, n.p. Forestry Commission, Countryside Agency.

Kirby, K. 1996. Importance of native woods for nature conservation in the National Parks of England and Wales. *Quarterly Journal of Forestry* 90 (1): 31–38.

Land Use Consultants. 2001. *Threats to Ancient Woodlands. Final Report*. London: Woodland Trust/WWF-UK.

Lankester, R. 1915. *Diversions of a Naturalist*. London: Methuen. Cited in Lowe, Philip. 1983. Values and institutions in the history of British nature conservation. In *Conservation in Perspective*, eds. A. Warren and F. Goldsmith, 329–352. Chichester: Wiley.

Lonsdale, D. 1999. *Tree Hazard Assessment and Management*. London: HMSO.

Lovell, N. 1998. *Locality and Belonging*. London: Routledge.

Lowe, P. 1983. Values and institutions in the history of British nature conservation. In *Conservation in Perspective*, eds. A. Warren and F. Goldsmith, 329–352. Chichester: Wiley.

Lowenthal, D. 1994. European and English landscapes as national symbols. In *Geography and National Identity*, ed. D. Hoosen, 15–38. Oxford: Blackwell.

Luhrmann, T. 1993. The resurgence of romanticism. In *Environmentalism – The View from Anthropology*, ed. K. Milton. London: Routledge.

Mitchell, J. 1994. Imperial landscape. In *Landscape and Power*, ed. J. Mitchell, 5–34. Chicago and London: Chicago University Press.

Morris, J. 1981. Barchester lives on. In *Places. An Anthology of Britain*, texts chosen by Ronald Blythe. Oxford: Oxford University Press.

Muir, R. 2005. *Ancient Trees, Living Landscapes*. Stroud: Tempus.

NHMF (National Heritage Memorial Fund) Heritage Lottery Fund. 2006. *Annual Report and Accounts 2006*. London: The Stationery Office.

Nottinghamshire Country Council. 2007. 'Robin Hood Festival', http://www.robinhood.co.uk/robin_hood_festival.php. Accessed 7 October 2007.

ODPM (Office of the Deputy Prime Minister). 2005. *Planning Policy Statement No 9 – Biodiversity and Geological Conservation*. London: The Stationery Office.

OPSI (Office of Public Sector Information). 2002. *National Heritage Act 2002*. Section 4: 'New trading functions of the Commission'. http://www.opsi.gov.uk/acts/acts2002/20020014.htm. Accessed 6 June 2007.

Pakenham, T. 1996. *Meetings with Remarkable Trees*. London: Weidenfeld and Nicolson.

Rackham, O. 2002. *Trees and Woodlands in the British Landscape. The Complete History of Britain's Trees, Woods and Hedgerows*. London: Phoenix.

Rackham, O. 2004. Pre-existing trees and woods in country-house parks. *Landscapes* 5 (2): 1–16.

Read, H. 2000. *Veteran Trees. A Guide to Good Management*. Peterborough: English Nature.

Southwood, T. 1961. The number of species of insect associated with various trees. *The Journal of Animal Ecology* 30 (1): 1–8.

Spencer, J. and K. Kirby. 1992. An inventory of ancient woodlands for England and Wales. *Biological Conservation* 62: 77–93.

Thomas, K. 1984. *Man and the Natural World. Changing Attitudes in England, 1500–1800*. Oxford: Oxford University Press.

Tsouvalis-Gerber, J. 1996. Making the invisible visible: ancient woodlands, British forest policy and the social construction of reality. In *European Woods and Forests: Studies in Cultural History*, ed. C. Watkins, 215–229. New York: CAB International.

Watkins, C. (ed.). 1998. 'A solemn and gloomy umbrage': changing interpretations of the ancient oaks of Sherwood Forest. In *European Woods and Forests: Studies in Cultural History*, 93–113. New York: CAB International.

Wright, P. 1992. The disenchanted forest. *Guardian Weekend* 7 November.

Concluding Remarks: Rebranding England Through Consensual Woodlands?

The purpose of this book was twofold. With the specific case of England, but with many possible applications to other countries, it aimed to show the uses and functions of 21st century woodlands in relation to their past roots and to present social trends, as well as to provide an analysis of the policies implemented to manage, promote and protect woodland.

Over the last decades, the priorities given to production and consumption have evolved in woodlands, with less of the former and much more of the latter. Nevertheless, these woodlands of consumption are not devoid of their own economic agenda. Behind the apparently benign 'quality of life' objectives of social and urban forestry, lie 'invisible earnings' and a socio-political agenda. Moreover, with derelict land resulting from both agricultural surpluses and defunct industries, woodlands have provided new land uses and new outlets for diversification. The introduction raised the question of whether 21st century forestry in England still hinged on the two tenets which have justified its pursuit since the Middle Ages: that there should be no better use of the land and that it should be profitable. After analysis, the answer is a resounding 'yes'. Forestry has to pay its way, whether the profits take the traditional form of cash for wood products or new forms of profits: increased tourism, improved image leading to inward investment, or positive consequences on community cohesion.

From the evidence presented in this necessarily brief English panorama, one is tempted to conclude that the 21st century opens on an unchallenged consensus on the worth of woodlands to provide solutions to complex issues in postmodern Western countries. Multi-purpose woodlands offer an aesthetic solution to postindustrial landscapes, and may contribute to inward investment as well as to a reevaluation of the landscapes. They provide practicable sources of diversification to the rural economy, whether in the form of short rotation coppice crops or amenity woodlands. In and around towns and cities, woodlands bring the countryside close to where people live, thus cumulating the advantages of recreation close to home, a tool against social exclusion and an element of the sustainable city equation. They also contribute to improving the individual and social health of the nation in more senses than one and suggest alternative paths in education. Last but not least, they build bridges between past and present and help construct the landscapes of the future.

Woodland policies have thus helped foster a positive discourse on the benefits of trees and woodlands in everyday life; conversely, they have benefited over the past 20 years from a favourable cultural context. Today's perceptions are the direct heirs of the changes brought about in the previous periods, in particular the replacement of the belief in a rationalist, progressive outlook, inherited from the Enlightenment, by a pessimistic assessment of 'progress' and of human intervention in nature, since the First World War in particular. With the backlash of technology since Rachel Carson's *Silent Spring* (1962) and fears linked to deforestation, globalisation and the loss of identity, trees and woodlands more than ever 'symbolise anti-modern (i.e. anti-urban, anti-consumerist and anti-industrial) values' [Rival 2000: 16]. Yet it is no longer believed that nature is best left to its own devices either, as Sir Ray Lankester did at the beginning of the century. Few people would now seriously assert that if 'one man is present in the neighbourhood, even at a long distance, he upsets the "balance of nature"' [Lowe 1983: 337], if only because there seems to be no such thing as balance in nature. The past 20 years have seen a clear statement that the best way to protect woodland is to manage it.

Tree-planting programmes to create or extend woodlands, numerous instances of which have been presented, are probably the most highly symbolical reaction to these fears and part of the 'feel-good' factor connected with trees. Paradoxically however, at least half of these newly-planted trees die for lack of care: the cathartic effect on the planters seems to prevail over possible long-term differences to the environment, which is one facet of the exploitation of the positive image of woodlands. Another one is the use of rural imagery to connote new urban projects (e.g. Thames Gateway) positively. There is considerable scope for the study of woodlands as post-modern pastiches in new urban settlements, their use as agents of consumption of an urban/rural lifestyle, their consequences on the conception of the countryside and their effects on planning policies like green belts. The appearance of oxymora like 'urban forestry', 'urban countryside', 'citification' of the forest – or 'forestification' of the city – point to a redefinition of the categories 'city' and 'country' and, beyond these, of Nature. Nature, it can be argued, has been externalised, objectified, and 'has become central to how urban spatial regulation and urban political regimes are reconstituted' [Keil and Graham 1998: 100].

Multi-purpose forestry has certainly striven to channel urbanites' access to 'countrysides' by providing everything to everyone, with the result that, instead of a rural/urban divide, a new divide operates between a consumerist approach and one linked to the 'discourse of extinction'[Harrison and Burgess 1994: 298], or between the 'commodity' and the 'ecology' narratives [Whatmore and Boucher 1993]. On the one hand, woodlands serve a specific agenda; on the other hand, they are to be protected for their own sake in the name of their intrinsic value. The consumerist approach dominates actions, if not discourse, down to the biodiversity benefits. It is striking to see that, in spite of the fact that 'it may be less cost-effective to create new semi-natural on open ground than to restore those ancient sites which have been converted to non-native plantations over the past 50 years' [Kirby 1996: 37], the grant system favours woodland creation to woodland restoration. What is significant is not so much the differential in the amounts granted, as the implications in terms of agenda: rural and industrial regeneration, as well as urban forestry, are preferred to biodiversity enhancement in public policies. Incidentally, there is no desire, either in public policies or in the population, to see the wildwood back in England – with its ugly, dangerous and unhealthy unkempt landscapes.

In this apparently consensual landscape, a lot of questions remain. One of them concerns what might happen should agricultural land come at a premium once again. This question may not be absurd and has been given particular resonance since 2000, with corn prices soaring due to ethanol production. If the past, and many farmers' reserved attitude to woodland-planting today, are anything to go by, today's woodland mania might well be affected. The Confederation of Forest Industries (UK) Ltd, created in 2006, whose purposes are political lobbying for the UK forestry sector and forest products sector, takes a rational approach, which is less emotional but maybe more appropriate to the woodlands interest in the long run:

> It is encouraging to see that the role that trees, woodlands and forests play in the UK is being recognised by an increasingly large number of people. There is renewed interest in trees and forests and the benefits they provide, but against this background there is a pressing need for increased commercial planting. Such planting must be appropriate in terms of species and silvicultural practice, so as to provide a quality crop. The old adage of 'the right tree, in the right place' has much to commend it [United Kingdom Forest Products Association 2007].

A second question has to do with the complexity of the sustainability agenda in forestry. As has been shown, reconciling its three poles is anything but easy, and at times they seem to exclude each other. Reaching all the objectives without sacrificing any one dimension is today's challenge. It will take time and a lot of debates about the desirable balance between the three poles. It will also require a lot of sensitive education, to avoid the risk of manipulation.

The last question that the present situation raises concerns the nature of the present political iconography of English woodlands. Both Prime Ministers Margaret Thatcher and Tony Blair had at heart a concern to restore national pride (whether through putting the '"Great" back in Great Britain', for the former, or 'rebranding' the nation, for the latter) and both saw during their time in office a lot done to enhance the visible signs of national identity. Values of stability and continuity were conveyed through the conservation of ancient woodlands and the creation of new, native woodlands. But do these correspond to everyone's idea of the national identity or do they cater for the taste of a dominant elite, who imposes it while pretending to consult the local populations? There is no straightforward answer, and more qualitative studies on the values of woodlands to particular groups (ethnic minorities, young people and women in particular) are required. This is a field which is just opening and the following thoughts are merely tentative and waiting to be confirmed or contradicted by fieldwork.

Concerning ethnic minorities, some references underline the multicultural dimension of woodlands as places for recreation and social mix: such is the case of the Sikh community and Bestwood Country Park or the Nottingham Inter Faith

Council trees in Nottingham, mentioned in the previous chapter. The same chapter also showed that the range of historical references to national identity included simultaneously almost all the influences exerted on the British Isles, to the point that the image of the palimpsest might be adequately replaced by that of overhead transparencies, to be viewed all at the same time if the correct light is provided. Indeed, variations on the theme of national identity as they appear in the discourse on woodland heritage alternatively take up, on top of the Celtic, Roman, Saxon, Danish and medieval influences already mentioned, the concepts of progress and reconciliation with urbanity typical of the 18th century and the rural idyll of the 19th century, the attraction for wilderness of the Romantics and the liking for homely landscapes and provincial life of the Victorians. More than any particular history, then, English woodlands evoke stability, timelessness, against rapid change with its unforeseeable consequences: 'England as she was: changeless in our fast-changing world', in the words of former Environment Minister Michael Heseltine in an article aptly entitled 'Is it as risk, this England?' [Heseltine 1990: 78–79]. Whether in Boscobel or elsewhere, veteran trees and ancient woodlands, under the care of their 21st century stewards, can hope to withstand the passing of time, globalisation and (possibly) climate change. They act as the instruments of the propagation of a consensual vision of England, as opposed to Scotland, Wales and Ireland and, even more so, to the rest of the world.

'Almost' all the influences are to be read in the wooded landscapes: the lacks are significant, and recent immigrants are conspicuous by their absence in the definition of new landscapes. From her study of ramblers, Wendy Darby goes as far as to say that ethnic minorities' perception of the English landscape is one of 'non-identity' [Darby 2000: 6]. Looking for connections, could one venture to make a parallel between the treatment of exotic trees and the perception of ethnic minorities on English ground? Mark Johnston and Lia Shimada remind us that the vocabulary used to qualify tree species is anything but neutral [Johnston and Shimada 2004: 188], and many occurrences during fieldwork confirm it: 'natives good, aliens bad' is still the dominant motto, unlikely to attract immigrants to tree-planting events. Unless, that is, the aliens have become part of the landscape with time, or unless they serve a specific purpose as adequately, or better, than the natives (by fulfilling priority jobs, to follow the latest trends in British immigration policies?). Such cases include the sycamore (*Acer pseudoplatanus*) or the London plane (*Platanus × hispanica*), both introduced from mainland Europe in the 17th century and which withstand pollution remarkably. J. Wilks has definitely been proved right when he asserted in 1972 that

> some years will pass before the immigrant pine cone is revered as emblematically as the acorn [Wilks 1972: 14],

and, we might add, this time is definitely not just around the corner as yet.

Some research leads to conclude that ethnic minorities do not easily relate to woodlands in England, for two main reasons: they prefer productive landscapes such as allotments to landscapes to be contemplated; and they find woodland landscapes too English. The rural idyll which so appealed to the urban Victorian middle class may be prolonged by walking in the ancient woodlands. Walking is a mostly middle-class pursuit. Wendy Darby gives the example of the Ramblers' Association project 'Let's Get Going', designed to bring into the countryside people with no tradition of walking, and which has not yielded the expected results, partly for practical reasons (difficulty of climbing over stiles with a sari) or cultural reasons (many Asian women do not go out without their menfolk, and landscape was experienced as being a bastion of 'Englishness' to these urban minorities). Furthermore, public participation, it has been shown, can be quite prescriptive.

A new hierarchy may be at work, inherited from the rural/urban divide, but transposed to ancient/new woodlands. In a new variant of the 'class-ridden society', the same object (woodland) might serve as a marker suited to social hierarchy, the 'polite society' of today taking to the ancient woodlands to cultivate Englishness in the countryside, leaving on the other side of the greenbelts the less privileged/less mobile urban dwellers, including ethnic minorities, to use the new community woodlands as venues for recreation. Who knows? They may in the long term, like the Sikh community of Khalsa Wood in Nottingham, build their own ethnic, urban identity into these new spaces so far virgin of national history.

Many of the trends described in 21st century English woodlands are not limited to England. On the one hand, they are typical of countries which have a lower than average woodland cover: thus, Denmark embarked in 1989 on a policy aimed at doubling its woodland cover within 80–100 years, bringing the forest cover to around 20%, essentially for recreational purposes. On the other hand, social forestry is the order of the day in many developed countries, Germany, Austria, Scandinavia, Switzerland and the United States among them.

Internally, these trends provide a striking echo of the preoccupations of society at large in England: the fear of the other, concern about the future, the rights of individuals versus the cohesion of the community are but a few of these. This interrelation is probably all the more perceptible as policies have become more reactive recently: the *England Forestry Strategy*, which came out in 1998, was reviewed in 2006; the report *Trees in Towns*, published in 1993, was followed by *Trees in Towns II* which, after much delaying, was published in 2008; Tree Preservation Orders (1962) were reviewed in 1999.

At the same time, however, and without being naïve about the potential of forestry to solve problems which have nothing to do with it, forests can also act as a piece of the jigsaw puzzle towards finding innovative solutions to the issues that plague postmodern societies. They may be part of the array of strategies to imagine solutions: in the form of voluntary work leading to paid employment, of an improved 'feel-good' factor likely to promote social cohesion and a social mix propitious to inclusion at all levels.

The last paradox of 21st century forestry could well be that it is less about planting/managing woodland than about being a vector to bring about social change. Indeed, forests straddle the public and the private spheres. They are the objects of policies concerning the environment in its widest sense, but they are also the realm of people's private lives, their relationships to nature and space, to leisure, to identity. Beyond the more or less successful attempts to commodify trees and woodlands, many people's reactions clearly express the permanence of the value of trees and woodlands as connections between people, but also between the earth and the stars, landscapes of the mind and soul.

References

Darby, W. 2000. *Landscape and Identity: Geographies of Nation and Class in England.* Oxford: Berg.

Harrison, C. and J. Burgess. 1994. Social constructions of nature: a case study of conflicts over the development of Rainham Marshes. *Transactions of the Institute of British Geographers* NS 19: 291–310.

Heseltine, M. 1990. Is is at risk, this England? *The Field* 272, May: 78–79.

Johnston, M. and L. Shimada. 2004. Urban forestry in a multicultural society. *Journal of Arboriculture* 30 (3): 185–192.

Keil, R. and J. Graham. 1998. Reasserting nature. Constructing urban environments after Fordism. In *Remaking Reality. Nature at the Millenium*, eds. B. Braun and N. Castree, 100–125. London: Routledge.

Kirby, K. 1996. Importance of native woods for nature conservation in the National Parks of England and Wales. *Quarterly Journal of Forestry* 90 (1): 31–38.

Lowe, P. 1983. Values and institutions in the history of British nature conservation. In *Conservation in Perspective*, eds. A. Warren and F. Goldsmith, 329–352. Chichester: Wiley.

Rival, L. Trees, from symbols of life and regeneration to political artifacts. In *The Social Life of Trees*, ed. L. Rival, p.16. Oxford: Berg, 2000.

United Kingdom Forest Products Association. 2007. *UK Softwood and Hardwood Sector Report.* http://www.ukfpa.co.uk/site/sawnews.html. Accessed 26 July 2007.

Whatmore, S. and S.n Boucher. 1993. Bargaining with nature: the discourse and practice of environmental gain. *Transactions of the Institute of British Geographers* 18: 166–178.

Wilks, J. 1972. *Trees of the British Isles in History and Legend.* London: Frederick Muller.

Bibliography

XII World Forestry Congress Bulletin 10 (12), 2003.

Addison, J. 1714. *Spectator* 583, 20 August. In *The Spectator. Joseph Addison and Richard Steele* (1711–1714), ed. Y. D. Bond. 1965. Oxford: Clarendon.

Agnoletti, M. and S. Anderson (eds.). 2000a. *Forest History: International Studies on Socio-economic and Forest Ecosystem Change*. IUFRO Research Series N°2. New York: CABI, in association with IUFRO.

Agnoletti, M. and S. Anderson (eds.). 2000b. *Methods and Approaches in Forest History*, 304p. IUFRO Research Series n°3. New York: CABI, in association with IUFRO.

Agyeman, J. 1996. *Involving Communities in Forestry... Through Community Participation*. Forestry Practice Guide 10. Edinburgh: The Forestry Authority and the Forestry Commission.

Aitchison, C. N. MacLeod and S. Shaw. 2000. *Leisure and Tourism Landscapes: Social and Cultural Geographies*. London: Routledge.

Alison, H. 2000. Learning the lessons from woods on your doorstep – Where next? In *Community Forestry – A Change for the Better. Conference Proceedings, 7–8 December 1999, The Guildhall, London*, Forestry Commission, n.p. Forestry Commission: Countryside Agency.

Amin, A. and S. Graham. 1997. The ordinary city. *Transactions of the Institute of British Geographers* 22 (4): 411–429.

Amin, A., D. Massey and N. Thrift. 2000. *Cities for the Many not the Few*. Bristol: Policy.

Andersen, F., C. Konijnendijk and T. Randrup. 2002. Higher education on urban forestry in Europe: An overview. *Forestry* 75 (5): 501–511.

Antonovsky, A. 1979. *Health, Stress and Coping*. San Francisco: Jossey-Bass.

Appelstrand, M. 2002. Participation and societal values: the challenge for lawmakers and policy practitioners. *Forest Policy and Economics* 4: 281–290.

Arnould, P., M. Hotyat and L. Simon. 1997. *Les forêts d'Europe*. Paris: Nathan.

Arnstein, S. 1969. A ladder of citizen participation. *Journal of American Institute Planners* 35: 216–224.

Audit Commission. 1995. *Listen Up! Effective Community Consultation*. London: The Audit Commission.

Baker, A. and G. Biger (eds.). 1992. *Ideology and Landscape in Historical Perspective*. Cambridge: Cambridge University Press.

Ball, R., S. Bussey, D. Patch, A. Simson and S. West. 1999. *Country Report on the United Kingdom*. In *COST Action E12 – Research and Development in Urban Forestry in Europe*, eds. Forrest, M., C. Konijnendijk and T. Randrup. Luxembourg: Office for Official Publications of the European Communities, EUR 19108 EN.

Banister, D. and R. Hickman. 2006. *Looking Over the Horizon: Visioning and Backcasting for UK Transport Policy*. London: Department for Transport.

Barnett, A. and R. Scruton (eds.). 1998. *Town and Country*. London: Jonathan Cape.

Barton, H. 2000. Conflicting visions of neighbourhood. In *Sustainable Communities – the Potential for Eco-neighbourhoods*, ed. H. Barton, 3–18. London: Earthscan.

Battershill, M. and A. Gilg. 1996. New approaches to creative conservation on farms in South-west England. *Journal of Environmental Management* 48: 321–340.

Beard, M. 1988. *English Landed Society in the Twentieth Century*. London: Routledge & Kegan Paul.

Beaverstock, J. et al. 1997. The National Forest and Global environmental sustainability. *European Environment* 7: 54–62.

Bechmann, R. 1990. *Trees and Man – The Forest in the Middle Ages*. New York: Paragon House.

Beckett, K., P. Freer-Smith and G. Taylor. 1998. Urban woodlands: their role in reducing the effects of particulate pollution. *Environmental Pollution* 99.

Bell, S., C. W. Thompson and P. Travlou. 2003. Contested views of freedom and control: children, teenagers and urban fringe woodlands in Central Scotland. *Urban Forestry and Urban Greening* 2: 87–100.

Bellringer, A. and S. Gillam. 1998. *Forest Visitor Surveys. 1994 and Earlier Years*. Edinburgh: Forestry Commission HQ Statistics.

Bending, N. and A. Moffat. 1997. *Tree Establishment on Landfill Sites: Research and Updated Guidance*. Edinburgh: Forestry Commission.

Bengston, D. 1994. Changing forest values and ecosystem management. *Social and Natural Resources* 7: 515–533.

Benson, J. and K. Willis. 1992. Valuing informal recreation on the Forestry Commission Estate. *Forestry Commission Bulletin* 104. London: HMSO.

Berger, P. and T. Luckmann. 1966. *The Social Construction of Reality. A Treatise in the Sociology of Knowledge*. New York: Anchor Books.

Best, R. 1973. *The Urban Countryside*. London: Faber & Faber.

Betts, A. and J. Claridge. 1994. *New Markets for Old Woods – Proceedings of a Conference at the University of Surrey*. Cambridge: Forestry Authority for England.

Binney, M. and E. Milne (eds.). 1982. *Vanishing Houses of England: A Pictorial Documentary of Lost Country Houses*. London: SAVE Britain's Heritage.

Biomass Task Force. 2005. *Report to Government*. London: DEFRA.

Bixler, R., C. Carlisle W. Hammitt and M. Floyd. 1994. Observed fears and discomforts among urban students on school field trips to wildland areas. *Journal of Environmental Education* 26: 24–35.

Bixler, R. and M. Floyd. 1994. Nature is scary, disgusting and uncomfortable. *Environment and Behavior* 29: 443–456.

Blackman, D. and R. Thackray. n.d. *The Green Infrastructure of Sustainable Communities*. Community Forests Partnership. http://www.communityforest.org.uk/resources/ECF_GI_Report.pdf. Accessed 6 August 2007.

Booker, C. *The Neophiliacs. A Study of the Revolution in English Life in the Fifties and Sixties*. London: Collins.

Bowler, I. and B. Ibery. 1999. Agricultural land-use and landscape change under the post-productivist transition – examples from the United Kingdom. In *Land Use Changes and their Environmental Impact in Rural Ares in Europe*, eds. R. Kronert, J. Baudry, I. Bowler and A. Reenberg, 121–139. Carnforth Lancashire: Parthenon.

Box, J. and C. Harrison. 1993. Natural spaces in urban places. *Town and Country Planning*: 231–235.

Bradley, G. (ed.). 1995. *Urban Forest Landscapes: Integrating Multidisciplinary Perspectives*. Seattle: University of Washington Press.

Bradshaw, A. and P. Gilbertson. 1985. Tree survival in cities. The extent and nature of the problem. *Arboricultural Journal* 9 (2): 131–142.

Bradshaw, A., B. Hunt and T. Walmsley. 1995. *Trees in the Urban Landscape: Principles and Practice*. London: E&FN Spon.

Braun, B. and N. Castree. 1998. *Remaking Reality. Nature at the Millenium*. London: Routledge.

Breese, S. 1998. In search of Englishness: in search of votes. In *History and Heritage: Consuming the Past in Contemporary Culture*, eds. J. Arnold, K. Davies and S. Ditchfield, 155–167. Shaftesbury: Donhead.

Brosse, J. 1989. *Mythologie des arbres*. Paris: Plon.
Brown, J. 1883. *The Forests of England and the Management of Them in Byegone Times*. Edinburgh: Oliver & Boyd.
Brown, N. 1997. Re-defining native woodland. *Forestry* 70 (3): 191–198.
Buchy, M. and S. Hoverman. 2000. Understanding public participation in forest planning: a review. *Forest Policy and Economics* 1: 15–25.
Burgess, J. 1995. *Growing in Confidence. Understanding People's Perceptions of Urban Fringe Woodlands*. Technical Report CCP457. Cheltenham: Countryside Commission.
Burgess, J. 2000. Situating knowledges, sharing values and reaching collective decisions: the cultural turn in environmental decision making. In *Cultural Turns, Geographical Turns: Perspectives on Cultural Geography*, eds. I. Cook, D. Crouch, S. Naylor and J. Ryan, 273–287. Harlow: Prentice Hall.
Burgess, J., C. Harrison and M. Limb. 1988. People, parks and the urban green: a study of popular meanings and values for open spaces in the city. *Urban Studies* 25: 455–473.
Burgess, J., C. Harrison and M. Limb. 1991. Contested meanings: the consumption of news about nature conservation. *Media, Culture and Society* 13 (4): 499–519.
Burgess, J., M. Limb and C. Harrison. 1988a. Exploring environmental values through the medium of small groups. 1. Theory and practice. *Environment and Planning A* 20: 309–326.
Burgess, J., M. Limb and C. Harrison. 1988b. Exploring environmental values through the medium of small groups. 2. Illustration of a group at work. *Environment and Planning A* 20: 457–476.
Burke, E. 1970 (1757). *A Philosophical Enquiry into the Origin of our Ideas of the Sublime and Beautiful*. Menston: Scolar.
Campbell, D. 1987. Landscape design in forestry. *Landscape Design* 166: 31–36.
Cannell, M. 1999. Growing trees to sequester carbon in the UK: answers to some common questions. *Forestry* 72 (3): 237–247.
Central Liverpool Primary Care Trust. 2004. *The REACT Project. A Celebration of Achievements to Date*. Liverpool: Central Liverpool Primary Care Trust.
Chambers, K. 1987. Urban forestry in the London borough of Tower Hamlets – an account of the European urban forestry project sponsored by the EEC. *Arboricultural Journal* 11 (1): 1–14.
Chief Medical Officer. 2002. *Health Check: On the State of Public Health*. London: Department of Health.
Church, A., N. Ravenscroft and G. Rodgers. 2005. *Woodland Owners' Attitudes to Public Access Provision in South-East England*. Edinburgh: Forestry Commission Information Note.
CJC Consulting. 2001. *Evaluation of Woodland Creation in England Under the WGS and the WGS/FWPS*. Edinburgh: Forestry Commission.
CJC Consulting et al. 2003. *Economic Analysis of Forestry Policy in England, Final Report*. London: DEFRA and the Treasury.
Clarke, G. 1985. Where did all the trees come from? An analysis of Bridgeman's planting at Stowe. *Journal of Garden History* 5 (1): 72–83.
Clarke, G., J. Darrall, R. Grove-White, P. MacNaghten and J. Urry. 1994. *Leisure Landscapes. Leisure, Culture and the English Countryside: Challenges and Conflicts*. Lancaster: Council for the Protection of Rural England.
Clegg & Co. 2002. *Evaluation of Woodland Creation in England under the Woodland Grant Scheme and the Farm Woodland Premium Scheme*. London: Department for Environment, Food and Rural Affairs.
Cloke, P., P. Milbourne and C. Thomas. 1996a. The English National Forest: local reactions to plans for renegotiated nature-society relations in the countryside. *Transactions of the Institute of British Geographers* NS21: 552–571.
Cloke, P., P. Milbourne and C. Thomas. 1996b. From wasteland to wonderland: opencast mining, regeneration and the English National Forest. *Geoforum* 27 (2): 159–174.
Coates, D. 1999. Sustainable rural development and the role of new farm woodlands in England. In *Farm Woodlands for the Future*, ed. P. Burgess, 1–10. London: BIOS Scientific Publishers.

Coates, P. 1998 (1st edition), 2005. *Nature: Western Attitudes since Ancient Times*. Berkeley: University of California Press.

Coates, P. 2005. *Nature: Western Attitudes Since Ancient Times*. Berkeley: University of California Press.

Cobham Resource Consultants. 1993. *London Tree Survey*. Abingdon, Oxfordshire: Prepared for Task Force Trees, Countryside Commission.

Coles, R. and S. Bussey. 2000. Urban forestry landscapes – urban lifestyles and urban living. In *Urban Lifestyles: Spaces. Places. People*, eds. J. Benson and M. Roe, 185–192. Rotterdam: Balkerna.

Coles, R. and M. Caserio. 2003. *Understanding and Facilitating the Social Outputs of Urban Green Spaces*. Paper read at the International Conference 'Open Space, People Space', Edinburgh. http://www.openspace.eca.ac.uk/conference/proceedings/PDF/Coles1.pdf. Accessed 16 May 2006.

Community Forest Partnership. 2005. *Community Forests Monitoring Report 2004–2005. Final Report*. Bury Saint Edmunds: Projects Partners Research. http://www.communityforest.org.uk/resources/monitoringreport_2004_05.pdf. Accessed 7 February 2007.

Conway, H. 1991. *People's Parks: The Design and Development of Victorian Parks in Britain*. Cambridge: Cambridge University Press.

Cooke, V. 2000. Social issues in community forestry. In *Community Forestry – A Change for the Better. Conference Proceedings, 7–8 December 1999, The Guildhall, London*, Forestry Commission, n.p. Forestry Commission, Countryside Agency.

Coppok, J. 1960. A decade of post-war forestry in Britain. *Economic Geography* 36 (2):127–138.

Corvol, A. 1997. La Forêt. In *Les lieux de mémoire*, ed. P. Nora, quarto 2, 2765–2816. Paris: Gallimard.

Cosgrove, D. and S. Daniels (eds.). 1988. *The Iconography of Landscape*. Cambridge: Cambridge University Press.

Cost Action E39. 2003. *Memorandum of Understanding for the implementation of a European Concerted Research Action designated as COST E39 "Forests, Trees, Human Health and Wellbeing"*. http://www.e39.com.ee/en/m-9/c-11/. Accessed 14 November 2007.

Countryside Agency. 2004. *Leisure Day Visits GB. Report of the 2002–2003 Great Britain Day Visits Survey*. London: Countryside Agency.

Countryside Agency. 2007. *The Countryside In and Around Towns*. Wetherby: Countryside Agency. http://www.countryside.gov.uk/Images/CAT_tcm2–22089.pdf. Accessed 11 February 2007.

Countryside Agency and Groundwork. 2005. *Countryside In and Around Towns*. Wetherby: Countryside Agency.

Countryside Commission. 1987. *Forestry in the Countryside*. CCP 245. Cheltenham: Countryside Commission.

Countryside Commission. 1994. *The National Forest. The Strategy*. CCP 468. Cheltenham: Countryside Commission.

Crabtree B. et al. 2000. *Economic Benefits of Community Forestry – Phase 1 Study*. Research contract for the Countryside Agency and Forestry Commission. Oxford: CJC Consulting with MLURI.

Crawford, C. and T. Cooper. 2004. *Sustainable Woodland Products. Review of Consumer Awareness and Best Marketing Practice*. Prepared for the Forestry Commission and the Countryside Agency. Guilford: Arkenford.

Crowe, S. 1978. *The Landscape of Forests and Woods*. Forestry Commission Booklet n°44. London: HMSO.

Cummins, J. 1988. *The Hound and the Hawk : The Art of Medieval Hunting*. London: Weidenfeld & Nicholson.

Curry, N. 2000. Community participation in outdoor recreation and the development of millenium greens in England. *Leisure Studies* 19 (1): 17–35.

Daniels, S. 1988. The Political iconography of woodland in later Georgian England. In *The Iconography of Landscape*, eds. D. Cosgrove and S. Daniels, 43–82. Cambridge: Cambridge University Press.

Daniels, S. 1993. *Fields of Vision. Landscape Imagery and National Identity in England and the United States*. Princeton: Princeton University Press.

Daniels, S. and D. Cosgrove. 1988. Introduction. Iconography and landscape. In *The Iconography of Landscape*, eds. D. Cosgrove and S. Daniels, 1–10. Cambridge: Cambridge University Press

Darby, H.C. 1951. The changing English landscape. *The Geographical Journal* CXVII (4): 377–398.

Darby, W. 2000. *Landscape and Identity: Geographies of Nation and Class in England*. Oxford: Berg.

Davies, C. 2001. Planning for community forests, 'the green lungs of European cities'. In *Partnership-Led Regeneration*. Papers of the 4th European Forum on Urban Forestry Held in Durham 15–19 May 2001: 25–32. http://www.vbv.be/efuf/_docs/EFUF4_Papers.PDF. Accessed 10 November 2007.

Davies, C. and J. Vaughan. 2002. Bridging the gap: the experience of England's Community Forests in engaging excluded communities with urban woodland and green spaces. Congress *Urban Woods for People*, 25–27 November 2002, Paris. http://www.onf.fr/europe/life-urbanwoods/pdf/davies_GB.pdf. Accessed 3 April 2006.

Deakin, R. 2007. *Wildwood. A Journey Through Trees*. London: Hamish Hamilton.

Decourt, N. 1999. *La forêt*. Paris : Presses Universitaires de France.

Delort, R. and F. Walter. 2001. *Histoire de l'environnement européen*. Paris: Presses Universitaires de France.

Department for Communities and Local Government. 1999. *Tree Preservation Orders: New Regulations*. London: DCLG. 14p. http://www.communities.gov.uk/pub/70/TreePreservationOrdersnewregulations_id1507070.pdf. Accessed 8 May 2007.

Department for Culture, Media and Sport Tourism Division. 2005. *National Sustainable Tourism Indicators*. London: Department for Culture, Media and Sport.

Department for Communities and Local Government. 2000. *Our Towns and Cities - the Future*. London: DCLG.

Department for Environment, Food and Rural Affairs. 2002. *Working with the Grain of Nature*. London: DEFRA.

Department for Environment, Food and Rural Affairs. 2004. Alun Michael announces two year extension in funding for community forests. *News Releases 2004*, 113/04, 29 March 2004. http:/ww.defra.gov.uk/news/2004/040329a.htm. Accessed 15 April 2004.

Department for Environment, Food and Rural Affairs. 2006a. *England Forestry Strategy Review. Topic Paper: Natural Resource Protection*. http://www.defra.gov.uk/corporate/consult/forestry-strategy/interactions.pdf. Accessed 12 April 2007.

Department for Environment, Food and Rural Affairs. 2006b. *Working with the Grain of Nature. Taking it Forward. Volume II*. London: DEFRA.

Department for Environment, Food and Rural Affairs and ERDP. 2001. *The Countryside Stewardship Scheme. Traditional Farming in the Modern Environment*. London: DEFRA.

Department for Environment, Food and Rural Affairs and Forestry Commission. 2005. *Keepers of Time. A Statement of Policy for England's Ancient and Native Woodland*. London: DEFRA/Forestry Commission.

Department for Environment, Food and Rural Affairs and Health and Safety Commission. 2006. *Pesticides. Code of Practice for Using Plant Protection Products*. London: DEFRA.

Department of Health. 1995a. *Committee on the Medical Effects of Air Pollutants: Health Effects of Non-biological Particles*. London: HMSO.

Department of Health. 1995b. *Committee on the Medical Effects of Air Pollutants: Asthma and Outdoor Air Pollutants*. London: HMSO.

Department of Health. 2004. *Choosing Health: Making Healthy Choices Easier*. London: Stationery Office.

Department of the Environment. 1986. *Waste Management Paper 26: Landfilling Wastes*. London: HMSO.

Department of the Environment. 1990. *This Common Inheritance: A Summary of the White Paper on the Environment*. Cm 1200. London: HMSO.

Department of the Environment. 1993. *Countryside Survey 1990*. London: Department of the Environment.

Department of the Environment/Welsh Office. 1992. *Planning Policy Guidance 21. Tourism*.

Department of the Environment, Transport and the Regions. 1998a. *Modern Local Government: In Touch with the People*. London: HMSO.

Department of the Environment, Transport and the Regions. 1998b. *Guidance on Enhancing Public Participation in Local Government*. London: DETR.

Department of Trade and Industry. 2006. *Renewable Energy: 2005/6 Review of the Renewables Obligation Statutory Consultation Document*. London: DTI.

Department of Trade and Industry and DEFRA. 2003. *Energy White Paper. Our Energy Future – Creating a Low Carbon Economy*. Cmnd 5761. London: Stationery Office.

Department of Trade and Industry and DEFRA. 2006. *The Government's Response to the Biomass Task Force*. London: DTI/DEFRA.

Department of Transport, Local Government and the Regions. 2002. *Green Spaces, Better Places*. Final Report of the Urban Green Spaces Task Force. London: DTLGR.

Des Arbres et des Hommes. 1979. Actes du colloque Forêt et Société – Lyon, 1979. Le Paradou: Bibliothèque des Ruralistes – Actes Sud.

De Vreese, R., K. Bonsen, G. Borgman, B. Van Alfen, J. van Rooijen and M. Borgman. 2004. *Educating the Urban Foresters. Proceedings of the 6th European Forum on Urban Forestry*, Arnhem (The Netherlands). Velp: University of Professional Education.

Dickinson, M., J. Mackay, A. Goodman and P. Putwain. 2000. Planting trees on contaminated soils: Issues and guidelines. *Land Contamination and Reclamation* 8 (2): 87–101.

Dobson, M. and A. Moffat. 1993. *The Potential for Woodland Establishment on Landfill Sites*. London : HMSO.

Donadieu, P. and M. Périgord. 2005. *Clés pour le paysage*. Paris: Ophrys.

Douglas, I. 2005. *Urban Greenspace and Mental Health*. Prepared for the UK MAB FORUM. http://ukmaburbanforum.org.uk/Publications/UF%20Publications%20and%20Reports/ukmabgrnspcepap2.htm. Accessed 16 May 2006.

Dower, M. 1965. *The Fourth Wave: The Challenge of Leisure*. London: The Civic Trust.

Dowson, D. 2001. Arboricultural education and qualification opportunities. *Arboricultural Journal* 25: 353–359.

Dumas, R. 2002. Traité de l'arbre: essai d'une philosophie occidentale. Arles: Actes Sud.

Duvigneaud, P. 1974. *La synthèse écologique*. Paris: Doin.

Elands, B. and K. Wiersum. 2001. Forestry and rural development in Europe: an exploration of socio-political discourses. *Forest Policy and Economics* 3: 5–16.

England Forestry Forum. 2005. *England Forestry Strategy Review. Topic Paper: Natural Resource Protection*. http://www.defra.gov.uk/corporate/consult/forestry-strategy/interactions.pdf. Accessed 12 April 2007

English Heritage. 1992. *Managing England's Heritage: Setting Our Priorities for the 1990s*. London: English Heritage.

English Nature. 2000. *The Future for Veteran Trees*. Peterborough: English Nature.

Etzioni, A. 2000. Creating good communities and good societies. *Contemporary Sociology* 29: 188–195.

Evans, D. 1992. *A History of Nature Conservation in Britain*. London: Routledge.

Evelyn, J. 1786. *Silva. Or, a Discourse on Forest Trees*. With notes by A. Hunter. York.

Fay, N. 2002. Environmental arboriculture, tree ecology and veteran tree management. *Arboricultural Journal* 26 (3): 213–238.

Finch, J. 2004. 'Grass, grass, grass': fox-hunting and the creation of the modern landscape. *Landscapes* 5 (2): 41–52.

Firn Crichton Roberts Ltd. and John Clegg & Co. 2000. *Monitoring the Health and Financial Viability of Forest Industry Businesses: Business Survey Report to the Forestry Commission & Forest Industry Associations*. Edinburgh: Forestry Commission.

Fischer, C. 1976. *The Urban Experience*. New York: Harcourt Brace Jovanovich.

Fitzgerald, B. 1966. *Portrait of the New Forest*. London: Hale.
Font, X. and J. Tribe (eds.). 2000. *Forest Tourism and Recreation: Case Studies in Environmental Management*. Wallingford: CABI.
Font, X., K. Yale and J. Tribe. 2001. Introducing environmental management systems in forest recreation: results from a consultation exercise. *Managing Leisure* 6: 154–167.
Forbes-Laird, J. 2003. National tree fatality database. *Arboricultural Newsletter*, No 123, December 2003.
Forest Enterprise. 2002. *Life in the Dead Wood. A Guide to Managing Deadwood in Forestry Commission Forests*. Edinburgh: Forestry Commission.
Forestry Commission. 1985a. *The Policy for Broadleaved Woodlands. Policy Paper N°5*. Edinburgh: Forestry Commission.
Forestry Commission. 1985b. *Guidelines for the Management of Broadleaved Woodlands*. Edinburgh: Forestry Commission.
Forestry Commission. 1998a. *England Forestry Strategy. A New Focus for England's Woodlands*. Cambridge: Forestry Commission.
Forestry Authority. 1998b. *The UK Forestry Standard. The Government's Approach to Sustainable Forestry*. Edinburgh: Forestry Commission.
Forestry Commission. 2001. *National Inventory of Woodlands and Trees – England*. Edinburgh: Forestry Commission.
Forestry Commission England. 2003a. *Public Opinion of Forestry 2003: England*. Edinburgh: Forestry Commission.
Forestry Commission. 2003b. *British Timber Statistics 2002*. Edinburgh: Forestry Commission.
Forestry Commission. 2004. *The UK Forestry Standard. The Government's Approach to Sustainable Forestry*. Edinburgh: Forestry Commission.
Forestry Commission. 2005a. *GB Public Opinion of Forestry Statistics 2005*. Edinburgh: Economics and Statistics Unit, Forestry Commission.
Forestry Commission. 2005b. *The London Trees and Woodland Framework*. http://www.forestry.gov.uk/forestry/infd-6qhdva. Accessed 11 April 2007.
Forestry Commission. 2006. *England Forestry Strategy. Implementation Progress Report. 1998–2006*. Cambridge: Forestry Commission.
Forestry Commission. 2007. *Public Opinion of Forestry 2007: England (preliminary report)*. Edinburgh: Economics and Statistics.
Forestry Commission England. 2006. *General Guide to EWGS*. Cambridge: Forestry Commission England.
Forestry Commission England. 2007. *English Woodland Grant Scheme*. http://www.forestry.gov.uk/ewgs. Accessed 13 November 2007.
Forestry Commission England and DEFRA. 2003. *The English Woodland Grant Scheme*. Edinburgh: Forestry Commission.
Freeman, M. 1996. Plebs or predators? Deer-stealing in Whichwood Forest, Oxfordshire in the eighteenth and nineteenth centuries. *Social History* 21 (1): 1–21.
Fraser, E. and A. Kenney. 2000. Cultural background and landscape history as factors affecting perceptions of the urban forest. *Journal of Arboriculture* 26 (2): 106–113.
Frumkin, H. 2002. Urban sprawl and public health. *Public Health Reports* 117: 201–217.
FSC (Forest Stewardship Council). 2004. *FSC Principles and Criteria for Forest Stewardship*. http://www.fsc.org/plantations/docs/FSC-STD-01–001%20FSC%20Principles%20and%20Criteria%20for%20 Forest%20Stewardship%202004–04.pdf
Gale, R. and S. Cordray. 1991. What should forests sustain? Eight answers. *Journal of Forestry* 89 (5): 31–36.
Gandrillon-Purcell. 2003. *'Albion's Vital Tree': Arbre, société et identité dans la peinture de portrait britannique de 1709 à 1792*. Unpublished doctor's thesis, Université de Paris VII – Denis Diderot.
Garner, A. 2004. Living history: trees and metaphors of identity in an English forest. *Journal of Material Culture* 9 (1): 87–100.

Gilpin, W. 1791. *Remarks on Forest Scenery and Other Woodland Views*. London.
Glicken, J. 2000. Getting stakeholder participation 'right': a discussion of participatory processes and possible pitfalls. *Environmental Science and Policy* 3 (6): 305–310.
Grahn, P. and U. Stigsdotter. 2003. Landscape planning and stress. *Urban Forestry and Urban Planning* 2: 1–18.
Gobster, P. 1995. Aldo Leopold's ecological esthetic: integrating esthetic and biodiversity values. *Journal of Forestry* 93 (2): 6–10.
Goldsmith, F. 1988. Threats to woodland in an urban landscape: a case study in Greater London. *Landscape and Urban Planning* 16: 221–228.
Goldsmith, F. and A. Warren (eds.). 1993. *Conservation in Progress*. Chichester: Wiley.
Grant, R. 1991. *The Royal Forests of England*. Stroud: Alan Sutton.
Greenhalgh, L. and K. Worpole. 1995. *Park Life. Urban Parks and Social Renewal*. London: Comedia/Demos.
Grey, G. and F. Deneke. 1978. *Urban Forestry*. New York: Wiley.
Grove-White, R. 1991. The emerging shape of environmental conflict in the 1990s. *RSA Journal* 139 (5419): 437–447.
Grove-White, R. 1997. The environmental 'valuation' controversy: observations on its recent history and significance. In *Valuing Nature. Economics, Ethics and Environment*, ed. J. Foster, 21–31. London: Routledge.
Guerrier, Y., N. Alexander, J. Chase and M. O'Brien (eds.). 1995. *Values and the Environment: A Social Science Perspective*. Chichester: Wiley.
Haines-Young et al. 2000. *Countryside Survey 2000. Accounting for Nature. Assessing Habitats in the UK Countryside. Executive Summary*. London: Department of the Environment, Transport And The Regions.
Handley, R. et al. 2003. *Accessible Natural Green Space Standards in Towns and Cities: A Review and Toolkit for their Implementation*. English Nature Research Reports, n°526. Peterborough: English Nature.
Harrison, C. 1981. A playground fro whom? Informal recreation in London's green belt. *Area* 13: 109–114.
Harrison, C. and J. Burgess. 1994. Social constructions of nature: a case study of conflicts over the development of Rainham Marshes. *Transactions of the Institute of British Geographers*. NS 19: 291–310.
Harrison, C., J. Burgess, A. Millward and W. Dawe. 1995. *Accessible Natural Greenspace in Towns and Cities. A Review of Appropriate Size and Distance Criteria*. English Nature report N° 153. Peterborough: English Nature.
Harrison, C., M. Limb and J. Burgess.1986. Recreation 2000: Views of the Country from the City. *Landscape Research* 11: 19–24.
Harrison, R. 1992. *Forests: The Shadow of Civilization*. Chicago: University of Chicago Press.
Hart, R. 1979. *Children's Experience of Place*. New York: Irvington Publishers.
Hartig, T. 2002. *Psychosocial Forestry and the Health of Urban Populations*. Paper presented at the 'Urban Woods for People' Congress. *La Perception de la Forêt*. Paris, November. http://www.onf.fr/europe/life-urbanwoods/pdf/Hartig_T.pdf. Accessed 10 November 2007.
Hastie, C. 2003. *The Benefits of Urban Trees. A Literature Review*. Warwick: Warwick District Council.
Hazucha, A. 2002. Neither deep nor shallow but national. Eco-nationalism in Wordsworth's *Guide to the Lakes. Interdisciplinary Studies in Literature and Environment* 9 (2): 61–73.
Henwood, K. and N. Pidgeon. 1998. *The Place of Forestry in Modern Welsh Culture and Life*. Report to the Forestry Commission. Bangor: University of Wales.
Hermeline, M. and G. Rey. 1994. *L'Europe et la forêt*. Paris: Eurofor; Luxembourg: Parlement européen/Office des publications des Communautés européennes.
Hibberd, B. 1989. *Urban Forestry Practice*. Forestry Commission Handbook 5. London: HMSO.
Hill, G., P. Courtney, R. Burton et al. 2003. *Forests' Role in Tourism: Phase 2. Final Report*. Aberdeen: Macaulay Institute.

Hill-Tout, P. 1999. Costed options for managing state forests in Britain. In *Non market Benefits of Forestry. International Symposium Held in Edinburgh*, ed. C. Roper and A. Park, 165–170. London: Stationery Office.
Hinde, T. 1985. *Forests of Britain*. London: Victor Gollancz.
HM Treasury. 1972. *Forestry in Great Britain: An Interdepartmental Cost/Benefit Study*. London: HMSO.
Hislop, M., M. Twery and H. Vihemäki. 2004. *Involving People in Forestry. Introduction*. Edinburgh: Forestry Commission.
Hodge, S. 1995. *Creating and Managing Woodlands Around Towns*. Forestry Commission Handbook 11. London: HMSO.
Hooke, D. (ed.). 2000. *Landscape: The Richest Historical Record*. Birmingham: Society for Landscape Studies, supplementary series 1.
Hoskins, W. 1975. *The Making of the English Landscape*. London: Pelican Books.
House of Commons Agriculture Committee. 1990. *Land Use and Forestry. Second Report, Vol. 1*. House of Commons Paper 16–1. London: HMSO.
House of Commons Environment Committee. 1993. Paper 257. *Forestry and the Environment. First Report, Session 1992–93*. London: HMSO.
Huddleston, J. *Climate Change Agreements – Results of the First Target Period Assessment*. Harwell: Future Energy Solutions.
Hughes, R. 1991. The role of urban forestry in environmental strategies and the economic regeneration of the post industrial town. *European Environment* 1 (2): 1–5.
Hull, R. 1992. Brief Encounters with urban forests produce moods that matter. *Journal of Arboriculture* 18 (6): 322–324.
Hull, B. and A. Harvey. 1989. Explaining the emotion people experience in suburban parks. *Environment and Behavior* 21 (3): 323–345.
Ineichen, B. 1993. *Homes and Health: How Housing and Health Interact*. London: Spon.
Jackson, R. et al. 2005. Trading water for carbon with biological carbon sequestration. *Science* 310 (5756): 1944–1947.
Jacobs, M. 1995. *Sustainability and Socialism*. London: Socialist Environment Resources Association.
Jacovelli, P. 2001. *The Bowland Initiative. Added Value Project 1999–2001. Final Report*. Edinburgh: Forestry Commission/Lancashire County Council.
James, N. 1981. *A History of English Forestry*. Oxford: Basil Blackwell.
Janse, G. and C. Konijnendijk. 2007. Communication between science, policy and citizens in public participation in urban forestry—Experiences from the Neighbourwoods project. *Urban Forestry and Urban Greening* 6 (1): 23–40.
Johnston, M. 1986. Community forestry: a question of priorities. *Arboricultural Journal* 10 (2): 95–100.
Johnston, M. 1996. A brief history of urban forestry in the United States. *Journal of Arboriculture* 20 (3): 257–278.
Johnston, M. 1997a. The early development of urban forestry in Britain, Part I. *Arboricultural Journal* 21: 107–126.
Johnston, M. 1997b. The early development of urban forestry in Britain, Part II. *Arboricultural Journal* 21: 317–330.
Johnston, M. 1999. The Springtime of Urban Forestry in Britain - Developments between the 1st and 3rd Conferences, 1988–1993, Part I. *Arboricultural Journal* 23: 233–260.
Johnston, M. 2000. The Springtime of Urban Forestry in Britain - Developments between the 1st and 3rd Conferences, 1988–1993, Part II. *Arboricultural Journal* 23: 313–341.
Johnston, M. 2001. A review of the Black Country Urban Forest Millennium Programme, 1995–2001. *Arboricultural Journal* 26:11–139.
Johnston, M. and B. Rushton. 1999. *A Survey of Urban Forestry in Britain*. Coleraine: University of Ulster.
Johnston, M. and L. Shimada. 2004. Urban forestry in a multicultural society. *Journal of Arboriculture* 30 (3): 185–192.

Jones, N. 2001. Political partnerships for regeneration. The Black Country Urban Forest: a UK case study. In *Partnership-Led Regeneration*. Papers of the 4th European Forum on Urban Forestry held in Durham 15–19 May 2001: 60–64. http://www.vbv.be/efuf/_docs/EFUF4_Papers.PDF.

Jones, O. and Paul C. 2002. *Tree Cultures. The Place of Trees and Trees in Their Place*. Oxford: Berg.

Jorgensen, E. 1970. Urban forestry in Canada. In *Proceedings of the 46th International Shade Tree Conference*, 43a–51a.

Kaplan, R. 1984. Human needs for renewable resources and supportive environments. In *Land Use and Forest Resources in a Changing Environment*, ed. G. Bradley, 133–140. Seattle and London: University of Washington Press.

Kennedy, J. M. Dombeck and N. Koch. 1998. Values, beliefs and management of public forests in the Western world at the close of the twentieth century. *Unasylva* 49 (192): 16–26.

King, M. 2000. Community involvement and community development – a review of lessons for the future. In *Community Forestry – A Change for the Better. Conference Proceedings, 7–8 December 1999, The Guildhall, London*, Forestry Commission, n.p. Forestry Commission, Countryside Agency.

Kirby, K. 1996. Importance of native woods for nature conservation in the National Parks of England and Wales. *Quarterly Journal of Forestry* 90 (1): 31–38.

Kitchen, L., T. Marsden and P. Milbourne. 2006. Community forests and regeneration in post-industrial landscapes. *Geoforum* 37: 831–843.

Kong, L. 2000. Nature's dangers, nature's pleasures: Urban children and the natural world. In *Children's Geographies*, eds. S. Holloway and G. Valentine, 257–271. London: Routledge.

Konijnendijk, C. 1997. A short history of urban forestry in Europe. *Journal of Arboriculture* 23 (1): 31–39.

Konijnendijk, C. 1999. *Urban Forestry in Europe: A Comparative Study of Concepts, Policies and Planning for Forest Conservation, Management and Development in and Around Major European Cities*. Doctoral dissertation. Research Notes No. 90. Faculty of Forestry, University of Joensuu.

Konijnendijk, C., M. Gauthier and R. van Veenhuizen. 2004. Editorial. *Urban Agriculture Magazine*, special issue 'Trees and Cities: Growing Together' 13: 3–5.

Konijnendijk, C., K. Nilsson T. Randrup and J. Schipperijn (eds.). 2005. *Urban Forests and Trees. A Reference Book*. Berlin: Springer.

Konijnendijk, C., R. Ricard, A. Kenney and T. Randrup. 2006. Defining urban forestry – a comparative perspective of North America and Europe. *Urban Forestry and Urban Greening* 4: 93–103.

Konijnendijk, C. and J. Schipperijn (eds.). 2004. *NeighbourWoods for Better Cities. Tools for Developing Multifunctional Community Woodlands in Europe*. Frederiksberg: Danish Center for Forest, landscape and Planning. http://www.vbv.be/neighbourwoods_leaflet.pdf. Accessed 2 November 2007.

Konijnendijk, C. and A. Simson. 2001. *Partnership-Led Regeneration*. Papers of the 4th European Forum on Urban Forestry held in Durham 15–19 May 2001. http://www.vbv.be/efuf/_docs/EFUF4_Papers.PDF. Accessed 7 June 2007.

Kumar, K. 2001. 'Englishness' and English National Identity. In *British Cultural Studies*, eds. D. Morley and K. Robins, 41–55. Oxford: Oxford University Press.

Kuo, F. and W. Sullivan. 2001. Aggression and violence in the inner-city. Effects of the environment via mental fatigue. *Environment and Behavior* 33 (4): 543–571.

Lambert, R. 2005. *Trees and People: Tracing the Early Provision of Forest and Woodland Recreation in Twentieth Century Britain*. http://www.nottingham.ac.uk/ttri/pdf/2005_3.pdf. Accessed 5 March 2006.

Land Use Consultants. 1993. *Trees in Towns: A Survey of Trees in 66 Towns and Villages in England*. London: HMSO.

Land Use Consultants. 2001. *Threats to Ancient Woodlands. Final Report*. London: Woodland Trust/ WWF-UK.

Land Use Consultants. 2006. *England's Trees, Woods and Forests. A Consultation Document. Analysis of Responses to the Consultation*. Bristol: DEFRA. http://www.defra.gov.uk/corporate/consult/forestry-strategy/responseanalysis-final.pdf. Accessed 7 February 2007.

Land Use Consultants and SQW Ltd. 2005. *Evaluation of the Community Forest Programme. Final Report*. Cheltenham: Countryside Agency.

Lankester, R. 1915. *Diversions of a Naturalist*. London: Methuen.

Large, P. 2002. From swanimote to disafforestation: Feckenham Forest in the early seventeenth century. In *The Estates of the English Crown 1558–1640*, ed. R. Hoyle, 389–417. Cambridge: Cambridge University Press.

Laverne, R. and K. Winson-Geideman. 2003. The influence of trees and landscaping on rental rates at office buildings'. *Journal of Arboriculture* 29 (5): 281–290.

Lawrence, H. 1993. The neoclassical origins of modern urban forests. *Forest and Conservation History* 37 (1): 26–36.

Lawrence, H. 2006. *City Trees: A Historical Geography from the Renaissance Through the Nineteenth Century*. Charlottesville: University of Virginia Press.

Learning Through Landscapes. 2003. *National School Grounds Survey 2003*. Winchester: LTT.

Lohr, V., C. Pearson-Mims, J. Tarnai and D. Dillman. 2004. How urban residents rate and rank the benefits and problems associated with trees in cities. *Journal of Arboriculture* 30 (1): 28–35.

Lonsdale, D. 1999. *Tree Hazard Assessment and Management*. London: HMSO.

Lovell, Nadia. 1998. *Locality and Belonging*. London: Routledge.

Low, S. and D. Lawrence-Zuniga. 2003. *The Anthropology of Space and Place: Locating Culture*. Oxford: Blackwell Publishing.

Lowe, P. and A. Flynn. 1989. Environmental politics and policy in the 1980s. In *The Political Geography of Contemporary Britain*, ed. J. Mohan, 255–279. London: Macmillan.

Lowe, P., J. Murdoch, T. Marsden, R. Munton and A. Flynn. 1993. Regulating the new rural spaces: the uneven development of land. *Journal Of Rural Studies* 9 (3): 205–222.

Lowe, P. and S. Ward (eds.). 1998. *British Environmental Policy and Europe*. London: Routledge.

Lowenthal, D. 1991. British national identity and the English landscape. *Rural History* 2 (2): 205–230.

Lowenthal, D. 1994. European and English landscapes as national symbols. In *Geography and National Identity*, ed. D. Hoosen, 15–38. Oxford: Blackwell.

Luhrmann, T. 1993. The resurgence of romanticism. In *Environmentalism – The View from Anthropology*, ed. K. Milton. London: Routledge.

McManus, P. 1999. Histories of forestry: ideas, networks and silences. *Environment and History* 5: 185–208.

McPherson, E. and C. Johnson. 1988. A community forest planning progress: case study of citizen participation. *Landscape and Urban Planning* 15: 1–2.

Mace, A., P. Hall and N. Gallent. 2007. New East Manchester: Urban renaissance or urban opportunism? *European Planning Studies* 15 (1): 51–65.

Mackenzie, A. and G. Paget. 1999. *Maximising Ethnic Minority Involvement in Environmental Activities*. Blackburn: Groundwork.

Maff/Forestry Commission. 1998. *A Guide to the Woodland Grant Scheme*. Edinburgh: Forestry Commission Grants and Licences Division.

Manwood, J. 1598. *A Treatise of the Laws of the Forest; Wherein Is Declared Not Only Those Laws as They Are Now in Force, but Also the Original and Beginning of Forests and What a Forest Is in Its Own Proper Nature*. London: Reprinted by William S. Hein & Co., Buffalo, 2002.

Martin, E. et al. 1957. Incidence of neurosis in a new housing estate. *British Journal of Preventive and Social Medicine* 11: 196–202.

Mather, A. 2001. Forests of consumption: postproductivism, postmaterialism and the postindustrial forest. *Environment and Planning C: Government and Policy* 19: 249–268.

Matless, D. 1996. Nature, the modern and the mythic: tales from early twentieth century geography. *Transactions of the Institute of British Geographers* NS 16: 272–286.

Matless, D. 1998. *Landscape and Englishness*. London: Reaktion Books.

Mayor of London. 2005. *Tree and Woodland Framework for London*. London: Greater London Authority.

Mental Health Foundation. 2006. *Up and Running. How Exercise Can Help Beat Depression*. London: Mental Health Foundation.

Mersey Forest. 2001. *Mersey Forest Plan*. Birchwood: Mersey Forest.

Mersey Forest and Red Rose Forest. 2004. *The Mersey and Red Rose Forests Timber Stations Project. September 2003-March 2004. Report to the Countryside Agency and Forestry Commission North West England Conservancy*. Forestry Commission England/Countryside Agency, 2004. http://www.merseyforest.org.uk/files/TSReport_v1.pdf. Accessed 30 October 2007.

Miles, R. 1967. *Forestry in the English Landscape*. London: Faber & Faber.

Miller, H. 1995. *Forest Policy: The International and British Dimensions*. Aberdeen: University of Aberdeen.

Miller, R. 1981. *State Forestry for the Axe: A Study of the Forestry Commission and De-Nationalisation by the Market,* Hobart Paper 91. London: Institute of Economic Affairs.

Millward, A. and B. Mostyn. 1989. *People and Nature in Cities. The Social Aspects of Planning and Managing Natural Parks in Urban Areas*. Peterborough: Nature Conservancy Council.

Ministry of Local Government and Planning. 1951. *Town and Country Planning 1943–1951: Progress Report*. HMSO: Cmnd 8204.

Mitchell, J. 1994. Imperial landscape. In *Landscape and Power*, ed. J. Mitchell, 5–34. Chicago and London: Chicago University Press.

Moffat, A., T. Hutchings and D. Elgy. 1997. *Tree Establishment on Landfill Sites. Interim Report*. Forest Research Report to DETR.

Moffat, A. and J. Laing. 2003. An audit of woodland performance on reclaimed land in England. *Arboricultural Journal* 27 (1):11–25.

Morris, J. 1981. Barchester lives on. In *Places. An Anthology of Britain*, texts chosen by Ronald Blythe. Oxford: Oxford University Press.

Mottet, J. (ed.). 2002. *L'arbre dans le paysage*. Seyssel: Champ Vallon.

Muir, R. 2005. *Ancient Trees, Living Landscapes*. Stroud: Tempus.

Murdoch, J. and S. Abram. 1998. Defining the limits of community governance', *Journal of Rural Studies* 14 (1): 41–50.

National Playing Fields Association. 2001. *The Six Acre Standard. Minimum Standards for Outdoor Playing Space*. London: National Playing Fields Association.

National Urban Forestry Unit. n.d. *Training the Long-term Unemployed to Manage the Urban Forest*. Case Study 48. Wolverhampton: NUFU.

National Urban Forestry Unit. n.d. *Trees and Woods in Towns and Cities. How to Develop Local Strategies for Urban Forestry: A Guide*. Wolverhampton: NUFU.

National Urban Forestry Unit. n.d. *Trees Matter. The Benefits of Trees and Woods in Towns: A Review*. Wolverhampton: NUFU.

New Forests for the 21st Century: The Scale, Nature and Location of Forestry Expansion in Britain. Speakers' papers at the Forestry Commission Conference, 26 and 27 March 1992, Forestry Commission.

Nicholson-Lord, D. 1991. Conference paper. In *A Seed in Time. The Second UK/international Conference in Urban Forestry*, eds. C. Field and M. Johnston, 311–315. Edinburgh: Seed in Time.

Nisbet, J. 1906. The history of the Forest of Dean, in Gloucestershire. *The English Historical Review* 21 (83): 445–459.

Nobles, R. 1980. Urban forestry/Arboriculture Programme. *Journal of Arboriculture* 6: 53–56.

Nowak, D. 2000. Impact of urban forest management on air pollution and greenhouse gases. In *Proceedings of the Society of American Foresters, 1999 National Convention, Portland, Oregon*, 143–148. Bethesda: Society of American Foresters.

Nowak, D. and E. McPherson. 1992. Quantifying the impact of trees: The Chicago urban forest climate project. *Unasylva* 173. http://www.fao.org/docrep/u9300e/u9300e08.htm. Accessed 12 November 2007.

Nowak, D., J. Stevens, S. Sisinni and C. Luley. 2002. Effects of urban tree management and species selection on atmospheric carbon dioxide. *Journal of Arboriculture* 28 (3): 113–122.
O'Brien, E. (ed.). 2001. *Social Forestry: Questions and Issues*. Farnham: Forest Research.
O'Brien, E. 2003. Human values and their importance to the development of forestry policy in Britain: a literature review. *Forestry* 76 (1): 3–17.
O'Brien, L. 2005. *Trees and Woodlands. Nature's Health Service*. Farnham: Forest Research.
O'Brien, L. and J. Claridge. 2002. *Trees Are Company. Social Science Research into Woodlands and the Natural Environment*. Edinburgh: Forestry Commission.
O'Brien, L. and P. Tabbush. 2005. *Accessibility of Woodlands and Natural Places. Addressing Crime and Safety Issues*. Farnham: Forest Research.
O'Callaghan, D. 2003. The challenges of planning for sustainable development in the UK. *Arboricultural Journal* 27 (2): 93–116.
Oakley, P. et al. 1991. *Projects with People. The Practice of Participation in Rural Development*.
Ode, Å. and G. Fry. 2002. Visual aspects in urban woodland management. *Urban Forestry and Urban Greening* 1 (1): 15–24.
Office of the Deputy Prime Minister. 2002. *Living Places: Cleaner, Safer, Greener*. London: ODPM.
Office of the Deputy Prime Minister. 2003. *Sustainable Communities. Building for the Future*. London: Her majesty's Stationery Office.
Office of the Deputy Prime Minister. 2004. *Making it Happen – The Northern Way*. London: ODPM.
Ong, D. 2001. The impact of environmental law on corporate governance: International and comparative perspectives. *European Journal of Environmental Law* 12 (4): 685–726.
Oosthoek, J.-W. 2000. *The Logic of British Forest-Policy, 1919–1970*. Paper presented at the 3rd Conference of the European Society for Ecological Economics. *Transitions Towards a Sustainable Europe. Ecology - Economy – Policy*. Vienna, Austria, May 3 - May 6, 2000. http://www.eh-resources.org/documents/esee_paper.pdf. Accessed 1 November 2007.
Orstein, R. and P. Ehrlich. 1989. *New World, New Mind. Changing the Way We Think to Change Our Future*. London: Methuen.
Pakenham, T. 1996. *Meetings with Remarkable Trees*. London: Weidenfeld & Nicolson.
Parkinson, J. 1995. *Corporate Power and Responsibility. Issues on the Theory of Company Laws*. Oxford: Oxford University Press.
Paris, J. 1972. The citification of the forest. *Canadian Pulp and Paper Magazine* 9: 119–122
Patmore, J. 1970. *Land and Leisure*. London: David and Charles.
Pauleit, S., N. Jones and G. Garcia-Martin et al. 2002. Tree establishment practice in towns and cities – results from a European survey. *Urban Forestry and Urban Greening* 1 (2): 83–96.
Paulson, R. 1989. *Breaking and Remaking : Aesthetic Practice in England, 1700–1820*. New Brunswick and London: Rutgers University Press.
Parlement Europeen. 1994. *L'Europe et la forêt – EUROFOR*. Coordonné par l'Office National des Forêts (France). 2 vols. Complete text available on http://www.europarl.eu.int/workingpapers/agri/default_fr.htm. Accessed 16 September 2007.
Penning-Rowsell, E. 1981. Fluctuating fortunes in gauging landscape value. *Progress in Human Geography* 5: 25–41.
Penning-Rowsell, E., G. Gullett, G. Searle and S. Witham. 1977. *Public Evaluation of Landscape Quality*. Polytechnic Planning Research Group Report N°13, Middlesex University. London: Middlesex
Penning-Rowsell, E. and D. Lowenthal (eds.). 1986. *Landscape Meanings and Values*. London: Allen & Unwin.
Perkins, J. 1920. A plea for the consideration of the aesthetic side in restocking our warfelled woods. *Quarterly Journal of Forestry* 13: 254–258.
Peterken, G. 1981. *Woodland Conservation and Management*. London: Chapman & Hall. Cited in J. Spencer and K. Kirby. 1992. An inventory of ancient woodlands for England and Wales. *Biological Conservation* 62: 77–93.
Peterken, G. 2001. Ecological effects of introduced tree species in Britain. *Forest Ecology and Management* 141: 31–42.

Petts, J. and B. Leach. 2000. *Evaluating Methods for Public Participation: Literature Review.* R&D Technical Report: E135. Bristol: Environment Agency.

Price, C. 1997. Twenty-five years of forestry cost-benefit analysis in Britain. *Forestry* 70 (3): 171–189.

Price, C. 2003. Quantifying the aesthetic benefits of urban forestry. *Urban Forestry and Urban Greening* 1 (3): 123–133.

Price, C. 2007. Do cultural landscapes have a sustainable future? In *Social and Cultural Values of Forests – Benefit for Today's Society. 12th IUFRO World Congress – Proceedings. News of Forest History* 4 (38): 12–24.

Pringle, D. 1994. *The First 75 Years. A Brief Account of the History of the Forestry Commission.* Edinburgh: Forestry Commission.

Public and Corporate Economic Consultants. 2000. *English Forestry Contribution to Rural Economies. Final Report.* Cambridge: PACEC.

Rackham, O. 1998. Trees and woodlands in a cultural landscape: the history of woods in England. *Forestry Sciences* 54: 139–147.

Rackham, O. 2000. Prospects for landscape history and historical ecology. *Landscapes* 1 (2): 3–15.

Rackham, O. 2002. *Trees and Woodlands in the British Landscape. The Complete History of Britain's Trees, Woods and Hedgerows.* London: Phoenix. '1st edition Dent 1976).

Rackham, O. 2004. Pre-existing trees and woods in country-house parks. *Landscapes* 5 (2): 1–16.

Rackham, O. 2006. *Woodlands.* London: Collins.

Randrup, T. and K. Nilsson. 1997. *Research on Urban Trees in Europe.* Third European Arboricultural Congress, Merano (Italy). http://www.gemeinde.meran.bz.it/giardinerie/congresso/atti/research.htm. Accessed 10 November 2007.

Rawlinson, H., N. Dickinson, P. Nolan and P. Putwain. 2004. Woodland establishment on closed old-style landfill sites in N.W. England. *Forest Ecology and Management* 202: 265–280.

Read, H. 2000. *Veteran Trees. A Guide to Good Management.* Peterborough: English Nature.

Readman P. 2001. Landscape preservation, 'advertising disfigurement' and the English National Identity c.1890–1914. *Rural History* 12 (1): 61–83.

Reed, M. 1990. *The Landscape of Britain from the Beginnings to 1914.* London: Routledge.

Repton, J. 1816. *Fragments on the Theory and Practice of Landscape Gardening.*

Rhode, C. 1983. *An Investigation of Reinforcement Mechanisms in Depressed Psychiatric Patients.* Unpublished PhD Thesis: University of London.

Rhode, C. and A. Kendle. 1994. *Human Well-Being, Natural Landscapes and Wildlife in Urban Areas. A Review.* English Nature Science report N°22. Peterborough: English Nature.

Rickman, R. 1991. *What's Good for Woods: Promoting the Industry, Protecting the Environment.* Policy Study N°129, London: Centre for Policy Studies.

Rival, L. 1998. *The Social Life of Trees.* Oxford: Berg.

Roberts, D., D. Eiser, B. Crabtree and G. Broom. 2000. *Forests' Role in Tourism: Phase 1.* Aberdeen: Macaulay Institute.

Roper, C. and A. Park. 1999. *Non Market Benefits of Forestry. International Symposium Held in Edinburgh.* London: Stationery Office.

Rowntree, R. and D. Nowak. 1991. Quantifying the role of urban forests in removing atmospheric carbon dioxide. *Journal of Arboriculture* 17 (10): 269–275.

Royle, E. 1987. *Modern Britain, a Social History 1750–1985.* London: Arnold.

Rural Innovation. 2004. *An Economic Study of the Proposals for the Development of Mountain Bike Trails and Associated Trailhead Facilities at Whinlatter Forest.* http://mtb-lakes.com/whin_pdfs/app_2.pdf.

Ryle, G. 1969. *Forest Service: The First Forty-Five Years of the Forestry Commission of Great Britain.* Newton Abbot: David & Charles.

Salter, B. 1998. Urban forestry: UK experience in practice. *Arboricultural Journal* 22 (1): 11–23.

Schama, S. 1995. *Landscape and Memory.* London: Harper Collins.

Schnellnhuber, H.J. (ed.). 2006. *Avoiding Dangerous Climate Change*. Cambridge: Cambridge University Press.

Sempik, J., J. Aldridge and S. Becker. 2005a. *Growing Together: A Practice Guide to Promoting Social Inclusion Through Gardening and Horticulture*. Bristol: Policy.

Sempik, J. J. Aldridge and S. Becker 2005b. *Health, Well-Being and Social Inclusion: Therapeutic Horticulture in the UK*. Bristol: Policy.

Sheail, J. 1998. The National Forest: an historical context. *East Midland Geographer* 21 (1): 12–22.

Sheppard, S. and H. Harshaw (eds.). 2000. *Forests and Landscapes. Linking Ecology, Sustainability and Aesthetics. IUFRO Research Series N°6*. New York: CABI, in association with IUFRO.

Shroeder, H. 1989. Environment, behavior, and design research on urban forests. In *Advances in Environment, Behavior and Design*, eds. E. Zube and G. Moore, 87–117. New York: Plenum.

Simmons, I. 2001. *An Environmental History of Great Britain*. Edinburgh: Edinburgh University Press.

Simson, A. and J. Ryan. 2003. *NeighbourWoods Case Study Report*. Leeds: Design Task Force/ Leeds Met University. http://www.sl.kvl.dk/euforic/docs/NeighbourWoods/D10d%20Leeds.pdf. Accessed 17 April 2007.

Slee, B. 2006. The socio-economic evaluation of the impact of forestry on rural development: a regional level analysis. *Forest Policy and Economics* 8: 542–554.

Slee, B., D. Jones and N. Curry. 2001. *Social Exclusion in Countryside Leisure in the United Kingdom*. Cardiff: Countryside Recreation Network.

Small Woods Association. 2001. *A Way into Woodland. Making Access Work Well for All Concerned*. Conference report of the National Small Woods Association. 2001. Small Woods Association, Forestry Commission.

Southwood, T. 1961. The number of species of insect associated with various trees. *The Journal of Animal Ecology* 30 (1): 1–8.

Spencer, J. and K. Kirby. 1992. An inventory of ancient woodlands for England and Wales. *Biological Conservation* 62: 77–93.

Stamper, P. 2002. The tree that hid a king: The Royal Oak at Boscobel, Shropshire. *Landscapes* 3 (1): 19–34.

Stewart, H., S. Owen, R. Donovan et al. 2002. *Trees and Sustainable Urban Air Quality*. Lancaster: University of Lancaster with Centre for Ecology and Hydrology.

Steward, T. V. Pickett, T. Parker and P. Feidler. 1992. The new paradigm in ecology: implications for conservation biology above the species level. In *Conservation Biology. The Theory and Practice of Nature Conservation, Preservation and Management*, eds. P. Feidler and S. Jain, 65–88. London: Chapman & Hall.

Stirrat, R., R. Scruton and N. Sargeant. 1984. *Urban Forestry and Job Creation*. Edinburgh: PEFS.

Stoker, G. 2004. *Transforming Local Governance: From Thatcherism to New Labour*. Beyond the Centre Series. Basingstoke: Palgrave Macmillan.

Sustainable Development Commission. 2005. *Leading by Example? Not exactly...* London: SDiG.

Sustainable Development Research Network. 2007. *Emerging Methods for Sustainability Valuation and Appraisal*. SDRN Briefing Four. http://admin.sd-research.org.uk/wp-content/uploads/2007/04/briefingprooffinal.pdf. Accessed 1 May 2007.

Taylor, W. 1945. *Forests and Forestry in Great Britain*. London: Lockwood.

Thirsk, J. (ed.). 2000. *Rural England: An Illustrated History of the Landscape*. Oxford: Oxford University Press.

Thomas, K. 1984. *Man and the Natural World. Changing Attitudes in England, 1500–1800*. Oxford: Oxford University Press.

Thompson, F. 1971. *English Landed Society in the Nineteenth Century*. London: Routledge & Kegan Paul.

Tsouvalis, J. 2000. Socialized nature: England's royal and plantation forests. In *Cultural Turns/ Geographical Turns: Perspectives on Cultural Geography*, eds. I. Cook, D. Crouch, S. Naylor and J. Ryan, 289–312. Harlow: Prentice Hall.

Tsouvalis, J. 2000. *A Critical Geography of Britain's State Forests*. Oxford: Oxford University Press.

Ulrich, R. 1984. View through a window may influence recovery from surgery. *Science* 224: 420–421.

United Nations Forum on Forests. 2003. *National Report to the Third Session of the United Nations Forum on Forests. United Kingdom of Great Britain and Northern Ireland*. http://www.un.org/esa/forests/pdf/national_reports/unff3/united_kingdom.pdf. Accessed 17 November 2006.

Valentine,G. and J. McKendrick. 1997. Children's outdoor play: Exploring parental concerns about children's safety and the changing nature of childhood. *Geoforum* 28 (2): 219–235.

Voisey, H. and T. O'riordan. 1998. Sustainable development: the UK national approach. In *The Transition to Sustainability*, eds. T. O'Riordan and H. Voisey, 156–173. London: Earthscan.

Walker, A., C. Kershaw and S. Nicholas. 2006. *Crime in England and Wales 2005–2006*. Home Office Statistical Bulletin 12/06. London: Home Office.

Walter, F. 2004. *Les figures paysagères de la nation. Territoire et paysage en Europe (16e - 20è siècle)*. Paris : Éditions de l'École des Hautes Études en Sciences Sociales.

Wanless, D. 2002. *Securing Our Future Health: Taking a Long-Term View*. London: HM Treasury.

Warburton, D. 1997. *Participatory Action in the Countryside. A Literature Review*. Cheltenham: Countryside Commission.

Warren, A. and F. Goldsmith (eds.). 1983. *Conservation in Perspective*. Chichester: Wiley.

Watford, M. and D. Sharpe. 1993. Green beliefs and religion. In *The Politics of Nature: Explorations in Green Political Theory*, eds. A. Dobson, and P. Lucardie, 210–228. London: Routledge.

Watkins, C. (ed.). 1998. *European Woods and Forests: Studies in Cultural History*. New York: CABI.

Watkins, C., D. Williams and T. Lloyd. 1996. Constraints on farm woodland planting in England: a study of Nottinghamshire farmers. *Forestry* 69 (2): 167–176.

Whatmore, S. and S. Boucher. 1993. Bargaining with nature: the discourse and practice of environmental gain. *Transactions of the Institute of British Geographers* 18: 166–178.

Whitaker, J. 1997. Get involved – working effectively with local people. In *Community Forestry – A Change for the Better. Conference Proceedings, 7–8 December 1999, The Guildhall, London*, Forestry Commission, n.p. Forestry Commission, Countryside Agency.

Whitelock, D. et al. 1965. *The Anglo-Saxon Chronicle. A revised Translation*. London: Eyre & Spottiswoode.

Whitt, L. and J. Slack. 1994. Communities, environments and cultural studies. *Cultural Studies* 8 (1): 5–31.

Wilks, J. 1972. *Trees of the British Isles in History and Legend*. London: Frederick Muller.

Williams, R. 1972. Ideas of nature. In *Problems in Materialism and Culture. Selected Essays*, Raymond Williams 1980: 67–85.

Williams, R. 1981. *Culture*. Glasgow: Fontana paperbacks.

Williams, M. and R. Williams. 1986. *John Clare, Selected Poetry and Prose*. London: Routledge.

Williamson, T. 1995. *Polite Landscapes – Gardens and Society in Eighteenth-Century England*. Stroud: Sutton Publishing.

Wiilis, K., G. Garrod, R. Scarpa et al. 2000. *Non Market Benefits of Forestry. Phase 1*. Report to the Forestry Commission. Newcastle: Centre for Research un Environmental Appraisal and Management, University of Newcastle.

Wilson, E. 1984. *Biophilia: The Human Bond with Other Species*. Cambridge: Harvard University Press.

Willoughby, I. and J. Dewar. 1995. *The Use of Herbicides in the Forest*. Field Book 8. Edinburgh: Forestry Commission.

Willoughby, I. et al. 2004. *Reducing Pesticide Use in Forestry. Practice Guide*. Edinburgh: Forestry Commission.

Wilson, B., G. Van Kooten, I. Vertinsky and L. Arthur (eds.). 1999. *Forest Policy: International Case Studies*. New York: CABI.

Winter, M. 1996. *Rural Politics: Policies for Agriculture, Forestry and the Environment*. London: Routledge.

Wirth, L. 1938. Urbanism as a way of life. *American Journal of Sociology* 44: 1–24.

Woodland Trust. 2002. *Reclaiming Our Forgotten Inheritance. The Race to Save Britain's Hidden Woodland Treasures*. Grantham: Woodland Trust.

Woodland Trust. 2004. *Space for People. Targeting Action for Woodland Access*. Grantham: The Woodland Trust.

Wolf, K. 1998a. *Urban Forest Values: Economic Benefits of Trees in Cities*. University of Washington College of Forest Resources, Factsheet n°29.

Wolf, K. 1998b. *Trees in Business Districts - Positive Effects on Consumer Behaviour*. University of Washington College of Forest Resources, Factsheet n°30.

Wolf, K. 1998c. *Trees in Business Districts - Comparing Values of Consumers and Business*. University of Washington College of Forest Resources, Factsheet n°31.

Wolf, K. 2000. *Community Image - Roadside Settings and Public Perceptions*. University of Washington College of Forest Resources, Factsheet #32

Wolf, K, 2003. *Public Response to the Urban Forest in Inner-City Business Districts. Journal of Arboriculture* 29 (3): 117–126.

World Commission on Environment and Development. 1987. *Our Common Future*. Oxford: Oxford University Press.

Worpole. 1997. *The Richness of Cities: Working Paper 2: Nothing to Fear? Trust and Respect in Urban Communities*. London: Comedia & Demos.

Worster, D. 1993. *The Wealth of Nature. Environmental History and the Ecological Imagination*. Oxford: Oxford University Press.

Zuckerman, S. 1957. *Enquiry into Forestry, Agriculture and Marginal Land*. London: HMSO.

Zukin, S. 1991. *Landscapes of Power*. Berkeley: University of California Press.

Zukin, S. 1995. *The Culture of Cities*. Cambridge: Blackwell.

Chronology

(international events are in bold type)

Around 8,000 B.C.: Glaciers retreated. Beginning of the Holocene
Around 6,000 B.C: Hunter-gatherers and fisherfolk of the late upper Palaeolithic and the Mesolithic gave way to farmers
Around 5,500 B.C.: Connection with the rest of Europe severed as the sea-level rose
5,500–3,000 B.C.: Forests of Britain at their maximum extent
Around 500 B.C.: Woodland clearance had caused forest to disappear on more than half the country
43 A.D.: Roman Conquest
409: Departure of the Romans from the British Isles
5th century A.D. onwards: Angles, Saxons and Jutes continued clearing the woods for agriculture
End of 8th century onwards: The Norse intensified wood clearance
1066: Battle of Hastings: William the Conqueror became king of England
1079: Creation of the New Forest
1086: *Domesday Book*
1100: King William II (Rufus) shot dead by an arrow in the New Forest
1184: Assize of Woodstock, defining the first Forest Laws (*Constitutiones de Foresta*, or *Charter of the Forest*)
1154–1189: Royal Forests reached their maximum extent
1215: *Magna Carta*
1217: *Carta de Foresta*, bringing Royal Forests back to their area before 1154
1258: Royal edicts against timber sales under Henry III
1290: Royal edicts against timber sales under Edward I
1290: Edward I said to have held his Parliament under an oak tree.
1348: Beginning of the Black Death
1482: *Enclosure Act* to allow natural regeneration after felling
1539: Creation of the last Royal Forest, the Forest of Honour, at Hampton Court
1543: *Statute of Woods*, the first timber preservation act
1558: Princess Elizabeth supposed to have learnt that she was queen under an oak tree at Hatfield
1558: Act prohibiting the use of timber in the iron industry
1580: First oak plantation in England, near Cranbourne Lodge, Windsor Great Park
1588: Routed Spanish *Invincible Armada* had allegedly targeted the Forest of Dean in order to destroy it
1598: Manwood's *Forest Laws*
1635–1638: Various portraits by Van Dyck depicted King Charles I under an oak tree
1641: *The Grand Remonstrance*
1649–1660: Abolition of the Royal Forests during the Commonwealth
1651: Charles II hid in the Boscobel Oak after the defeat of his army at Worcester
1664: Publication of John Evelyn's *Silva, or a Discourse of Forest Trees and the Propagation of Timber in His Majesty's Dominions*

1668: *Act for the Increase and Preservation of Timber within the Forest of Dean*
1669: Colbert's Ordonnance de Saint-Germain en Laye to manage woodland resources in France
1671: *Game Law* made hunting the privilege of propertied gentlemen, enshrining the principle of private property
1722: *Black Act*, making poaching and cutting trees a capital offence
1758–1835: Royal Society for the Encouragement of Arts, Manufactures and Commerce awarded prizes for tree-planting
1766: Legislation made it criminal to gather dead wood from plantations, copses and hedges
1789: First school of forestry created in Hunden (Hesse)
1811: John Nash designed Marylebone Park, London
1816: School of forestry founded in Tharandt (Saxony)
1827: Douglas fir (*Pseudotsuga menziesii*) introduced from North America
1831: Sitka spruce (*Picea sitchensis*) introduced from North America
1833: Parliamentary Select Committee on Public Walks
1836, 1840, 1845: *General Enclosure Acts*
1844: Paxton designs Birkenhead Park
1842: Chadwick's *Report on the Sanitary Condition of the Labouring Population of Great Britain*
1845: *Second Report on the State of Large Towns and Populous Districts*
1845: Creation of Victoria Park in the East End of London
1851: *Deer Removal Act*
1853: Lodgepole pine (*Pinus contorta*) introduced from North America
1859: Term 'forestry' appeared for the first time in the Oxford Dictionary
1862: Battle of Hampton Roads, during the American Civil War, proved the superiority of ironclad over wooden-hulled ships
1865: Creation of the Commons, Open Spaces and Footpaths Preservation Society
1877: *New Forest Act*
1878: *Epping Forest Act*
1881: Creation of the Royal Arboricultural Society
1885: Creation of a forestry department at the Royal Indian Engineering College at Coopers Hill, Surrey
1885: Select Committee on Forestry set up
1892: Foundation of IUFRO
1894: Creation of the National Trust
1907: Readership in forestry opened at Cambridge University
1909: Final report of the Royal Commission on Coast Erosion, Reclamation of Tidal Lands and Afforestation
1909: *Development and Road Improvement Fund Act*
1915: Board of Agriculture appointed a Home-Grown Timber Committee
1916: Creation of a subcommittee on forestry (Acland Committee) within the Reconstruction Committee
1917: Creation of a Directorate of Timber Supplies, later Timber Supply Department of the Board of Trade
1919: *Forestry Act* 1919, creating the Forestry Commission
1921: Creation of the Empire Forestry Association
1923: *Forestry (Transfer of Woods) Act*
1925: National Playing Fields Association (NPFA) founded
1926: First World Forestry Congress, Rome
1926: Creation of the Council for the Protection of Rural England
1930s: Creation of National Forest Parks
1936: Second World Forestry Congress, Budapest
1943: Institute of Forest History created at the University of Freiburg
1943: White Paper *Post-War Forest Policy*

1944: White Paper *Post-War Forest Policy: Private Woodlands*
1945: Creation of the FAO
1945: *Forestry Act*
1946: Foundation of the Forest History Society in the United States
1946: First *New Towns Act*
1949: Third World Forestry Congress, Helsinki
1949: *National Parks and Access to the Countryside Act*
1951: *Forestry Act*
1954: Fourth World Forestry Congress, Dehra-Dun (India)
1957: *Report on Forestry, Agriculture and Marginal Land* (Zuckerman Report)
1960: Fifth World Forestry Congress, Seattle
1962: Landscape architect Sylvia Crowe appointed landscape consultant to the Forestry Commission
1965: First occurrence of the term 'urban forestry' (Eric Jorgensen, Toronto)
1966: Sixth World Forestry Congress, Madrid
1967: *Forestry Act*
1968: *Countryside Act*
1972: Seventh World Forestry Congress, Buenos Aires
1972: Club of Rome report *The Limits to Growth*
1972: First 'Earth Summit' in Stockholm
1973: 'Plant a tree in' 73' campaign, giving birth to National tree Week in 1975
1973: The United Kingdom joins the EEC
1973–1976: First European Environmental Action Programme
1975: White Paper *Sport and Leisure*
1978: Eighth World Forestry Congress, Jakarta
1981: Creation of the Groupe d'Histoire des Forêts Françaises, France
1981: *Ancient Woodland Inventory* started by the Conservancy Council
1981: *Forestry Act*
1981: Creation of the Environment Directorate General (DG XI) within the European Commission
1982: Loughborough conference jointly organised by the Forestry Commission and the Institute of Chartered Foresters leading to Broadleaves policy
1983: *National Heritage Act* leading to the creation of English Heritage
1985: Ninth World Forestry Congress, Mexico
1985: *Broadleaves in Britain*
1985: *Wildlife and Countryside (Amendment) Act*
1986: International Conference on Trees and Forests *(SILVA Conference), Paris*
1986: *Agriculture Act*: end of the productivist era of British agriculture
1987: Brundtland report
1987: Storm in the South-East of England, reportedly the worst one since 1703
1988: Creation of charitable association Tree Register of the British Isles (TROBI)
1987: 'Forest of London' urban forestry project launched
1988: *Farm Land and Rural Development Act*, British farmers encouraged to take land out of arable production, introduction of the first Farm Woodland Scheme
1988: M. Thatcher's 'Green Speech' to the Royal Society
1989: Introduction of the first Woodland Grant Scheme
1990: White Paper *This Common Inheritance*
1990: First Ministerial Conference on the Protection of the Forests of Europe (MCPFE), Strasbourg, France
1990: Creation of the National Forest, promoted by the Government as the largest UK environmental project of the 20th century.
1990: Foundation of the national school grounds charity 'Learning Through Landscapes'
1990: Black Country Urban Forest (BCUF) launched
1991: Community Forests programme, the largest national urban forestry programme to date, launched by Prime Minister John Major

1991: *Tourism and the Environment. Maintaining the Balance*, by the Department of Employment and the English Tourist Board's Tourism and Environment Task Force
1991: Tenth World Forestry Congress, Paris
1991: *Ancient Woodland Inventory* taken up by English Nature in 1991
1991: *Forestry Policy for Great Britain*, policy statement by the Forestry Commission
1992: *Statement of Forest Principles*, issued at the United Nations Conference on Environment and Development, Rio
1992: Introduction of the Farm Woodland Premium Scheme
1992: Restructuring of the Forestry Commission
1993: Second Ministerial Conference on the Protection of Forests in Europe, Helsinki
1993: Forest Stewardship Council (FSC), an international network, set up
1993: Foundation of the European Forest Institute (EFI)
1993: Publication of the *Survey of Derelict Land in England* by the Department of Environment
1993: First eco-cemetery in Britain, Carlisle in 1993
1993: *Accessible Natural Greenspace Standard* by English Nature
1994: European Parliament report on the forests of Europe
1994: Publication of *Sustainable Development. The UK Strategy* and of *Biodiversity. The UK Action Plan*
1994: Publication of *Sustainable Forestry: the UK Programme*
1994: *Position Statement on Environmentally Sustainable Forestry and Woodland Management* by English Nature
1994: Forest Enterprise set up as an Agency
1995: Policy Studies Division of the Forestry Commission became Forest Research, an Executive Agency
1995: Intergovernmental Panel on Forests (IPF) appointed in 1995 under the United Nations Commission on Sustainable Development
1995: Creation of the National Urban Forestry Unit, disbanded in 2005
1995: First survey of *GB Public Opinion of Forestry Statistics*, then biennial
1995: Research project 'Urban Forestry: overview and analysis of European urban forest policies' launched by the European Forest Institute
1996–2001: 250 'Woods on Your Doorstep' created to celebrate the new millennium
1997: Forest Research, the Forestry Commission's research branch, became an Executive Agency
1997: Eleventh World Forestry Congress, Antalya
1997: Tree-ROUTE (Research On Urban Trees in Europe) Network set up by the Danish Forest and Landscape Research Institute
1997–2000: Programme TOURFOR (European Award for Tourism in Forest Area Project), under the auspices of DGXI
1997–2000: Intergovernmental Forum on forests
1997–2002: COST (European Cooperation in the field of Scientific and Technical Research) Action E12 'Urban Forests and Trees', funded by the European Commission
1998: Third Ministerial Conference on the Protection of Forests in Europe, Lisbon
1998: European Forum on Urban Forestry (EFUF), a network of urban foresters, set up
1998: Convention On Access To Information, Public Participation in Decision-Making And Access To Justice In Environmental Matters (Aarhus Convention) by the United Nations Economic Commission for Europe, ratified by the UK in 2005.
1998: *UK Forestry Standard*
1998: *The England Forestry Strategy. A New Focus for England's Woodlands*
1999: *UK Woodland Assurance Scheme*
1999: Devolution to Scotland, Wales and Northern Ireland, leading to a reorganisation of the Forestry Commission
1999: Report of the Urban Task Force *Towards an Urban Renaissance*
1999: Creation of the Commission for Architecture and the Built Environment (CABE) to advise government on architecture, urban design, and more recently, on public space

1999–2000: 'Millenium Yews' distributed in 7,000 parishes
2000: Urban White Paper *Our Towns and Cities: the Future*
2000: Rural White Paper *Our Countryside: the Future*
2000: *Countryside and Rights of Way Act*
2000–2003: Establishment of the United Nations Forum on Forests (UNFF)
2000: Social Research Unit of the Forestry Commission created
2000–2006: Energy Crops Scheme, run by DEFRA and the Forestry Commission
2001–2002: *Ancient Woodland Inventory* integrated into the National Digital Archive of Datasets
2001: European Forestry Research and Information Centre (EUFORIC) created by the European Forestry Institute
2001–2004: European programme NeighbourWoods aimed at producing a comprehensive toolbox for socially-inclusive planning and design of urban woodlands
2002: World Summit on Sustainable Development, Johannesburg
2002: *England Biodiversity Strategy*
2002: Office of the Deputy Prime Minister report *Living Places: Cleaner, Safer, Greener*
2002: Report of the Urban Task Force, *Green Spaces, Better Places*
2002: Forest School England Network created
2002: *National Heritage Act 2002*
2002: *Working With the Grain of Nature*, England's Biodiversity Strategy, published
2002: *Forestry Devolution Review*
2003: Newlands (New Economic Environments – through Woodlands) programme launched
2003: *Sustainable Communities: Building for the Future* launched
2003: Twelfth World Forestry Congress, in Quebec City
2003: Fourth Ministerial Conference on the Protection of Forests in Europe, Vienna
2003: *UK National Forest Programme*
2003: *UK Energy White Paper*
2004: Second edition of the *UK Forestry Standard* published
2004: Integration of the Baltic countries into the European Union
2004: *Government's Rural Strategy*
2004: *Framework for Public Involvement in Forest Planning* developed jointly by Forest Research, Forest Enterprise and the USDA Forest Service
2005: English Woodland Grant Scheme launched by the Forestry Commission
2005: Report of the Biomass Task Force published
2006: *England Forestry Strategy Implementation Progress Report* published
2007: MSc course in urban forestry created at Myerscough College (Preston, Lancashire)

Index

World Forests (Series)

I. M. Palo and J. Uusivuori (eds.), *World Forests, Society and Environment.* 1999. ISBN 978-0-7923-5301-3.
II. M. Palo and H. Vanhanen (eds.), *World Forests from Deforestation to Transition?* 2001. ISBN 978-0-7923-6683-2.
III. M. Palo, J. Uusivuori, and G. Mery, *World Forests, Markets and Policies.* 2001. ISBN 978-0-7923-7170-4.
IV. J.-A. Lamberg, J. Näsi, J. Ojala, and P. Sajasalo (eds.), T*he Evolution of Competitive Strategies in Global Forestry Industries.* 2006. ISBN 978-1-4020-4015-6.
V. W. De Jong, D. Donovan, and Ken-ichi Abe (eds.), *Extreme Conflict and Tropical Forest.* 2007. ISBN 978-1-4020-5461-7.
VI. S. Nail, *Forest Policies and Social Change in England.* 2008. ISBN 978-1-4020-8364-8.